General Motors
Chevrolet Lumina APV, Oldsmobile Silhouette and Pontiac Trans Sport
Automotive Repair Manual

by J J Haynes
and John H Haynes
Member of the Guild of Motoring Writers

Models covered:
Chevrolet Lumina APV,
Oldsmobile Silhouette and Pontiac Trans Sport
1990 through 1995

(5E10 - 38035) ABCD
(2035)

Haynes Publishing Group
Sparkford Nr Yeovil
Somerset BA22 7JJ England

Haynes North America, Inc
861 Lawrence Drive
Newbury Park
California 91320 USA

Acknowledgements
Technical writers who contributed to this project include Robert Maddox, Mike Stubblefield and Larry Warren

© **Haynes North America, Inc. 1992, 1995, 1996**
With permission from J.H. Haynes & Co. Ltd.

A book in the Haynes Automotive Repair Manual Series

Printed in the U.S.A.

All rights reserved. No part of this book may be reproduced or transmitted in any form or by any means, electronic or mechanical, including photocopying, recording or by any information storage or retrieval system, without permission in writing from the copyright holder.

ISBN 1 56392 208 8

Library of Congress Catalog Card Number 96-75325

While every attempt is made to ensure that the information in this manual is correct, no liability can be accepted by the authors or publishers for loss, damage or injury caused by any errors in, or omissions from, the information given.

Contents

Introductory pages

About this manual	0-5
Introduction to the Chevrolet Lumina APV, Oldsmobile Silhouette and Pontiac Trans Sport	0-5
Vehicle identification numbers	0-6
Buying parts	0-7
Maintenance techniques, tools and working facilities	0-7
Booster battery (jump) starting	0-13
Jacking and towing	0-13
Automotive chemicals and lubricants	0-14
Conversion factors	0-15
Safety first!	0-16
Troubleshooting	0-17

Chapter 1
Tune-up and routine maintenance — 1-1

Chapter 2 Part A
3.1L V6 engine — 2A-1

Chapter 2 Part B
3800 V6 engine — 2B-1

Chapter 2 Part C
General engine overhaul procedures — 2C-1

Chapter 3
Cooling, heating and air conditioning systems — 3-1

Chapter 4
Fuel and exhaust systems — 4-1

Chapter 5
Engine electrical systems — 5-1

Chapter 6
Emissions control systems — 6-1

Chapter 7
Automatic transaxle — 7-1

Chapter 8
Driveaxles — 8-1

Chapter 9
Brakes — 9-1

Chapter 10
Suspension and steering systems — 10-1

Chapter 11
Body — 11-1

Chapter 12
Chassis electrical system — 12-1

Wiring diagrams — 12-16

Index — IND-1

1990 Pontiac Trans Sport

Haynes author, mechanic and photographer with Pontiac Trans Sport

About this manual

Its purpose

The purpose of this manual is to help you get the best value from your vehicle. It can do so in several ways. It can help you decide what work must be done, even if you choose to have it done by a dealer service department or a repair shop; it provides information and procedures for routine maintenance and servicing; and it offers diagnostic and repair procedures to follow when trouble occurs.

We hope you use the manual to tackle the work yourself. For many simpler jobs, doing it yourself may be quicker than arranging an appointment to get the vehicle into a shop and making the trips to leave it and pick it up. More importantly, a lot of money can be saved by avoiding the expense the shop must pass on to you to cover its labor and overhead costs. An added benefit is the sense of satisfaction and accomplishment that you feel after doing the job yourself.

Using the manual

The manual is divided into Chapters. Each Chapter is divided into numbered Sections, which are headed in bold type between horizontal lines. Each Section consists of consecutively numbered paragraphs.

At the beginning of each numbered Section you will be referred to any illustrations which apply to the procedures in that Section. The reference numbers used in illustration captions pinpoint the pertinent Section and the Step within that Section. That is, illustration 3.2 means the illustration refers to Section 3 and Step (or paragraph) 2 within that Section.

Procedures, once described in the text, are not normally repeated. When it's necessary to refer to another Chapter, the reference will be given as Chapter and Section number. Cross references given without use of the word "Chapter" apply to Sections and/or paragraphs in the same Chapter. For example, "see Section 8" means in the same Chapter.

References to the left or right side of the vehicle assume you are sitting in the driver's seat, facing forward.

Even though we have prepared this manual with extreme care, neither the publisher nor the author can accept responsibility for any errors in, or omissions from, the information given.

NOTE

A **Note** provides information necessary to properly complete a procedure or information which will make the procedure easier to understand.

CAUTION

A **Caution** provides a special procedure or special steps which must be taken while completing the procedure where the Caution is found. Not heeding a Caution can result in damage to the assembly being worked on.

WARNING

A **Warning** provides a special procedure or special steps which must be taken while completing the procedure where the Warning is found. Not heeding a Warning can result in personal injury.

Introduction to the Chevrolet Lumina, Oldsmobile Silhouette and Pontiac Trans Sport

These front-wheel drive mid-size All Purpose Vehicle (APV) or Multi-purpose (MPV) models are available in van body styles only. The body construction is of the space frame type with plastic body panels.

Engines used in these vehicles include the 60-degree 3.1 liter V6 and 90-degree 3800 V6.

Throttle body injection (TBI) is used on the 3.1 liter engine while the 3800 engine is equipped with multi-port fuel injection.

The engine drives the front wheels through either a three- or four-speed automatic transaxle via driveaxles equipped with Constant Velocity (CV) joints. The power assisted rack and pinion steering is mounted behind the engine.

The front suspension is composed of MacPherson strut with lower control arms and a stabilizer bar. The rear suspension is beam axle, with trailing arms, shock absorber units and a track bar.

The brakes are disc on the front and drum on the rear wheels. Power assist is standard equipment.

Vehicle identification numbers

Vehicle identification numbers

Modifications are a continuing and unpublicized part of vehicle manufacturing. Since spare parts manuals and lists are compiled on a numerical basis, the individual vehicle numbers are essential to correctly identify the component required.

Vehicle Identification Number (VIN)

This very important identification number is stamped on a plate attached to the left side of the dashboard and is visible through the driver's side of the windshield (see illustration). The VIN also appears on the Vehicle Certificate of Title and Registration. It contains valuable information such as where and when the vehicle was manufactured, the model year and the body style.

Service parts identification label

This label is located on the glove compartment door (see illustration). It lists the VIN number, wheelbase, paint number, options and other information specific to the vehicle it's attached to. Always refer to this label when ordering parts.

Engine identification numbers

This number is found on a pad on the left end of the block, just above the starter motor (see illustration). An optional location for the number on 3800 engines is on the right end of the engine block, to the rear of the water pump.

Automatic transaxle number

The nameplate/ID number on the 3T40 transaxle is attached to the upper surface of the case, near the rear (see illustration). On the 4T60 transaxle, the transaxle VIN number is stamped into the horizontal casting rib on the right rear of the housing; an identification label is mounted above.

Vehicle Emissions Control Information label

This label is found in the engine compartment. See Chapter 6 for more information on this label.

The Vehicle Identification Number (VIN) is on a plate attached to the top of the dashboard on the driver's side of the vehicle - it can be seen from outside the vehicle, looking through the windshield

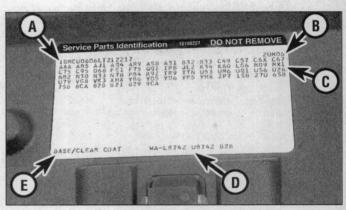

The service parts identification label is located on the glove box door and contains much important information

A Vehicle Identification Number
B Body type and style
C Options
D Paint codes
E Paint type

3.1L V6 engine number locations

A Engine code
B VIN number

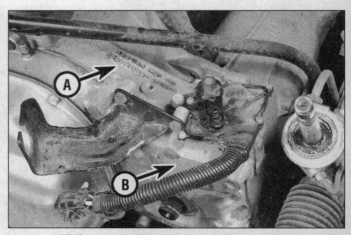

3T40 transaxle identification number locations

A VIN number
B Transaxle identification number

Buying parts

Replacement parts are available from many sources, which generally fall into one of two categories - authorized dealer parts departments and independent retail auto parts stores. Our advice concerning these parts is as follows:

Retail auto parts stores: Good auto parts stores will stock frequently needed components which wear out relatively fast, such as clutch components, exhaust systems, brake parts, tune-up parts, etc. These stores often supply new or reconditioned parts on an exchange basis, which can save a considerable amount of money. Discount auto parts stores are often very good places to buy materials and parts needed for general vehicle maintenance such as oil, grease, filters, spark plugs, belts, touch-up paint, bulbs, etc. They also usually sell tools and general accessories, have convenient hours, charge lower prices and can often be found not far from home.

Authorized dealer parts department: This is the best source for parts which are unique to the vehicle and not generally available elsewhere (such as major engine parts, transmission parts, trim pieces, etc.).

Warranty information: If the vehicle is still covered under warranty, be sure that any replacement parts purchased - regardless of the source - do not invalidate the warranty!

To be sure of obtaining the correct parts, have engine and chassis numbers available and, if possible, take the old parts along for positive identification.

Maintenance techniques, tools and working facilities

Maintenance techniques

There are a number of techniques involved in maintenance and repair that will be referred to throughout this manual. Application of these techniques will enable the home mechanic to be more efficient, better organized and capable of performing the various tasks properly, which will ensure that the repair job is thorough and complete.

Fasteners

Fasteners are nuts, bolts, studs and screws used to hold two or more parts together. There are a few things to keep in mind when working with fasteners. Almost all of them use a locking device of some type, either a lockwasher, locknut, locking tab or thread adhesive. All threaded fasteners should be clean and straight, with undamaged threads and undamaged corners on the hex head where the wrench fits. Develop the habit of replacing all damaged nuts and bolts with new ones. Special locknuts with nylon or fiber inserts can only be used once. If they are removed, they lose their locking ability and must be replaced with new ones.

Rusted nuts and bolts should be treated with a penetrating fluid to ease removal and prevent breakage. Some mechanics use turpentine in a spout-type oil can, which works quite well. After applying the rust penetrant, let it work for a few minutes before trying to loosen the nut or bolt. Badly rusted fasteners may have to be chiseled or sawed off or removed with a special nut breaker, available at tool stores.

If a bolt or stud breaks off in an assembly, it can be drilled and removed with a special tool commonly available for this purpose. Most automotive machine shops can perform this task, as well as other repair procedures, such as the repair of threaded holes that have been stripped out.

Flat washers and lockwashers, when removed from an assembly, should always be replaced exactly as removed. Replace any damaged washers with new ones. Never use a lockwasher on any soft metal surface (such as aluminum), thin sheet metal or plastic.

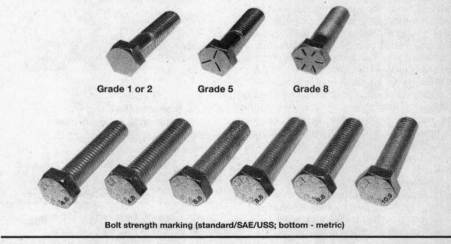

Bolt strength marking (standard/SAE/USS; bottom - metric)

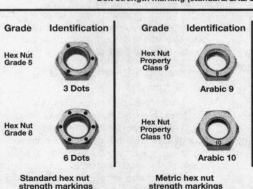

Standard hex nut strength markings

Metric hex nut strength markings

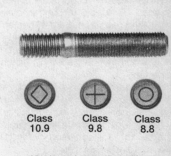

Metric stud strength markings

Fastener sizes

For a number of reasons, automobile manufacturers are making wider and wider use of metric fasteners. Therefore, it is important to be able to tell the difference between standard (sometimes called U.S. or SAE) and metric hardware, since they cannot be interchanged.

All bolts, whether standard or metric, are sized according to diameter, thread pitch and length. For example, a standard 1/2 - 13 x 1 bolt is 1/2 inch in diameter, has 13 threads per inch and is 1 inch long. An M12 - 1.75 x 25 metric bolt is 12 mm in diameter, has a thread pitch of 1.75 mm (the distance between threads) and is 25 mm long. The two bolts are nearly identical, and easily confused, but they are not interchangeable.

In addition to the differences in diameter, thread pitch and length, metric and standard bolts can also be distinguished by examining the bolt heads. To begin with, the distance across the flats on a standard bolt head is measured in inches, while the same dimension on a metric bolt is sized in millimeters (the same is true for nuts). As a result, a standard wrench should not be used on a metric bolt and a metric wrench should not be used on a standard bolt. Also, most standard bolts have slashes radiating out from the center of the head to denote the grade or strength of the bolt, which is an indication of the amount of torque that can be applied to it. The greater the number of slashes, the greater the strength of the bolt. Grades 0 through 5 are commonly used on automobiles. Metric bolts have a property class (grade) number, rather than a slash, molded into their heads to indicate bolt strength. In this case, the higher the number, the stronger the bolt. Property class numbers 8.8, 9.8 and 10.9 are commonly used on automobiles.

Strength markings can also be used to distinguish standard hex nuts from metric hex nuts. Many standard nuts have dots stamped into one side, while metric nuts are marked with a number. The greater the number of dots, or the higher the number, the greater the strength of the nut.

Metric studs are also marked on their ends according to property class (grade). Larger studs are numbered (the same as metric bolts), while smaller studs carry a geometric code to denote grade.

It should be noted that many fasteners, especially Grades 0 through 2, have no distinguishing marks on them. When such is the case, the only way to determine whether it is standard or metric is to measure the thread pitch or compare it to a known fastener of the same size.

Standard fasteners are often referred to as SAE, as opposed to metric. However, it should be noted that SAE technically refers to a non-metric fine thread fastener only. Coarse thread non-metric fasteners are referred to as USS sizes.

Since fasteners of the same size (both standard and metric) may have different strength ratings, be sure to reinstall any bolts, studs or nuts removed from your vehicle in their original locations. Also, when replacing a fastener with a new one, make sure that the new one has a strength rating equal to or greater than the original.

Metric thread sizes	Ft-lbs	Nm
M-6	6 to 9	9 to 12
M-8	14 to 21	19 to 28
M-10	28 to 40	38 to 54
M-12	50 to 71	68 to 96
M-14	80 to 140	109 to 154
Pipe thread sizes		
1/8	5 to 8	7 to 10
1/4	12 to 18	17 to 24
3/8	22 to 33	30 to 44
1/2	25 to 35	34 to 47
U.S. thread sizes		
1/4 - 20	6 to 9	9 to 12
5/16 - 18	12 to 18	17 to 24
5/16 - 24	14 to 20	19 to 27
3/8 - 16	22 to 32	30 to 43
3/8 - 24	27 to 38	37 to 51
7/16 - 14	40 to 55	55 to 74
7/16 - 20	40 to 60	55 to 81
1/2 - 13	55 to 80	75 to 108

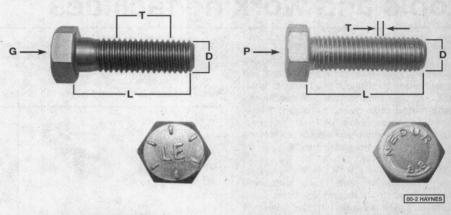

Standard (SAE and USS) bolt dimensions/grade marks

- G Grade marks (bolt strength)
- L Length (in inches)
- T Thread pitch (number of threads per inch)
- D Nominal diameter (in inches)

Metric bolt dimensions/grade marks

- P Property class (bolt strength)
- L Length (in millimeters)
- T Thread pitch (distance between threads in millimeters)
- D Diameter

Tightening sequences and procedures

Most threaded fasteners should be tightened to a specific torque value (torque is the twisting force applied to a threaded component such as a nut or bolt). Overtightening the fastener can weaken it and cause it to break, while undertightening can cause it to eventually come loose. Bolts, screws and studs, depending on the material they are made of and their thread diameters, have specific torque values, many of which are noted in the Specifications at the beginning of each Chapter. Be sure to follow the torque recommendations closely. For fasteners not assigned a specific torque, a general torque value chart is presented here as a guide. These torque values are for dry (unlubricated) fasteners threaded into steel or cast iron (not aluminum). As was previously mentioned, the size and grade of a fastener determine the amount of torque that can safely be applied to it. The figures listed here are approximate for Grade 2 and Grade 3 fasteners. Higher grades can tolerate higher torque values.

Fasteners laid out in a pattern, such as cylinder head bolts, oil pan bolts, differential cover bolts, etc., must be loosened or tightened in sequence to avoid warping the com-

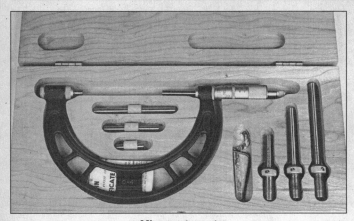
Micrometer set

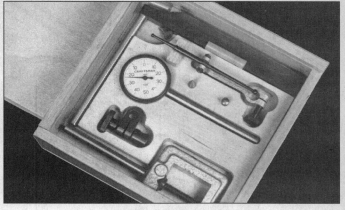
Dial indicator set

ponent. This sequence will normally be shown in the appropriate Chapter. If a specific pattern is not given, the following procedures can be used to prevent warping.

Initially, the bolts or nuts should be assembled finger-tight only. Next, they should be tightened one full turn each, in a criss-cross or diagonal pattern. After each one has been tightened one full turn, return to the first one and tighten them all one-half turn, following the same pattern. Finally, tighten each of them one-quarter turn at a time until each fastener has been tightened to the proper torque. To loosen and remove the fasteners, the procedure would be reversed.

Component disassembly

Component disassembly should be done with care and purpose to help ensure that the parts go back together properly. Always keep track of the sequence in which parts are removed. Make note of special characteristics or marks on parts that can be installed more than one way, such as a grooved thrust washer on a shaft. It is a good idea to lay the disassembled parts out on a clean surface in the order that they were removed. It may also be helpful to make sketches or take instant photos of components before removal.

When removing fasteners from a component, keep track of their locations. Sometimes threading a bolt back in a part, or putting the washers and nut back on a stud, can prevent mix-ups later. If nuts and bolts cannot be returned to their original locations, they should be kept in a compartmented box or a series of small boxes. A cupcake or muffin tin is ideal for this purpose, since each cavity can hold the bolts and nuts from a particular area (i.e. oil pan bolts, valve cover bolts, engine mount bolts, etc.). A pan of this type is especially helpful when working on assemblies with very small parts, such as the carburetor, alternator, valve train or interior dash and trim pieces. The cavities can be marked with paint or tape to identify the contents.

Whenever wiring looms, harnesses or connectors are separated, it is a good idea to identify the two halves with numbered pieces of masking tape so they can be easily reconnected.

Gasket sealing surfaces

Throughout any vehicle, gaskets are used to seal the mating surfaces between two parts and keep lubricants, fluids, vacuum or pressure contained in an assembly.

Many times these gaskets are coated with a liquid or paste-type gasket sealing compound before assembly. Age, heat and pressure can sometimes cause the two parts to stick together so tightly that they are very difficult to separate. Often, the assembly can be loosened by striking it with a soft-face hammer near the mating surfaces. A regular hammer can be used if a block of wood is placed between the hammer and the part. Do not hammer on cast parts or parts that could be easily damaged. With any particularly stubborn part, always recheck to make sure that every fastener has been removed.

Avoid using a screwdriver or bar to pry apart an assembly, as they can easily mar the gasket sealing surfaces of the parts, which must remain smooth. If prying is absolutely necessary, use an old broom handle, but keep in mind that extra clean up will be necessary if the wood splinters.

After the parts are separated, the old gasket must be carefully scraped off and the gasket surfaces cleaned. Stubborn gasket material can be soaked with rust penetrant or treated with a special chemical to soften it so it can be easily scraped off. A scraper can be fashioned from a piece of copper tubing by flattening and sharpening one end. Copper is recommended because it is usually softer than the surfaces to be scraped, which reduces the chance of gouging the part. Some gaskets can be removed with a wire brush, but regardless of the method used, the mating surfaces must be left clean and smooth. If for some reason the gasket surface is gouged, then a gasket sealer thick enough to fill scratches will have to be used during reassembly of the components. For most applications, a non-drying (or semi-drying) gasket sealer should be used.

Hose removal tips

Warning: *If the vehicle is equipped with air conditioning, do not disconnect any of the A/C hoses without first having the system depressurized by a dealer service department or a service station.*

Hose removal precautions closely parallel gasket removal precautions. Avoid scratching or gouging the surface that the hose mates against or the connection may leak. This is especially true for radiator hoses. Because of various chemical reactions, the rubber in hoses can bond itself to the metal spigot that the hose fits over. To remove a hose, first loosen the hose clamps that secure it to the spigot. Then, with slip-joint pliers, grab the hose at the clamp and rotate it around the spigot. Work it back and forth until it is completely free, then pull it off. Silicone or other lubricants will ease removal if they can be applied between the hose and the outside of the spigot. Apply the same lubricant to the inside of the hose and the outside of the spigot to simplify installation.

As a last resort (and if the hose is to be replaced with a new one anyway), the rubber can be slit with a knife and the hose peeled from the spigot. If this must be done, be careful that the metal connection is not damaged.

If a hose clamp is broken or damaged, do not reuse it. Wire-type clamps usually weaken with age, so it is a good idea to replace them with screw-type clamps whenever a hose is removed.

Tools

A selection of good tools is a basic requirement for anyone who plans to maintain and repair his or her own vehicle. For the owner who has few tools, the initial investment might seem high, but when compared to the spiraling costs of professional auto maintenance and repair, it is a wise one.

To help the owner decide which tools are needed to perform the tasks detailed in this manual, the following tool lists are offered: *Maintenance and minor repair, Repair/overhaul* and *Special*.

The newcomer to practical mechanics

0-10 Maintenance techniques, tools and working facilities

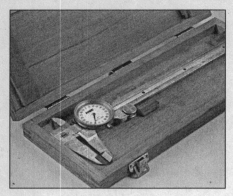
Dial caliper

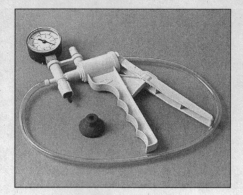
Hand-operated vacuum pump

Timing light

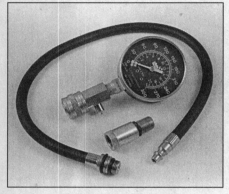
Compression gauge with spark plug hole adapter

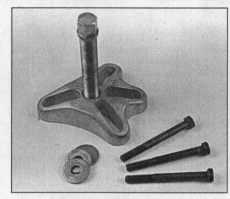
Damper/steering wheel puller

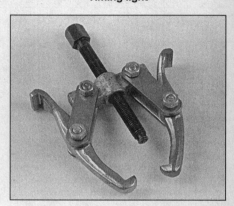
General purpose puller

Hydraulic lifter removal tool

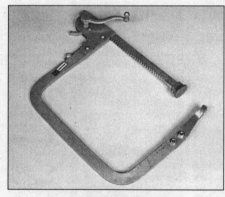
Valve spring compressor

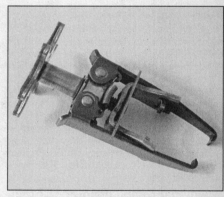
Valve spring compressor

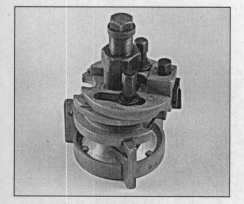
Ridge reamer

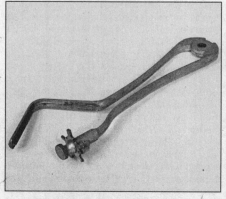
Piston ring groove cleaning tool

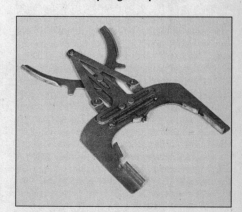
Ring removal/installation tool

Maintenance techniques, tools and working facilities 0-11

Ring compressor

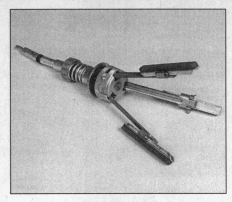
Cylinder hone

Brake hold-down spring tool

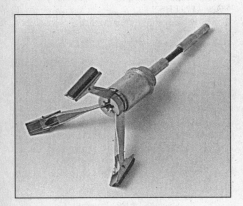
Brake cylinder hone

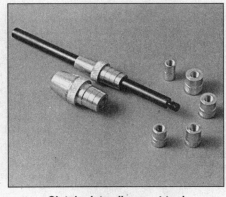
Clutch plate alignment tool

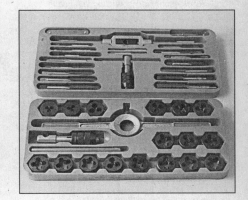
Tap and die set

should start off with the *maintenance and minor repair* tool kit, which is adequate for the simpler jobs performed on a vehicle. Then, as confidence and experience grow, the owner can tackle more difficult tasks, buying additional tools as they are needed. Eventually the basic kit will be expanded into the *repair and overhaul* tool set. Over a period of time, the experienced do-it-yourselfer will assemble a tool set complete enough for most repair and overhaul procedures and will add tools from the special category when it is felt that the expense is justified by the frequency of use.

Maintenance and minor repair tool kit

The tools in this list should be considered the minimum required for performance of routine maintenance, servicing and minor repair work. We recommend the purchase of combination wrenches (box-end and open-end combined in one wrench). While more expensive than open end wrenches, they offer the advantages of both types of wrench.

Combination wrench set (1/4-inch to 1 inch or 6 mm to 19 mm)
Adjustable wrench, 8 inch
Spark plug wrench with rubber insert
Spark plug gap adjusting tool
Feeler gauge set
Brake bleeder wrench
Standard screwdriver (5/16-inch x 6 inch)
Phillips screwdriver (No. 2 x 6 inch)
Combination pliers - 6 inch
Hacksaw and assortment of blades
Tire pressure gauge
Grease gun
Oil can
Fine emery cloth
Wire brush
Battery post and cable cleaning tool
Oil filter wrench
Funnel (medium size)
Safety goggles
Jackstands (2)
Drain pan

Note: *If basic tune-ups are going to be part of routine maintenance, it will be necessary to purchase a good quality stroboscopic timing light and combination tachometer/dwell meter. Although they are included in the list of special tools, it is mentioned here because they are absolutely necessary for tuning most vehicles properly.*

Repair and overhaul tool set

These tools are essential for anyone who plans to perform major repairs and are in addition to those in the maintenance and minor repair tool kit. Included is a comprehensive set of sockets which, though expensive, are invaluable because of their versatility, especially when various extensions and drives are available. We recommend the 1/2-inch drive over the 3/8-inch drive. Although the larger drive is bulky and more expensive, it has the capacity of accepting a very wide range of large sockets. Ideally, however, the mechanic should have a 3/8-inch drive set and a 1/2-inch drive set.

Socket set(s)
Reversible ratchet
Extension - 10 inch
Universal joint
Torque wrench (same size drive as sockets)
Ball peen hammer - 8 ounce
Soft-face hammer (plastic/rubber)
Standard screwdriver (1/4-inch x 6 inch)
Standard screwdriver (stubby - 5/16-inch)
Phillips screwdriver (No. 3 x 8 inch)
Phillips screwdriver (stubby - No. 2)
Pliers - vise grip
Pliers - lineman's
Pliers - needle nose
Pliers - snap-ring (internal and external)
Cold chisel - 1/2-inch
Scribe
Scraper (made from flattened copper tubing)
Centerpunch
Pin punches (1/16, 1/8, 3/16-inch)
Steel rule/straightedge - 12 inch
Allen wrench set (1/8 to 3/8-inch or 4 mm to 10 mm)
A selection of files
Wire brush (large)
Jackstands (second set)
Jack (scissor or hydraulic type)

Maintenance techniques, tools and working facilities

Note: *Another tool which is often useful is an electric drill with a chuck capacity of 3/8-inch and a set of good quality drill bits.*

Special tools

The tools in this list include those which are not used regularly, are expensive to buy, or which need to be used in accordance with their manufacturer's instructions. Unless these tools will be used frequently, it is not very economical to purchase many of them. A consideration would be to split the cost and use between yourself and a friend or friends. In addition, most of these tools can be obtained from a tool rental shop on a temporary basis.

This list primarily contains only those tools and instruments widely available to the public, and not those special tools produced by the vehicle manufacturer for distribution to dealer service departments. Occasionally, references to the manufacturer's special tools are included in the text of this manual. Generally, an alternative method of doing the job without the special tool is offered. However, sometimes there is no alternative to their use. Where this is the case, and the tool cannot be purchased or borrowed, the work should be turned over to the dealer service department or an automotive repair shop.

Valve spring compressor
Piston ring groove cleaning tool
Piston ring compressor
Piston ring installation tool
Cylinder compression gauge
Cylinder ridge reamer
Cylinder surfacing hone
Cylinder bore gauge
Micrometers and/or dial calipers
Hydraulic lifter removal tool
Balljoint separator
Universal-type puller
Impact screwdriver
Dial indicator set
Stroboscopic timing light (inductive pick-up)
Hand operated vacuum/pressure pump
Tachometer/dwell meter
Universal electrical multimeter
Cable hoist
Brake spring removal and installation tools
Floor jack

Buying tools

For the do-it-yourselfer who is just starting to get involved in vehicle maintenance and repair, there are a number of options available when purchasing tools. If maintenance and minor repair is the extent of the work to be done, the purchase of individual tools is satisfactory. If, on the other hand, extensive work is planned, it would be a good idea to purchase a modest tool set from one of the large retail chain stores. A set can usually be bought at a substantial savings over the individual tool prices, and they often come with a tool box. As additional tools are needed, add-on sets, individual tools and a larger tool box can be purchased to expand the tool selection. Building a tool set gradually allows the cost of the tools to be spread over a longer period of time and gives the mechanic the freedom to choose only those tools that will actually be used.

Tool stores will often be the only source of some of the special tools that are needed, but regardless of where tools are bought, try to avoid cheap ones, especially when buying screwdrivers and sockets, because they won't last very long. The expense involved in replacing cheap tools will eventually be greater than the initial cost of quality tools.

Care and maintenance of tools

Good tools are expensive, so it makes sense to treat them with respect. Keep them clean and in usable condition and store them properly when not in use. Always wipe off any dirt, grease or metal chips before putting them away. Never leave tools lying around in the work area. Upon completion of a job, always check closely under the hood for tools that may have been left there so they won't get lost during a test drive.

Some tools, such as screwdrivers, pliers, wrenches and sockets, can be hung on a panel mounted on the garage or workshop wall, while others should be kept in a tool box or tray. Measuring instruments, gauges, meters, etc. must be carefully stored where they cannot be damaged by weather or impact from other tools.

When tools are used with care and stored properly, they will last a very long time. Even with the best of care, though, tools will wear out if used frequently. When a tool is damaged or worn out, replace it. Subsequent jobs will be safer and more enjoyable if you do.

How to repair damaged threads

Sometimes, the internal threads of a nut or bolt hole can become stripped, usually from overtightening. Stripping threads is an all-too-common occurrence, especially when working with aluminum parts, because aluminum is so soft that it easily strips out.

Usually, external or internal threads are only partially stripped. After they've been cleaned up with a tap or die, they'll still work. Sometimes, however, threads are badly damaged. When this happens, you've got three choices:

1) Drill and tap the hole to the next suitable oversize and install a larger diameter bolt, screw or stud.
2) Drill and tap the hole to accept a threaded plug, then drill and tap the plug to the original screw size. You can also buy a plug already threaded to the original size. Then you simply drill a hole to the specified size, then run the threaded plug into the hole with a bolt and jam nut. Once the plug is fully seated, remove the jam nut and bolt.
3) The third method uses a patented thread repair kit like Heli-Coil or Slimsert. These easy-to-use kits are designed to repair damaged threads in straight-through holes and blind holes. Both are available as kits which can handle a variety of sizes and thread patterns. Drill the hole, then tap it with the special included tap. Install the Heli-Coil and the hole is back to its original diameter and thread pitch.

Regardless of which method you use, be sure to proceed calmly and carefully. A little impatience or carelessness during one of these relatively simple procedures can ruin your whole day's work and cost you a bundle if you wreck an expensive part.

Working facilities

Not to be overlooked when discussing tools is the workshop. If anything more than routine maintenance is to be carried out, some sort of suitable work area is essential.

It is understood, and appreciated, that many home mechanics do not have a good workshop or garage available, and end up removing an engine or doing major repairs outside. It is recommended, however, that the overhaul or repair be completed under the cover of a roof.

A clean, flat workbench or table of comfortable working height is an absolute necessity. The workbench should be equipped with a vise that has a jaw opening of at least four inches.

As mentioned previously, some clean, dry storage space is also required for tools, as well as the lubricants, fluids, cleaning solvents, etc. which soon become necessary.

Sometimes waste oil and fluids, drained from the engine or cooling system during normal maintenance or repairs, present a disposal problem. To avoid pouring them on the ground or into a sewage system, pour the used fluids into large containers, seal them with caps and take them to an authorized disposal site or recycling center. Plastic jugs, such as old antifreeze containers, are ideal for this purpose.

Always keep a supply of old newspapers and clean rags available. Old towels are excellent for mopping up spills. Many mechanics use rolls of paper towels for most work because they are readily available and disposable. To help keep the area under the vehicle clean, a large cardboard box can be cut open and flattened to protect the garage or shop floor.

Whenever working over a painted surface, such as when leaning over a fender to service something under the hood, always cover it with an old blanket or bedspread to protect the finish. Vinyl covered pads, made especially for this purpose, are available at auto parts stores.

Booster battery (jump) starting

Observe these precautions when using a booster battery to start a vehicle:
a) Before connecting the booster battery, make sure the ignition switch is in the Off position.
b) Turn off the lights, heater and other electrical loads.
c) Your eyes should be shielded. Safety goggles are a good idea.
d) Make sure the booster battery is the same voltage as the dead one in the vehicle.
e) The two vehicles MUST NOT TOUCH each other!
f) Make sure the transmission is in Neutral (manual) or Park (automatic).
g) If the booster battery is not a maintenance-free type, remove the vent caps and lay a cloth over the vent holes.

Connect the red jumper cable to the positive (+) terminals of each battery.

Connect one end of the black jumper cable to the negative (-) terminal of the booster battery. The other end of this cable should be connected to a good ground on the vehicle to be started, such as a bolt or bracket on the engine block **(see illustration)**. Make sure the cable will not come into contact with the fan, drivebelts or other moving parts of the engine.

Start the engine using the booster battery, then, with the engine running at idle speed, disconnect the jumper cables in the reverse order of connection.

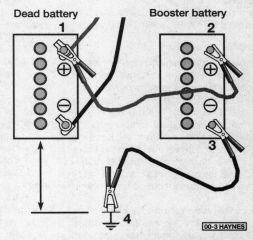

Make the booster battery cable connections in the numerical order shown (note that the negative cable of the booster battery is NOT attached to the negative terminal of the dead battery)

Jacking and towing

Jacking

Warning: *The jack supplied with the vehicle should only be used for raising the vehicle when changing a tire or placing jackstands under the frame. Never work under the vehicle or start the engine while the jack is being used as the only means of support.*

The vehicle must be on a level surface with the wheels blocked and the transaxle in Park. Apply the parking brake if the front of the vehicle must be raised. Make sure no one is in the vehicle as it's being raised with the jack.

Remove the jack and lug nut wrench from the compartment in the right rear corner of the passenger compartment. If needed, lower the spare tire from under the rear of the vehicle by inserting the tapered end of the jack handle into the hole above the rear bumper and turning it counterclockwise with the ratchet handle.

To replace the tire, use the tapered end of the lug wrench to pry loose the wheel cover. **Note:** *If the vehicle is equipped with aluminum wheels, it may be necessary to pry out the special lug nut covers. Also, aluminum wheels normally have anti-theft lug nuts (one per wheel) which require using a special "key" between the lug wrench and lug nut. The key is usually in the glove compartment.* Loosen the lug nuts one-half turn, but leave them in place until the tire is raised off the ground.

Position the jack under the side of the vehicle at the indicated jacking point. There's a front and rear jacking point on each side of the vehicle.

Turn the jack handle clockwise (the lug wrench also serves as the jack handle) until the tire clears the ground. Remove the lug nuts and pull the tire off. Clean the mating surfaces of the hub and wheel, then install the spare. Replace the lug nuts with the bevelled edges facing in and tighten them snugly. Don't attempt to tighten them completely until the vehicle is lowered or it could slip off the jack.

Turn the jack handle counterclockwise to lower the vehicle. Remove the jack and tighten the lug nuts in a criss-cross pattern. If possible, tighten the nuts with a torque wrench (see Chapter 1 for the torque figures). If you don't have access to a torque wrench, have the nuts checked by a service station or repair shop as soon as possible. **Caution:** *The compact spare included with these vehicles is intended for temporary use only. Have the tire repaired and reinstall it on the vehicle at the earliest opportunity and don't exceed 50 mph with the spare tire on the car.*

Install the wheel cover, then stow the tire, jack and wrench and unblock the wheels.

Towing

We recommend these vehicles be towed from the front, with the front wheels off the ground. If it's absolutely necessary, these vehicles can be towed from the rear with the rear wheels off the ground, provided that speeds don't exceed 35 mph and the distance is less than 50 miles; the transaxle can be damaged if these mileage/speed limitations are exceeded. **Caution:** *Do not tow these vehicles with all four wheels on the ground.*

Equipment specifically designed for towing should be used. It must be attached to the main structural members of the vehicle, not the bumpers or brackets.

Safety is a major consideration when towing and all applicable state and local laws must be obeyed. A safety chain must be used at all times.

The parking brake must be released and the transaxle must be in Neutral. The steering must be unlocked (ignition switch in the Off position). Remember that power steering and power brakes won't work with the engine off.

Automotive chemicals and lubricants

A number of automotive chemicals and lubricants are available for use during vehicle maintenance and repair. They include a wide variety of products ranging from cleaning solvents and degreasers to lubricants and protective sprays for rubber, plastic and vinyl.

Cleaners

Carburetor cleaner and choke cleaner is a strong solvent for gum, varnish and carbon. Most carburetor cleaners leave a dry-type lubricant film which will not harden or gum up. Because of this film it is not recommended for use on electrical components.

Brake system cleaner is used to remove grease and brake fluid from the brake system, where clean surfaces are absolutely necessary. It leaves no residue and often eliminates brake squeal caused by contaminants.

Electrical cleaner removes oxidation, corrosion and carbon deposits from electrical contacts, restoring full current flow. It can also be used to clean spark plugs, carburetor jets, voltage regulators and other parts where an oil-free surface is desired.

Demoisturants remove water and moisture from electrical components such as alternators, voltage regulators, electrical connectors and fuse blocks. They are non-conductive, non-corrosive and non-flammable.

Degreasers are heavy-duty solvents used to remove grease from the outside of the engine and from chassis components. They can be sprayed or brushed on and, depending on the type, are rinsed off either with water or solvent.

Lubricants

Motor oil is the lubricant formulated for use in engines. It normally contains a wide variety of additives to prevent corrosion and reduce foaming and wear. Motor oil comes in various weights (viscosity ratings) from 0 to 50. The recommended weight of the oil depends on the season, temperature and the demands on the engine. Light oil is used in cold climates and under light load conditions. Heavy oil is used in hot climates and where high loads are encountered. Multi-viscosity oils are designed to have characteristics of both light and heavy oils and are available in a number of weights from 5W-20 to 20W-50.

Gear oil is designed to be used in differentials, manual transmissions and other areas where high-temperature lubrication is required.

Chassis and wheel bearing grease is a heavy grease used where increased loads and friction are encountered, such as for wheel bearings, balljoints, tie-rod ends and universal joints.

High-temperature wheel bearing grease is designed to withstand the extreme temperatures encountered by wheel bearings in disc brake equipped vehicles. It usually contains molybdenum disulfide (moly), which is a dry-type lubricant.

White grease is a heavy grease for metal-to-metal applications where water is a problem. White grease stays soft under both low and high temperatures (usually from -100 to +190-degrees F), and will not wash off or dilute in the presence of water.

Assembly lube is a special extreme pressure lubricant, usually containing moly, used to lubricate high-load parts (such as main and rod bearings and cam lobes) for initial start-up of a new engine. The assembly lube lubricates the parts without being squeezed out or washed away until the engine oiling system begins to function.

Silicone lubricants are used to protect rubber, plastic, vinyl and nylon parts.

Graphite lubricants are used where oils cannot be used due to contamination problems, such as in locks. The dry graphite will lubricate metal parts while remaining uncontaminated by dirt, water, oil or acids. It is electrically conductive and will not foul electrical contacts in locks such as the ignition switch.

Moly penetrants loosen and lubricate frozen, rusted and corroded fasteners and prevent future rusting or freezing.

Heat-sink grease is a special electrically non-conductive grease that is used for mounting electronic ignition modules where it is essential that heat is transferred away from the module.

Sealants

RTV sealant is one of the most widely used gasket compounds. Made from silicone, RTV is air curing, it seals, bonds, waterproofs, fills surface irregularities, remains flexible, doesn't shrink, is relatively easy to remove, and is used as a supplementary sealer with almost all low and medium temperature gaskets.

Anaerobic sealant is much like RTV in that it can be used either to seal gaskets or to form gaskets by itself. It remains flexible, is solvent resistant and fills surface imperfections. The difference between an anaerobic sealant and an RTV-type sealant is in the curing. RTV cures when exposed to air, while an anaerobic sealant cures only in the absence of air. This means that an anaerobic sealant cures only after the assembly of parts, sealing them together.

Thread and pipe sealant is used for sealing hydraulic and pneumatic fittings and vacuum lines. It is usually made from a Teflon compound, and comes in a spray, a paint-on liquid and as a wrap-around tape.

Chemicals

Anti-seize compound prevents seizing, galling, cold welding, rust and corrosion in fasteners. High-temperature anti-seize, usually made with copper and graphite lubricants, is used for exhaust system and exhaust manifold bolts.

Anaerobic locking compounds are used to keep fasteners from vibrating or working loose and cure only after installation, in the absence of air. Medium strength locking compound is used for small nuts, bolts and screws that may be removed later. High-strength locking compound is for large nuts, bolts and studs which aren't removed on a regular basis.

Oil additives range from viscosity index improvers to chemical treatments that claim to reduce internal engine friction. It should be noted that most oil manufacturers caution against using additives with their oils.

Gas additives perform several functions, depending on their chemical makeup. They usually contain solvents that help dissolve gum and varnish that build up on carburetor, fuel injection and intake parts. They also serve to break down carbon deposits that form on the inside surfaces of the combustion chambers. Some additives contain upper cylinder lubricants for valves and piston rings, and others contain chemicals to remove condensation from the gas tank.

Miscellaneous

Brake fluid is specially formulated hydraulic fluid that can withstand the heat and pressure encountered in brake systems. Care must be taken so this fluid does not come in contact with painted surfaces or plastics. An opened container should always be resealed to prevent contamination by water or dirt.

Weatherstrip adhesive is used to bond weatherstripping around doors, windows and trunk lids. It is sometimes used to attach trim pieces.

Undercoating is a petroleum-based, tar-like substance that is designed to protect metal surfaces on the underside of the vehicle from corrosion. It also acts as a sound-deadening agent by insulating the bottom of the vehicle.

Waxes and polishes are used to help protect painted and plated surfaces from the weather. Different types of paint may require the use of different types of wax and polish. Some polishes utilize a chemical or abrasive cleaner to help remove the top layer of oxidized (dull) paint on older vehicles. In recent years many non-wax polishes that contain a wide variety of chemicals such as polymers and silicones have been introduced. These non-wax polishes are usually easier to apply and last longer than conventional waxes and polishes.

Conversion factors

Length (distance)
Inches (in)	X 25.4	= Millimetres (mm)	X 0.0394	= Inches (in)
Feet (ft)	X 0.305	= Metres (m)	X 3.281	= Feet (ft)
Miles	X 1.609	= Kilometres (km)	X 0.621	= Miles

Volume (capacity)
Cubic inches (cu in; in^3)	X 16.387	= Cubic centimetres (cc; cm^3)	X 0.061	= Cubic inches (cu in; in^3)
Imperial pints (Imp pt)	X 0.568	= Litres (l)	X 1.76	= Imperial pints (Imp pt)
Imperial quarts (Imp qt)	X 1.137	= Litres (l)	X 0.88	= Imperial quarts (Imp qt)
Imperial quarts (Imp qt)	X 1.201	= US quarts (US qt)	X 0.833	= Imperial quarts (Imp qt)
US quarts (US qt)	X 0.946	= Litres (l)	X 1.057	= US quarts (US qt)
Imperial gallons (Imp gal)	X 4.546	= Litres (l)	X 0.22	= Imperial gallons (Imp gal)
Imperial gallons (Imp gal)	X 1.201	= US gallons (US gal)	X 0.833	= Imperial gallons (Imp gal)
US gallons (US gal)	X 3.785	= Litres (l)	X 0.264	= US gallons (US gal)

Mass (weight)
Ounces (oz)	X 28.35	= Grams (g)	X 0.035	= Ounces (oz)
Pounds (lb)	X 0.454	= Kilograms (kg)	X 2.205	= Pounds (lb)

Force
Ounces-force (ozf; oz)	X 0.278	= Newtons (N)	X 3.6	= Ounces-force (ozf; oz)
Pounds-force (lbf; lb)	X 4.448	= Newtons (N)	X 0.225	= Pounds-force (lbf; lb)
Newtons (N)	X 0.1	= Kilograms-force (kgf; kg)	X 9.81	= Newtons (N)

Pressure
Pounds-force per square inch (psi; lbf/in^2; lb/in^2)	X 0.070	= Kilograms-force per square centimetre (kgf/cm^2; kg/cm^2)	X 14.223	= Pounds-force per square inch (psi; lbf/in^2; lb/in^2)
Pounds-force per square inch (psi; lbf/in^2; lb/in^2)	X 0.068	= Atmospheres (atm)	X 14.696	= Pounds-force per square inch (psi; lbf/in^2; lb/in^2)
Pounds-force per square inch (psi; lbf/in^2; lb/in^2)	X 0.069	= Bars	X 14.5	= Pounds-force per square inch (psi; lbf/in^2; lb/in^2)
Pounds-force per square inch (psi; lbf/in^2; lb/in^2)	X 6.895	= Kilopascals (kPa)	X 0.145	= Pounds-force per square inch (psi; lbf/in^2; lb/in^2)
Kilopascals (kPa)	X 0.01	= Kilograms-force per square centimetre (kgf/cm^2; kg/cm^2)	X 98.1	= Kilopascals (kPa)

Torque (moment of force)
Pounds-force inches (lbf in; lb in)	X 1.152	= Kilograms-force centimetre (kgf cm; kg cm)	X 0.868	= Pounds-force inches (lbf in; lb in)
Pounds-force inches (lbf in; lb in)	X 0.113	= Newton metres (Nm)	X 8.85	= Pounds-force inches (lbf in; lb in)
Pounds-force inches (lbf in; lb in)	X 0.083	= Pounds-force feet (lbf ft; lb ft)	X 12	= Pounds-force inches (lbf in; lb in)
Pounds-force feet (lbf ft; lb ft)	X 0.138	= Kilograms-force metres (kgf m; kg m)	X 7.233	= Pounds-force feet (lbf ft; lb ft)
Pounds-force feet (lbf ft; lb ft)	X 1.356	= Newton metres (Nm)	X 0.738	= Pounds-force feet (lbf ft; lb ft)
Newton metres (Nm)	X 0.102	= Kilograms-force metres (kgf m; kg m)	X 9.804	= Newton metres (Nm)

Vacuum
Inches mercury (in. Hg)	X 3.377	= Kilopascals (kPa)	X 0.2961	= Inches mercury
Inches mercury (in. Hg)	X 25.4	= Millimeters mercury (mm Hg)	X 0.0394	= Inches mercury

Power
Horsepower (hp)	X 745.7	= Watts (W)	X 0.0013	= Horsepower (hp)

Velocity (speed)
Miles per hour (miles/hr; mph)	X 1.609	= Kilometres per hour (km/hr; kph)	X 0.621	= Miles per hour (miles/hr; mph)

*Fuel consumption**
Miles per gallon, Imperial (mpg)	X 0.354	= Kilometres per litre (km/l)	X 2.825	= Miles per gallon, Imperial (mpg)
Miles per gallon, US (mpg)	X 0.425	= Kilometres per litre (km/l)	X 2.352	= Miles per gallon, US (mpg)

Temperature
Degrees Fahrenheit = (°C x 1.8) + 32   Degrees Celsius (Degrees Centigrade; °C) = (°F - 32) x 0.56

It is common practice to convert from miles per gallon (mpg) to litres/100 kilometres (l/100km), where mpg (Imperial) x l/100 km = 282 and mpg (US) x l/100 km = 235

Safety first!

Regardless of how enthusiastic you may be about getting on with the job at hand, take the time to ensure that your safety is not jeopardized. A moment's lack of attention can result in an accident, as can failure to observe certain simple safety precautions. The possibility of an accident will always exist, and the following points should not be considered a comprehensive list of all dangers. Rather, they are intended to make you aware of the risks and to encourage a safety conscious approach to all work you carry out on your vehicle.

Essential DOs and DON'Ts

DON'T rely on a jack when working under the vehicle. Always use approved jackstands to support the weight of the vehicle and place them under the recommended lift or support points.

DON'T attempt to loosen extremely tight fasteners (i.e. wheel lug nuts) while the vehicle is on a jack - it may fall.

DON'T start the engine without first making sure that the transmission is in Neutral (or Park where applicable) and the parking brake is set.

DON'T remove the radiator cap from a hot cooling system - let it cool or cover it with a cloth and release the pressure gradually.

DON'T attempt to drain the engine oil until you are sure it has cooled to the point that it will not burn you.

DON'T touch any part of the engine or exhaust system until it has cooled sufficiently to avoid burns.

DON'T siphon toxic liquids such as gasoline, antifreeze and brake fluid by mouth, or allow them to remain on your skin.

DON'T inhale brake lining dust - it is potentially hazardous (see *Asbestos* below).

DON'T allow spilled oil or grease to remain on the floor - wipe it up before someone slips on it.

DON'T use loose fitting wrenches or other tools which may slip and cause injury.

DON'T push on wrenches when loosening or tightening nuts or bolts. Always try to pull the wrench toward you. If the situation calls for pushing the wrench away, push with an open hand to avoid scraped knuckles if the wrench should slip.

DON'T attempt to lift a heavy component alone - get someone to help you.

DON'T rush or take unsafe shortcuts to finish a job.

DON'T allow children or animals in or around the vehicle while you are working on it.

DO wear eye protection when using power tools such as a drill, sander, bench grinder, etc. and when working under a vehicle.

DO keep loose clothing and long hair well out of the way of moving parts.

DO make sure that any hoist used has a safe working load rating adequate for the job.

DO get someone to check on you periodically when working alone on a vehicle.

DO carry out work in a logical sequence and make sure that everything is correctly assembled and tightened.

DO keep chemicals and fluids tightly capped and out of the reach of children and pets.

DO remember that your vehicle's safety affects that of yourself and others. If in doubt on any point, get professional advice.

Asbestos

Certain friction, insulating, sealing, and other products - such as brake linings, brake bands, clutch linings, torque converters, gaskets, etc. - may contain asbestos. Extreme care must be taken to avoid inhalation of dust from such products, since it is hazardous to health. If in doubt, assume that they do contain asbestos.

Fire

Remember at all times that gasoline is highly flammable. Never smoke or have any kind of open flame around when working on a vehicle. But the risk does not end there. A spark caused by an electrical short circuit, by two metal surfaces contacting each other, or even by static electricity built up in your body under certain conditions, can ignite gasoline vapors, which in a confined space are highly explosive. Do not, under any circumstances, use gasoline for cleaning parts. Use an approved safety solvent.

Always disconnect the battery ground (-) cable at the battery before working on any part of the fuel system or electrical system. Never risk spilling fuel on a hot engine or exhaust component. It is strongly recommended that a fire extinguisher suitable for use on fuel and electrical fires be kept handy in the garage or workshop at all times. Never try to extinguish a fuel or electrical fire with water.

Fumes

Certain fumes are highly toxic and can quickly cause unconsciousness and even death if inhaled to any extent. Gasoline vapor falls into this category, as do the vapors from some cleaning solvents. Any draining or pouring of such volatile fluids should be done in a well ventilated area.

When using cleaning fluids and solvents, read the instructions on the container carefully. Never use materials from unmarked containers.

Never run the engine in an enclosed space, such as a garage. Exhaust fumes contain carbon monoxide, which is extremely poisonous. If you need to run the engine, always do so in the open air, or at least have the rear of the vehicle outside the work area.

If you are fortunate enough to have the use of an inspection pit, never drain or pour gasoline and never run the engine while the vehicle is over the pit. The fumes, being heavier than air, will concentrate in the pit with possibly lethal results.

The battery

Never create a spark or allow a bare light bulb near a battery. They normally give off a certain amount of hydrogen gas, which is highly explosive.

Always disconnect the battery ground (-) cable at the battery before working on the fuel or electrical systems.

If possible, loosen the filler caps or cover when charging the battery from an external source (this does not apply to sealed or maintenance-free batteries). Do not charge at an excessive rate or the battery may burst.

Take care when adding water to a non maintenance-free battery and when carrying a battery. The electrolyte, even when diluted, is very corrosive and should not be allowed to contact clothing or skin.

Always wear eye protection when cleaning the battery to prevent the caustic deposits from entering your eyes.

Household current

When using an electric power tool, inspection light, etc., which operates on household current, always make sure that the tool is correctly connected to its plug and that, where necessary, it is properly grounded. Do not use such items in damp conditions and, again, do not create a spark or apply excessive heat in the vicinity of fuel or fuel vapor.

Secondary ignition system voltage

A severe electric shock can result from touching certain parts of the ignition system (such as the spark plug wires) when the engine is running or being cranked, particularly if components are damp or the insulation is defective. In the case of an electronic ignition system, the secondary system voltage is much higher and could prove fatal.

Troubleshooting

Contents

Symptom	Section
Engine and performance	
Engine backfires	15
Engine diesels (continues to run) after switching off	18
Engine hard to start when cold	3
Engine hard to start when hot	4
Engine lacks power	14
Engine lopes while idling or idles erratically	8
Engine misses at idle speed	9
Engine misses throughout driving speed range	10
Engine rotates but will not start	2
Engine runs with oil pressure light on	17
Engine stalls	13
Engine starts but stops immediately	6
Engine stumbles on acceleration	11
Engine surges while holding accelerator steady	12
Engine will not rotate when attempting to start	1
Oil puddle under engine	7
Pinging or knocking engine sounds during acceleration or uphill	16
Starter motor noisy or excessively rough in engagement	5
Engine electrical system	
Battery will not hold a charge	19
Voltage warning light fails to come on when key is turned on	21
Voltage warning light fails to go out	20
Fuel system	
Excessive fuel consumption	22
Fuel leakage and/or fuel odor	23
Cooling system	
Coolant loss	28
External coolant leakage	26
Internal coolant leakage	27
Overcooling	25
Overheating	24
Poor coolant circulation	29
Automatic transaxle	
Engine will start in gears other than Park or Neutral	34
Fluid leakage	30
General shift mechanism problems	32
Transaxle fluid brown or has a burned smell	31
Transaxle slips, shifts roughly, is noisy or has no drive in forward or reverse gears	35
Transaxle will not downshift with accelerator pedal pressed to the floor	33
Driveaxles	
Clicking noise in turns	36
Knock or clunk when accelerating after coasting	37
Shudder or vibration during acceleration	38
Brakes	
Brake pedal feels spongy when depressed	46
Brake pedal travels to the floor with little resistance	47
Brake roughness or chatter (pedal pulsates)	41
Dragging brakes	44
Excessive brake pedal travel	43
Excessive pedal effort required to stop vehicle	42
Grabbing or uneven braking action	45
Noise (high-pitched squeal when the brakes are applied)	40
Parking brake does not hold	48
Vehicle pulls to one side during braking	39
Suspension and steering systems	
Abnormal or excessive tire wear	50
Abnormal noise at the front end	55
Cupped tires	60
Erratic steering when braking	57
Excessive pitching and/or rolling around corners or during braking	58
Excessive play or looseness in steering system	64
Excessive tire wear on inside edge	62
Excessive tire wear on outside edge	61
Hard steering	53
Steering wheel does not return to center position correctly	54
Rattling or clicking noise in rack and pinion	65
Shimmy, shake or vibration	52
Suspension bottoms	59
Tire tread worn in one place	63
Vehicle pulls to one side	49
Wander or poor steering stability	56
Wheel makes a "thumping" noise	51

Troubleshooting

This section provides an easy reference guide to the more common problems which may occur during the operation of your vehicle. Various symptoms and their possible causes are grouped under headings denoting components or systems, such as Engine, Cooling system, etc. They also refer to the Chapter and/or Section that deals with the problem.

Remember that successful troubleshooting isn't a mysterious "black art" practiced only by professional mechanics. It's simply the result of knowledge combined with an intelligent, systematic approach to a problem. Always use a process of elimination, starting with the simplest solution and working through to the most complex - and never overlook the obvious. Anyone can run the gas tank dry or leave the lights on overnight, so don't assume that you're exempt from such oversights.

Finally, always establish a clear idea why a problem has occurred and take steps to ensure that it doesn't happen again. If the electrical system fails because of a poor connection, check all other connections in the system to make sure they don't fail as well. If a particular fuse continues to blow, find out why - don't just go on replacing fuses. Remember, failure of a small component can often be indicative of potential failure or incorrect functioning of a more important component or system.

Engine and performance

1 Engine will not rotate when attempting to start

1 Battery terminal connections loose or corroded (Chapter 1).
2 Battery discharged or faulty (Chapter 1).
3 Automatic transaxle not completely engaged in Park (Chapter 7). Also, the neutral start switch could be faulty (Chapter 7).
4 Broken, loose or disconnected wiring in the starting circuit (Chapters 5 and 12).
5 Starter motor pinion jammed in flywheel ring gear (Chapter 5).
6 Starter solenoid faulty (Chapter 5).
7 Starter motor faulty (Chapter 5).
8 Ignition switch faulty (Chapter 12).
9 Starter pinion or flywheel teeth worn or broken (Chapter 5).

2 Engine rotates but will not start

1 Fuel tank empty.
2 Battery discharged (engine rotates slowly) (Chapter 5).
3 Battery terminal connections loose or corroded (Chapter 1).
4 Leaking fuel injector(s), fuel pump, pressure regulator, etc. (Chapter 4).
5 Fuel not reaching fuel injection system (Chapter 4).
6 Ignition components damp or damaged (Chapter 5).
7 Worn, faulty or incorrectly gapped spark plugs (Chapter 1).
8 Broken, loose or disconnected wiring in the starting circuit (Chapter 5).
9 Broken, loose or disconnected wires at the ignition coil(s) or faulty coil(s) (Chapter 5).
10 Distributor loose, causing ignition timing to change (3.1L models only). Turn distributor as necessary to start engine, then set the ignition timing as soon as possible (Chapter 1).

3 Engine hard to start when cold

1 Battery discharged or low (Chapter 1).
2 Fuel system malfunctioning (Chapter 4).
3 Injector(s) leaking (Chapter 4).

4 Engine hard to start when hot

1 Air filter clogged (Chapter 1).
2 Fuel not reaching the fuel injection system (Chapter 4).
3 Corroded battery connections, especially ground (Chapter 1).

5 Starter motor noisy or excessively rough in engagement

1 Pinion or flywheel gear teeth worn or broken (Chapter 5).
2 Starter motor mounting bolts loose or missing (Chapter 5).

6 Engine starts but stops immediately

1 Loose or faulty electrical connections at distributor, coil, coil pack or alternator (Chapter 5).
2 Insufficient fuel reaching the fuel injectors (Chapter 4).
3 Vacuum leak at the gasket between the intake manifold/plenum and throttle body (Chapters 1 and 4).

7 Oil puddle under engine

1 Oil pan gasket and/or oil pan drain bolt seal leaking (Chapters 1 and 2).
2 Oil pressure sending unit leaking (Chapter 2).
3 Valve cover gaskets leaking (Chapter 2).
4 Engine oil seals leaking (Chapter 2).
5 Timing cover sealant or sealing flange leaking (Chapter 2).

8 Engine lopes while idling or idles erratically

1 Vacuum leakage (Chapter 4).
2 Leaking EGR valve or plugged PCV valve (Chapters 1 and 6).
3 Air filter clogged (Chapter 1).
4 Fuel pump not delivering sufficient fuel to the fuel injection system (Chapter 4).
5 Leaking head gasket (Chapter 2).
6 Timing chain worn (Chapter 2).
7 Camshaft lobes worn (Chapter 2).

9 Engine misses at idle speed

1 Spark plugs worn or not gapped properly (Chapter 1).
2 Faulty spark plug wires (Chapter 1).
3 Vacuum leaks (Chapters 1 and 4).
4 Ignition system malfunctioning (Chapter 5).
5 Uneven or low compression (Chapter 2).
6 Ignition timing incorrect (3.1L engine only) (Chapter 1).

10 Engine misses throughout driving speed range

1 Fuel filter clogged and/or impurities in the fuel system (Chapters 1 and 4).
2 Low fuel output at the injector (Chapter 4).
3 Faulty or incorrectly gapped spark plugs (Chapter 1).
4 Leaking spark plug wires (Chapter 1).
5 Other fault in the ignition system (Chapter 5).
6 Faulty emission system components (Chapter 6).
7 Low or uneven cylinder compression pressures (Chapter 2).
8 Weak or faulty ignition system (Chapter 5).
9 Vacuum leak in fuel injection system, intake manifold or vacuum hoses (Chapter 4).
10 Ignition timing incorrect (3.1L engine only) (Chapter 1).

11 Engine stumbles on acceleration

1 Spark plugs fouled (Chapter 1).
2 Fuel injection system needs adjustment or repair (Chapter 4).
3 Fuel filter clogged (Chapter 1).
4 Faulty spark plug wires (Chapter 1).
5 Intake manifold air leak (Chapter 4).

12 Engine surges while holding accelerator steady

1 Intake air leak (Chapter 4).
2 Fuel pump faulty (Chapter 4).

Troubleshooting 0-19

3 Loose fuel injector harness connections (Chapter 4).
4 Defective ECM (Chapter 6).

13 Engine stalls

1 Fuel filter clogged and/or water and impurities in the fuel system (Chapters 1 and 4).
2 Ignition components damp or damaged (Chapter 5).
3 Faulty emissions system components (Chapter 6).
4 Faulty or incorrectly gapped spark plugs (Chapter 1).
5 Faulty spark plug wires (Chapter 1).
6 Vacuum leak in the fuel injection system, intake manifold or vacuum hoses (Chapter 4).

14 Engine lacks power

1 Faulty or incorrectly gapped spark plugs (Chapter 1).
2 Fuel injection system out of adjustment or malfunctioning (Chapter 4).
3 Faulty coil(s) (Chapter 5).
4 Brakes binding (Chapter 1).
5 Automatic transaxle fluid level incorrect (Chapter 1).
6 Ignition timing incorrect (3.1L engine only) (Chapter 1).
7 Fuel filter clogged and/or impurities in the fuel system (Chapter 1).
8 Emission control system not functioning properly (Chapter 6).
9 Low or uneven cylinder compression pressures (Chapter 2).

15 Engine backfires

1 Emissions system not functioning properly (Chapter 6).
2 Spark plug wires not routed correctly or faulty wires or, if equipped, distributor cap or rotor (Chapter 1).
3 Faulty secondary ignition system (Chapter 5).
4 Fuel injection system in need of adjustment or worn excessively (Chapter 4).
5 Vacuum leak at fuel injectors, intake manifold or vacuum hoses (Chapter 4).
6 Valves sticking (Chapter 2).
7 Ignition timing incorrect (3.1L engine only) (Chapter 1).

16 Pinging or knocking engine sounds during acceleration or uphill

1 Incorrect grade of fuel.
2 Excessive carbon build-up in combustion chambers.

3 Fuel injection system in need of repair (Chapter 4).
4 Improper or damaged spark plugs or wires (Chapter 1).
5 Worn or damaged ignition components (Chapter 5).
6 Faulty emissions system (Chapter 6).
7 Vacuum leak (Chapter 4).
8 Ignition timing incorrect (3.1L engine only) (Chapter 1).

17 Engine runs with oil pressure light on

1 Low oil level (Chapter 1).
2 Short in wiring circuit (Chapter 12).
3 Faulty oil pressure sender (Chapter 2).
4 Worn engine bearings and/or oil pump (Chapter 2).

18 Engine diesels (continues to run) after switching off

1 Excessive engine operating temperature (Chapter 3).
2 Excessive carbon build-up in combustion chambers (Chapter 2).
3 Fuel injection system malfunctioning (Chapter 4).
4 Ignition system malfunctioning (Chapter 5).
5 Ignition timing incorrect (3.1L engine only) (Chapter 1).

Engine electrical system

19 Battery will not hold a charge

1 Alternator drivebelt defective or not adjusted properly (Chapter 1).
2 Battery terminals loose or corroded (Chapter 1).
3 Alternator not charging properly (Chapter 5).
4 Loose, broken or faulty wiring in the charging circuit (Chapter 5).
5 Short in vehicle wiring (Chapters 5 and 12).
6 Internally defective battery (Chapters 1 and 5).

20 Voltage warning light fails to go out

1 Faulty alternator or charging circuit (Chapter 5).
2 Alternator drivebelt defective or out of adjustment (Chapter 1).
3 Alternator voltage regulator inoperative (Chapter 5).

21 Voltage warning light fails to come on when key is turned on

1 Warning light bulb defective (Chapter 12).
2 Fault in the printed circuit, dash wiring or bulb holder (Chapter 12).

Fuel system

22 Excessive fuel consumption

1 Dirty or clogged air filter element (Chapter 1).
2 Fuel leak (usually accompanied by a fuel odor) (see the next Section).
3 Emissions system not functioning properly (Chapter 6).
4 Fuel injection internal parts excessively worn or damaged (Chapter 4).
5 Low tire pressure or incorrect tire size (Chapter 1).
6 Automatic transaxle (Chapter 7) slipping (normally you will also hear excessive engine revving during acceleration).

23 Fuel leakage and/or fuel odor

1 Leak in a fuel feed or vent line (Chapter 4).
2 Tank overfilled.
3 Evaporative canister filter clogged (Chapters 1 and 6).
4 Fuel injector internal parts excessively worn (Chapter 4).

Cooling system

24 Overheating

1 Insufficient coolant in system (Chapter 1).
2 Water pump drivebelt defective or out of adjustment (Chapter 1).
3 Radiator core blocked or grille restricted (Chapter 3).
4 Thermostat faulty (Chapter 3).
5 Electric cooling fan blades broken or cracked (Chapter 3).
6 Radiator cap not maintaining proper pressure (Chapter 3).

25 Overcooling

Faulty thermostat (Chapter 3).

26 External coolant leakage

1 Deteriorated/damaged hoses or loose

clamps (Chapters 1 and 3).
2   Water pump seal defective (Chapters 1 and 3).
3   Leakage from radiator core or header tank (Chapter 3).
4   Engine drain or water jacket core plugs leaking (Chapter 2).

27   Internal coolant leakage

1   Leaking cylinder head gasket (Chapter 2).
2   Cracked cylinder bore or cylinder head (Chapter 2).

28   Coolant loss

1   Too much coolant in system (Chapter 1).
2   Coolant boiling away because of overheating (Chapter 3).
3   Internal or external leakage (Chapter 3).
4   Faulty radiator cap (Chapter 3).

29   Poor coolant circulation

1   Inoperative water pump (Chapter 3).
2   Restriction in cooling system (Chapters 1 and 3).
3   Water pump drivebelt defective or out of adjustment (Chapter 1).
4   Thermostat sticking (Chapter 3).

Automatic transaxle

Note: *Due to the complexity of the automatic transaxle, it's difficult for the home mechanic to properly diagnose and service this component. For problems other than the following, the vehicle should be taken to a dealer service department or a transmission shop.*

30   Fluid leakage

1   Automatic transmission fluid is a deep red color. Fluid leaks should not be confused with engine oil, which can easily be blown by air flow to the transaxle.
2   To pinpoint a leak, first remove all built-up dirt and grime from the transaxle housing with degreasing agents and/or steam cleaning. Drive the vehicle at low speeds so air flow will not blow the leak far from its source. Raise the vehicle and determine where the leak is coming from. Common areas of leakage are:
   a) Pan (Chapters 1 and 7)
   b) Filler pipe (Chapter 7)
   c) Transaxle oil lines (Chapter 7)
   d) Speedometer gear or sensor (Chapter 7)
   e) Modulator

31   Transaxle fluid brown or has a burned smell

Transaxle overheated. Change fluid (Chapter 1).

32   General shift mechanism problems

1   Chapter 7 Part B deals with checking and adjusting the shift linkage on automatic transaxles. Common problems which may be attributed to poorly adjusted linkage are:
   a) *Engine starting in gears other than Park or Neutral.*
   b) *Indicator on shifter pointing to a gear other than the one actually being used.*
   c) *Vehicle moves when in Park.*
2   Refer to Chapter 7 Part B for the shift linkage adjustment procedure.

33   Transaxle will not downshift with accelerator pedal pressed to the floor

Throttle valve (TV) cable out of adjustment (Chapter 7).

34   Engine will start in gears other than Park or Neutral

Starter safety switch malfunctioning (Chapter 7).

35   Transaxle slips, shifts roughly, is noisy or has no drive in forward or reverse gears

There are many probable causes for the above problems, but the home mechanic should be concerned with only one possibility - fluid level. Before taking the vehicle to a repair shop, check the level and condition of the fluid as described in Chapter 1. Correct the fluid level as necessary or change the fluid and filter if needed. If the problem persists, have a professional diagnose the probable cause.

Driveaxles

36   Clicking noise in turns

Worn or damaged outer CV joint. Check for cut or damaged boots (Chapter 1). Repair as necessary (Chapter 8).

37   Knock or clunk when accelerating after coasting

Worn or damaged outer CV joint. Check for cut or damaged boots (Chapter 1). Repair as necessary (Chapter 8).

38   Shudder or vibration during acceleration

1   Excessive inner CV joint angle. Check and correct as necessary (Chapter 8).
2   Worn or damaged CV joints. Repair or replace as necessary (Chapter 8).
3   Sticking inboard joint assembly. Correct or replace as necessary (Chapter 8).

Brakes

Note: *Before assuming that a brake problem exists, make sure . . .*
   a) *The tires are in good condition and properly inflated (Chapter 1).*
   b) *The front end alignment is correct (Chapter 10).*
   c) *The vehicle isn't loaded with weight in an unequal manner.*

39   Vehicle pulls to one side during braking

1   Incorrect tire pressures (Chapter 1).
2   Front end out of line (have the front end aligned).
3   Unmatched tires on same axle.
4   Restricted brake lines or hoses (Chapter 9).
5   Malfunctioning brake assembly (Chapter 9).
6   Loose suspension parts (Chapter 10).
7   Loose brake calipers (Chapter 9).

40   Noise (high-pitched squeal when the brakes are applied)

Front disc brake pads worn out. The noise comes from the wear sensor rubbing against the disc. Replace pads with new ones immediately (Chapter 9).

41   Brake roughness or chatter (pedal pulsates)

1   Excessive front brake disc lateral runout (Chapter 9).
2   Disc or drum defective. Remove (Chapter 9) and check for excessive lateral runout or parallelism. Have the disc or drum resurfaces or replace it with a new one.
3   Uneven pad wear caused by caliper not sliding due to improper clearance or dirt (Chapter 9).

42 Excessive pedal effort required to stop vehicle

1 Malfunctioning power brake booster (Chapter 9).
2 Partial system failure (Chapter 9).
3 Excessively worn pads (Chapter 9).
4 One or more caliper pistons or wheel cylinders seized or sticking (Chapter 9).
5 Brake linings or pads contaminated with oil or grease (Chapter 9).
6 New pads or shoes installed and not yet seated. It will take a while for the new material to seat.

43 Excessive brake pedal travel

1 Partial brake system failure (Chapter 9).
2 Insufficient fluid in master cylinder (Chapters 1 and 9).
3 Air trapped in system (Chapters 1 and 9).
4 Rear brakes not adjusting properly. Make a series of starts and stops while the vehicle is in Reverse. If this does not correct the situation, remove the drums and inspect the self-adjusters (Chapter 9).

44 Dragging brakes

1 Master cylinder pistons not returning correctly (Chapter 9).
2 Restricted brakes lines or hoses (Chapters 1 and 9).
3 Incorrect parking brake adjustment (Chapter 9).

45 Grabbing or uneven braking action

1 Malfunction of proportioner valves (Chapter 9).
2 Malfunction of power brake booster unit (Chapter 9).
3 Malfunction of the Anti-lock Brake System (ABS) (Chapter 9).
4 Binding brake pedal mechanism (Chapter 9).

46 Brake pedal feels spongy when depressed

1 Air in hydraulic lines. Bleed brake system (Chapter 9).
2 Master cylinder mounting bolts loose (Chapter 9).
3 Master cylinder defective (Chapter 9).

47 Brake pedal travels to the floor with little resistance

Little or no fluid in the master cylinder reservoir caused by leaking caliper or wheel cylinder pistons, loose, damaged or disconnected brake lines (Chapter 9).

48 Parking brake does not hold

Check the parking brake. Adjust if necessary (Chapter 9).

Suspension and steering systems

Note: *Before attempting to diagnose the suspension and steering systems, perform the following preliminary checks:*
a) *Check the tire pressures and look for uneven wear.*
b) *Check the steering universal joints or coupling from the column to the steering gear for loose fasteners and wear.*
c) *Check the front and rear suspension and the steering gear assembly for loose and damaged parts.*
d) *Look for out-of-round or out-of-balance tires, bent rims and loose and/or rough wheel bearings.*

49 Vehicle pulls to one side

1 Mismatched or uneven tires (Chapter 10).
2 Broken or sagging springs (Chapter 10).
3 Front wheel alignment incorrect (Chapter 10).
4 Front brakes dragging (Chapter 9).

50 Abnormal or excessive tire wear

1 Front wheel alignment incorrect (Chapter 10).
2 Sagging or broken springs (Chapter 10).
3 Tire out-of-balance (Chapter 10).
4 Worn shock absorber (Chapter 10).
5 Overloaded vehicle.
6 Tires not rotated regularly.

51 Wheel makes a "thumping" noise

1 Blister or bump on tire (Chapter 1).
2 Improper shock absorber action (Chapter 10).

52 Shimmy, shake or vibration

1 Tire or wheel out-of-balance or out-of-round (Chapter 10).
2 Loose or worn wheel bearings (Chapter 10).
3 Worn tie-rod ends (Chapter 10).
4 Worn balljoints (Chapter 10).
5 Excessive wheel runout (Chapter 10).
6 Blister or bump on tire (Chapter 1).

53 Hard steering

1 Lack of lubrication at balljoints, tie-rod ends and steering gear assembly (Chapter 10).
2 Front wheel alignment incorrect (Chapter 10).
3 Low tire pressure (Chapter 1).

54 Steering wheel does not return to center position correctly

1 Lack of lubrication at balljoints and tie-rod ends (Chapters 1 and 10).
2 Binding in steering column (Chapter 10).
3 Defective rack-and-pinion assembly (Chapter 10).
4 Front wheel alignment problem (Chapter 10).

55 Abnormal noise at the front end

1 Lack of lubrication at balljoints and tie-rod ends (Chapter 1).
2 Loose upper strut mount (Chapter 10).
3 Worn tie-rod ends (Chapter 10).
4 Loose stabilizer bar (Chapter 10).
5 Loose wheel lug nuts (Chapter 1).
6 Loose suspension bolts (Chapter 10).

56 Wander or poor steering stability

1 Mismatched or uneven tires (Chapter 10).
2 Lack of lubrication at balljoints or tie-rod ends (Chapters 1 and 10).
3 Worn shock absorbers (Chapter 10).
4 Loose stabilizer bar (Chapter 10).
5 Broken or sagging springs (Chapter 10).
6 Front wheel alignment incorrect (Chapter 10).
7 Worn steering gear clamp bushings (Chapter 10).

57 Erratic steering when braking

1 Wheel bearings worn (Chapters 8 and 10).
2 Broken or sagging springs (Chapter 10).
3 Leaking wheel cylinder or caliper (Chapter 9).
4 Warped brake discs (rotors) (Chapter 9).
5 Worn steering gear clamp bushings (Chapter 10).

58 Excessive pitching and/or rolling around corners or during braking

1 Loose stabilizer bar (Chapter 10).
2 Worn shock absorbers or mounts

(Chapter 10).
3   Broken or sagging springs (Chapter 10).
4   Overloaded vehicle.

59 Suspension bottoms

1   Overloaded vehicle.
2   Worn shock absorbers (Chapter 10).
3   Incorrect, broken or sagging springs (Chapter 10).

60 Cupped tires

1   Front wheel alignment incorrect (Chapter 10).
2   Worn shock absorbers (Chapter 10).
3   Wheel bearings worn (Chapters 8 and 10).
4   Excessive tire or wheel runout (Chapter 10).
5   Worn balljoints (Chapter 10).

61 Excessive tire wear on outside edge

1   Inflation pressures incorrect (Chapter 1).
2   Excessive speed in turns.
3   Front end alignment incorrect (excessive toe-in or positive camber). Have professionally aligned.
4   Suspension arm bent or twisted (Chapter 10).

62 Excessive tire wear on inside edge

1   Inflation pressures incorrect (Chapter 1).
2   Front end alignment incorrect (toe-out or excessive negative camber). Have professionally aligned.
3   Loose or damaged steering components (Chapter 10).

63 Tire tread worn in one place

1   Tires out-of-balance.
2   Damaged or buckled wheel. Inspect and replace if necessary.
3   Defective tire (Chapter 1).

64 Excessive play or looseness in steering system

1   Wheel bearings worn (Chapter 10).
2   Tie-rod end loose or worn (Chapter 10).
3   Steering gear loose (Chapter 10).

65 Rattling or clicking noise in rack and pinion

Steering gear clamps loose (Chapter 10).

Chapter 1
Tune-up and routine maintenance

Contents

Section		Section	
Air filter replacement	19	Positive Crankcase Ventilation (PCV) valve check and replacement	30
Automatic transaxle fluid and filter change	27	Power steering fluid level check	7
Automatic transaxle fluid level check	6	Seat belt check	22
Battery check, maintenance and charging	8	Spare tire and jack check	24
Brake check	17	Spark plug replacement	32
Chassis lubrication	26	Spark plug wire, distributor cap and rotor check and replacement	33
Cooling system check	9	Starter safety switch check	23
Cooling system servicing (draining, flushing and refilling)	29	Suspension and steering check	14
Driveaxle boot check	13	Thermostatic air cleaner check (3.1L engine only)	28
Drivebelt check and replacement	21	Throttle Body Injection (TBI) mounting bolt torque check (3.1L engine only)	20
Engine oil and filter change	12	Tire and tire pressure checks	5
Evaporative emissions control system check	31	Tire rotation	16
Exhaust system check	15	Tune-up general information	3
Fluid level checks	4	Underhood hose check and replacement	10
Fuel filter replacement	25	Wiper blade inspection and replacement	11
Fuel system check	18		
Ignition timing check and adjustment (3.1L engine only)	34		
Introduction	1		
Maintenance schedule	2		

Specifications

Recommended lubricants and fluids

Note: *Listed here are manufacturer recommendations at the time this manual was written. Manufacturers occasionally upgrade their fluid and lubricant specifications, so check with your auto parts stores for current recommendations.*

Engine oil
    Type .... API grade SG, SG/CC or SG/CD multigrade and fuel-efficient oil
    Viscosity .... See accompanying chart
Automatic transaxle fluid .... Dexron II, IIE or III Automatic Transmission Fluid (ATF)
Engine coolant .... 50/50 mixture of water and ethylene glycol-based antifreeze
Brake fluid .... Delco Supreme II or DOT 3 fluid
Power steering fluid .... GM power steering fluid or equivalent
Engine oil (with filter change) .... 4.0 qts
Fuel tank .... 20 gals
Cooling system .... 12 qts

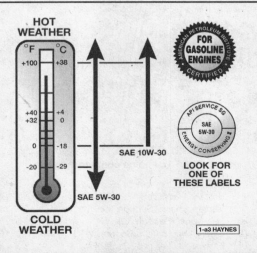

Engine oil viscosity chart – For best fuel economy and cold starting, select the lowest SAE viscosity grade for the expected temperature range

Chapter 1 Tune-up and routine maintenance

Capacities*

Automatic transaxle (fluid and filter replacement)
- Three speed .... 4 qts
- Four speed .... 6 qts

*All capacities approximate. Add as necessary to bring to appropriate level.

Ignition system

Spark plug type
- 3.1L engine .... AC type R43TS or equivalent
- 3800 engine .... AC type R43LTS6 or equivalent

Spark plug gap
- 3.1L engine .... 0.045 inch
- 3800 engine .... 0.060 inch

Firing order
- 3.1L engine 1-2-3-4-5-6
- 3800 engine 1-6-5-4-3-2

Cylinder locations .... See Chapter 2
Radiator cap pressure rating .... 15 psi
Brake pad wear limit .... 1/8 inch

Torque specifications   Ft-lbs

Throttle body nuts/bolts 18
Spark plugs
- 3.1L engine 18
- 3800 engine 12
Engine oil drain plug .... 15 to 20
Automatic transaxle pan bolts .... 8 to 10
Wheel lug nuts 100

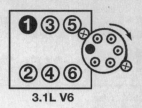
3.1L V6

FRONT

3800 V6

Cylinder location and distributor rotation

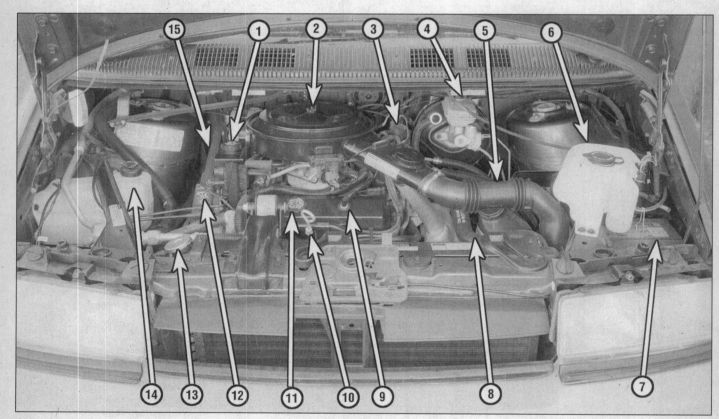

A typical 3.1L engine compartment

1. Power steering fluid reservoir (Section 7)
2. Air cleaner housing (Section 19)
3. Distributor (Section 38)
4. Brake fluid reservoir (Section 4)
5. Automatic transmission fluid dipstick location (Section 6)
6. Washer fluid reservoir (Section 4)
7. Battery (Section 8)
8. Radiator hose (Section 10)
9. PCV valve (Section 30)
10. Engine oil dipstick (Section 4)
11. Engine oil filler cap (Section 4)
12. Drivebelt routing decal (Section 21)
13. Radiator cap (Sections 9 and 29)
14. Radiator coolant reservoir (sections 4 and 29)
15. Serpentine drivebelt (Section 21)

Chapter 1 Tune-up and routine maintenance

1-3

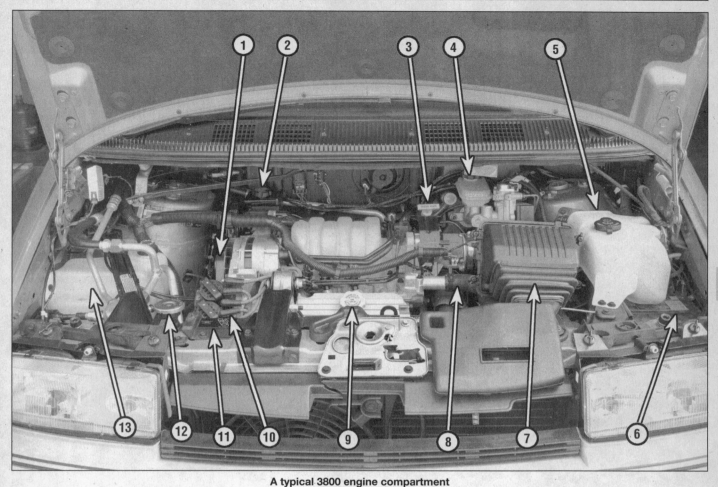

A typical 3800 engine compartment

1 Serpentine drivebelt (Section 21)
2 Power steering fluid reservoir (Section 7)
3 Automatic transmission fluid dipstick location (Section 6)
4 Brake fluid reservoir (Section 4)
5 Washer fluid reservoir (Section 4)
6 Battery (Section 8)
7 Air cleaner housing (Section 19)
8 Radiator hose (Section 10)
9 Engine oil filler cap (Section 4)
10 Direct Ignition System (DIS) coil and wires (Section 33)
11 Drivebelt routing decal (Section 21)
12 Radiator cap (Sections 9 and 29)
13 Radiator coolant reservoir (sections 4 and 29)

1 Introduction

This Chapter is designed to help the home mechanic maintain the Chevrolet Lumina, Pontiac Trans Sport and Oldsmobile Silhouette APV models with the goals of maximum performance, economy, safety and reliability in mind.

Included is a master maintenance schedule, followed by procedures dealing specifically with each item on the schedule. Visual checks, adjustments, component replacement and other helpful items are included. Refer to the accompanying illustrations of the engine compartment and the underside of the vehicle for the locations of various components.

Servicing your vehicle in accordance with the mileage/time maintenance schedule and the step-by-step procedures will result in a planned maintenance program that should produce a long and reliable service life. Keep in mind that it's a comprehensive plan, so maintaining some items but not others at the specified intervals will not produce the same results.

As you service your vehicle, you'll discover that many of the procedures can - and should - be grouped together because of the nature of the particular procedure you're performing or because of the close proximity of two otherwise unrelated components to one another. For example, if the vehicle is raised, you should inspect the exhaust, suspension, steering and fuel systems while you're under the vehicle. When you're rotating the tires, it makes good sense to check the brakes since the wheels are already removed. Finally, let's suppose you have to borrow or rent a torque wrench. Even if you only need it to tighten the spark plugs, you might as well check the torque of as many critical fasteners as time allows.

The first step in this maintenance program is to prepare yourself before the actual work begins. Read through all the procedures you're planning to do, then gather up all the parts and tools needed. If it looks like you might run into problems during a particular job, seek advice from a mechanic or an experienced do-it-yourselfer.

1-4 Chapter 1 Tune-up and routine maintenance

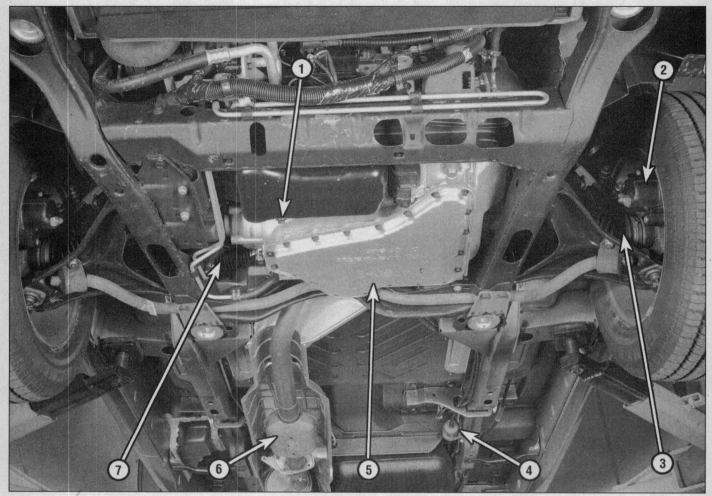

An underside view of a typical engine transaxle

1. Engine oil drain plug (Section 12)
2. Front disc brake caliper (Section 17)
3. Driveaxle boot (Section 13)
4. Fuel filter (Section 25)
5. Automatic transaxle fluid pan (Section 27)
6. Exhaust system (Section 15)
7. Steering gear boot (Section 14)

Chapter 1 Tune-up and routine maintenance

A typical rear underside view

1. Fuel filter hose (Section 18)
2. Spare tire (Section 24)
3. Exhaust system hanger (Section 15)
4. Muffler (Section 15)
5. Rear shock absorber (Section 14)
6. Fuel tank (Section 18)
7. Rear drum brake (Section 17)

2 GM APV Maintenance schedule

The following maintenance intervals are based on the assumption that the vehicle owner will be doing the maintenance or service work, as opposed to having a dealer service department do the work. Although the time/mileage intervals are loosely based on factory recommendations, most have been shortened to ensure, for example, that such items as lubricants and fluids are checked/changed at intervals that promote maximum engine/driveline service life. Also, subject to the preference of the individual owner interested in keeping his or her vehicle in peak condition at all times, and with the vehicle's ultimate resale in mind, many of the maintenance procedures may be performed more often than recommended in the following schedule. We encourage such owner initiative.

When the vehicle is new it should be serviced initially by a factory authorized dealer service department to protect the factory warranty. In many cases the initial maintenance check is done at no cost to the owner (check with your dealer service department for more information).

Every 250 miles or weekly, whichever comes first

Check the engine oil level (Section 4)
Check the engine coolant level (Section 4)
Check the windshield washer fluid level (Section 4)
Check the brake and clutch fluid levels (Section 4)
Check the tires and tire pressures (Section 5)

Every 3000 miles or 3 months, whichever comes first

All items listed above plus:
Check the automatic transaxle fluid level (Section 6)
Check the power steering fluid level (Section 7)
Check and service the battery (Section 8)
Check the cooling system (Section 9)
Inspect and replace, if necessary, all underhood hoses (Section 10)
Inspect and replace, if necessary, the windshield wiper blades (Section 11)
Change the engine oil and filter (Section 12)
Check the driveaxle boots (Section 13)
Inspect the suspension and steering components (Section 14)
Inspect the exhaust system (Section 15)

Every 7500 miles or 12 months, whichever comes first

All items listed above plus:
Rotate the tires (Section 16)
Check the brakes (Section 17)*
Inspect the fuel system (Section 18)
Replace the air filter and PCV filter (Section 19)
Check the throttle body mounting bolt torque (3.1L engine only) (Section 20)
Check the engine drivebelts (Section 21)
Check the seat belts (Section 22)
Check the starter safety switch (Section 23)
Check the spare tire and jack (Section 24)

Every 30,000 miles or 24 months, whichever comes first

All items listed above plus:
Replace the fuel filter (Section 25)
Lubricate the chassis (Section 26)
Change the automatic transaxle fluid (Section 27)**
Check the thermostatic air cleaner operation (3.1L engine only) (Section 28)
Service the cooling system (drain, flush and refill) (Section 29)
Inspect and replace, if necessary, the PCV valve (Section 30)
Inspect the evaporative emissions control system (Section 31)
Replace the spark plugs (Section 32)
Inspect the spark plug wires, distributor cap, rotor and wires (Section 33)
Check and adjust if necessary, the ignition timing (3.1L engine only) (Section 34)

Every 60,000 miles or 36 months, whichever comes first

Replace the engine drivebelt (Section 21)

*If the vehicle frequently tows a trailer, is operated primarily in stop-and-go conditions or its brakes receive severe usage for any other reason, check the brakes every 3000 miles or three months.

**If operated under one or more of the following conditions, change the automatic transaxle fluid every 15,000 miles:

In heavy city traffic where the outside temperature regularly reaches 90-degrees F (32-degrees C) or higher
In hilly or mountainous terrain
Frequent trailer pulling

Chapter 1 Tune-up and routine maintenance

4.2 The engine oil dipstick is mounted on the front side of the engine

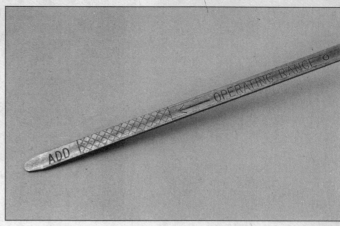

4.4 The oil level should be in the cross-hatched area – if it's below the ADD line, add enough oil to bring the level into the cross-hatched area

3 Tune-up general information

Note: *On models equipped with the Delco Loc II audio system, be sure the lockout feature is turned off before performing any procedure which requires disconnecting the battery.*

The term tune-up is used in this manual to represent a combination of individual operations rather than one specific procedure.

If, from the time the vehicle is new, the routine maintenance schedule is followed closely and frequent checks are made of fluid levels and high wear items, as suggested throughout this manual, the engine will be kept in relatively good running condition and the need for additional work will be minimized.

More likely than not, however, there will be times when the engine is running poorly due to lack of regular maintenance. This is even more likely if a used vehicle, which has not received regular and frequent maintenance checks, is purchased. In such cases, an engine tune-up will be needed outside of the regular routine maintenance intervals.

The first step in any tune-up or diagnostic procedure to help correct a poor running engine is a cylinder compression check. A compression check (see Chapter 2, Part C) will help determine the condition of internal engine components and should be used as a guide for tune-up and repair procedures. If, for instance, a compression check indicates serious internal engine wear, a conventional tune-up won't improve the performance of the engine and would be a waste of time and money. Because of its importance, the compression check should be done by someone with the right equipment and the knowledge to use it properly.

The following procedures are those most often needed to bring a generally poor running engine back into a proper state of tune.

Minor tune-up

Check all engine related fluids (Section 4)
Clean, inspect and test the battery (Section 8)
Check and adjust the drivebelts (Section 21)
Replace the spark plugs (Section 32)
Inspect the spark plug wires (Section 33)
Check the PCV valve (Section 30)
Check the air filter (Section 19)
Check the cooling system (Section 9)
Check all underhood hoses (Section 10)

Major tune-up

All items listed under Minor tune-up plus . . .
Check the EGR system (Chapter 6)
Check the ignition timing (Section 34)
Check the charging system (Chapter 5)
Check the fuel system (Section 18)
Replace the air filter (Section 19)
Replace the spark plug wires (Section 33)

4 Fluid level checks (every 250 miles or weekly)

Note: *The following are fluid level checks to be done on a 250 mile or weekly basis. Additional fluid level checks can be found in specific maintenance procedures which follow. Regardless of intervals, be alert to fluid leaks under the vehicle which would indicate a problem to be corrected immediately.*

1 Fluids are an essential part of the lubrication, cooling, brake and windshield washer systems. Because the fluids gradually become depleted and/or contaminated during normal operation of the vehicle, they must be periodically replenished. See *Recommended lubricants and fluids* at the beginning of this Chapter before adding fluid to any of the following components. **Note:** *The vehicle must be on level ground when fluid levels are checked.*

Engine oil

Refer to illustrations 4.2, 4.4 and 4.6

2 The engine oil level is checked with a dipstick **(see illustration)**. The dipstick extends through a metal tube down into the oil pan.

3 The oil level should be checked before the vehicle has been driven, or about 15 minutes after the engine has been shut off. If the oil is checked immediately after driving the vehicle, some of the oil will remain in the upper part of the engine, resulting in an inaccurate reading on the dipstick.

4 Pull the dipstick from the tube and wipe all the oil from the end with a clean rag or paper towel. Insert the clean dipstick all the way back into the tube and pull it out again. Note the oil at the end of the dipstick. Add oil as necessary to keep the level above the ADD mark in the cross-hatched area of the dipstick **(see illustration)**.

5 Do not overfill the engine by adding too much oil since this may result in oil fouled spark plugs, oil leaks or oil seal failures.

6 Oil is added to the engine after removing a twist-off cap located on the engine (see

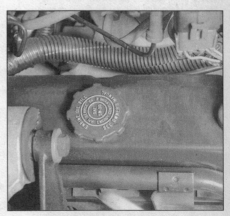

4.6 The engine oil filter cap is clearly marked and threads into the valve cover

4.9 The coolant reservoir is located on the right (passenger) side of the engine compartment – coolant can be added to the reservoir after unscrewing the cap

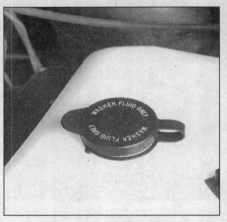

4.14 Flip the windshield washer fluid cap up to add fluid

4.17 The level in the translucent plastic brake reservoir can be checked without removing the cap, although it can be removed to check the precise level and fluid condition

illustration). A funnel may help to reduce spills.

7   Checking the oil level is an important preventive maintenance step. A consistently low oil level indicates oil leakage through damaged seals, defective gaskets or past worn rings or valve guides. If the oil looks milky in color or has water droplets in it, the cylinder head gasket may be blown or the head or block may be cracked. The engine should be checked immediately. The condition of the oil should also be checked. Whenever you check the oil level, slide your thumb and index finger up the dipstick before wiping off the oil. If you see small dirt or metal particles clinging to the dipstick, the oil should be changed (see Section 12).

Engine coolant

Refer to illustration 4.9

Warning: *Do not allow antifreeze to come in contact with your skin or painted surfaces of the vehicle. Flush contaminated areas immediately with plenty of water. Do not store new coolant or leave old coolant lying around where it's accessible to children or pets - they're attracted by its sweet smell. Ingestion of even a small amount of coolant can be fatal! Wipe up garage floor and drip pan coolant spills immediately. Keep antifreeze containers covered and repair leaks in the cooling system immediately.*

8   All vehicles covered by this manual are equipped with a pressurized coolant recovery system. A white plastic coolant reservoir located at the front of the engine compartment is connected by a hose to the radiator filler neck. As the engine warms up and the coolant expands, it escapes through a valve in the radiator cap and travels through the hose into the reservoir. As the engine cools, the coolant is automatically drawn back into the cooling system to maintain the correct level.

9   The coolant level in the reservoir should be checked regularly. **Warning:** *Do not remove the radiator cap to check the coolant level when the engine is warm.* The level in the reservoir varies with the temperature of the engine. When the engine is cold, the coolant level should be at or slightly above the FULL COLD mark on the reservoir **(see illustration)**. Once the engine has warmed up, the level should be at or near the FULL HOT mark. If it isn't, allow the engine to cool, then unscrew the cap from the reservoir and add a 50/50 mixture of ethylene glycol based antifreeze and water.

10   Drive the vehicle and recheck the coolant level. If only a small amount of coolant is required to bring the system up to the proper level, water can be used. However, repeated additions of water will dilute the antifreeze and water solution. In order to maintain the proper ratio of antifreeze and water, always top up the coolant level with the correct mixture. An empty plastic milk jug or bleach bottle makes an excellent container for mixing coolant. Do not use rust inhibitors or additives.

11   If the coolant level drops consistently, there may be a leak in the system. Inspect the radiator, hoses, filler cap, drain plugs and water pump (see Section 9). If no leaks are noted, have the radiator cap pressure tested by a service station.

12   If you have to remove the radiator cap, wait until the engine has cooled completely, then wrap a thick cloth around the cap and turn it to the first stop. If coolant or steam escapes, let the engine cool down longer, then remove the cap.

13   Check the condition of the coolant as well. It should be relatively clear. If it is brown or rust colored, the system should be drained, flushed and refilled. Even if the coolant appears to be normal, the corrosion inhibitors wear out, so it must be replaced at the specified intervals.

Windshield washer fluid

Refer to illustration 4.14

14   Fluid for the windshield washer system is located in a plastic reservoir on the left side of the engine compartment **(see illustration)**. In milder climates, plain water can be used in the reservoir, but it should be kept no more than two-thirds full to allow for expansion if the water freezes. In colder climates, use windshield washer system antifreeze, available at any auto parts store, to lower the freezing point of the fluid. Mix the antifreeze with water in accordance with the manufacturer's directions on the container. **Caution:** *Do not use cooling system antifreeze - it will damage the vehicle's paint.*

15   To help prevent icing in cold weather, warm the windshield with the defroster before using the washer.

Battery electrolyte

16   All vehicles covered by this manual are equipped with a battery which is permanently sealed (except for vent holes) and has no filler caps. Water does not have to be added to these batteries at any time; however, if a maintenance-type battery has been installed on the vehicle since it was new, remove all the cell caps on top of the battery (usually there are two caps that cover three cells each). If the electrolyte level is low, add distilled water until the level is above the plates. There is usually a split-ring indicator in each cell to help you judge when enough water has been added. Add water until the electrolyte level is just up to the bottom of the split ring indicator. Do not overfill the battery or it will spew out electrolyte when it is charging.

Brake fluid

Refer to illustration 4.17

17   The brake master cylinder is mounted on the front of the power booster unit in the engine compartment **(see illustration)**.

18   Unscrew the cap and make sure the fluid level is even with the bottom of the filler neck opening.

19   When adding fluid, pour it carefully into the reservoir to avoid spilling it on surrounding painted surfaces. Be sure the specified fluid is used, since mixing different types of

Chapter 1 Tune-up and routine maintenance

brake fluid can cause damage to the system. See *Recommended lubricants and fluids* at the front of this Chapter or your owner's manual. **Warning:** *Brake fluid can harm your eyes and damage painted surfaces, so use extreme caution when handling or pouring it. Do not use brake fluid that has been standing open or is more than one year old. Brake fluid absorbs moisture from the air. Excess moisture can cause a dangerous loss of braking effectiveness.*

20 At this time the fluid and master cylinder can be inspected for contamination. The system should be drained and refilled if deposits, dirt particles or water droplets are seen in the fluid.

21 After filling the reservoir to the proper level, make sure the cap is on tight to prevent fluid leakage.

22 The brake fluid level in the master cylinder will drop slightly as the pads at each wheel wear down during normal operation. If the master cylinder requires repeated replenishing to keep it at the proper level, this is an indication of leakage in the brake system, which should be corrected immediately. Check all brake lines and connections (see Section 17 for more information).

23 If, when checking the master cylinder fluid level, you discover one or both reservoirs empty or nearly empty, the brake system should be bled (see Chapter 9).

5 Tire and tire pressure checks (every 250 miles or weekly)

Refer to illustrations 5.2, 5.3, 5.4a, 5.4b and 5.8

1 Periodic inspection of the tires may spare you the inconvenience of being stranded with a flat tire. It can also provide you with vital information regarding possible problems in the steering and suspension systems before major damage occurs.

2 The original tires on this vehicle are equipped with 1/2-inch wide bands that appear when tread depth reaches 1/16-inch, indicating the tires are worn out. Tread wear can be monitored with a simple, inexpensive device known as a tread depth indicator **(see illustration)**.

3 Note any abnormal tread wear **(see illustration)**. Tread pattern irregularities such as cupping, flat spots and more wear on one side than the other are indications of front end alignment and/or balance problems. If any of these conditions are noted, take the vehicle to a tire shop or service station to cor-

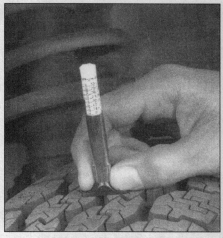

5.2 Use a tire tread depth indicator to monitor tire wear – they are available at auto parts stores and service stations and cost very little

rect the problem.

4 Look closely for cuts, punctures and embedded nails or tacks. Sometimes a tire will hold air pressure for short time or leak down very slowly after a nail has embedded

UNDERINFLATION

CUPPING

Cupping may be caused by:
- Underinflation and/or mechanical irregularities such as out-of-balance condition of wheel and/or tire, and bent or damaged wheel.
- Loose or worn steering tie-rod or steering idler arm.
- Loose, damaged or worn front suspension parts.

OVERINFLATION

INCORRECT TOE-IN OR EXTREME CAMBER

FEATHERING DUE TO MISALIGNMENT

5.3 This chart will help you determine the condition of the tires, the probable cause(s) of abnormal wear and the corrective action necessary

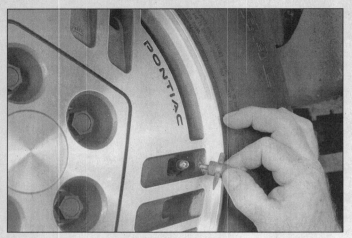

5.4a If a tires loses air on a steady basis, check the valve core first to make sure it's snug (special inexpensive wrenches are commonly available at auto parts stores)

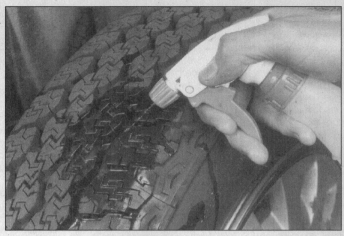

5.4b If the valve core is tight, raise the corner of the vehicle with the low tire and spray soapy water solution onto the tread as the tire is turned slowly – leaks will cause small bubbles to appear

itself in the tread. If a slow leak persists, check the valve stem core to make sure it's tight **(see illustration)**. Examine the tread for an object that may have embedded itself in the tire or for a "plug" that may have begun to leak (radial tire punctures are repaired with a plug that's installed in a puncture). If a puncture is suspected, it can be easily verified by spraying a solution of soapy water onto the suspected area **(see illustration)**. The soapy solution will bubble if there's a leak. Unless the puncture is unusually large, a tire shop or service station can usually repair the tire.

5 Carefully inspect the inner sidewall of each tire for evidence of brake fluid. If you see any, inspect the brakes immediately.

6 Correct air pressure adds miles to the lifespan of the tires, improves mileage and enhances overall ride quality. Tire pressure cannot be accurately estimated by looking at a tire, especially if it's a radial. A tire pressure gauge is essential. Keep an accurate gauge in the vehicle. The pressure gauges attached to the nozzles of air hoses at gas stations are often inaccurate.

7 Always check tire pressure when the tires are cold. Cold, in this case, means the vehicle has not been driven over a mile in the three hours preceding a tire pressure check. A pressure rise of four to eight pounds is not uncommon once the tires are warm.

8 Unscrew the valve cap protruding from the wheel or hubcap and push the gauge firmly onto the valve stem **(see illustration)**. Note the reading on the gauge and compare the figure to the recommended tire pressure shown on the label attached to the inside of the glove compartment door.

Be sure to reinstall the valve cap to keep dirt and moisture out of the valve stem mechanism. Check all four tires and, if necessary, add enough air to bring them up to the recommended pressure.

9 Don't forget to keep the spare tire inflated to the specified pressure (refer to your owner's manual or the tire sidewall).

5.8 To extend the life of the tires, check the air pressure at least once a week with an accurate gauge (don't forget the spare!)

6 Automatic transaxle fluid level check (every 3000 miles or 3 months)

Refer to illustrations 6.3 and 6.6

1 The automatic transaxle fluid level should be carefully maintained. Low fluid level can lead to slipping or loss of drive, while overfilling can cause foaming and loss of fluid.

2 With the parking brake set, start the engine, then move the shift lever through all the gear ranges, ending in Park. The fluid level must be checked with the vehicle level and the engine running at idle. **Note:** *Incorrect fluid level readings will result if the vehicle has just been driven at high speeds for an extended period, in hot weather in city traffic, or if it has been pulling a trailer. If any of these conditions apply, wait until the fluid has cooled (about 30 minutes).*

3 With the transaxle at normal operating

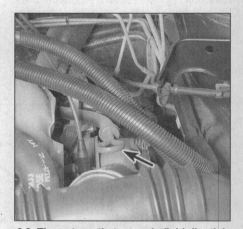

6.3 The automatic transaxle fluid dipstick (arrow) is located at the rear of the engine compartment

temperature, remove the dipstick from the filler tube. The dipstick is located at the rear of the engine compartment **(see illustration)**.

4 Carefully touch the fluid at the end of the dipstick to determine if the fluid is cool, warm or hot. Wipe the fluid from the dipstick with a clean rag and push it back into the filler tube until the cap seats.

5 Pull the dipstick out again and note the fluid level.

6 If the fluid felt cool, the level should be about 1/8-to-3/8 inch below the "ADD 1 PT" mark **(see illustration)**. If it felt warm, the level should be close to the "ADD 1 PT" mark. If the fluid was hot, the level should be within the cross-hatched area. If additional fluid is required, pour it directly into the tube using a funnel. It takes about one pint to raise the level from the ADD mark to the upper edge of the cross-hatched area with a hot transaxle, so add the fluid a little at a time and keep checking the level until it's correct.

7 The condition of the fluid should also be checked along with the level. If the fluid at the

Chapter 1 Tune-up and routine maintenance

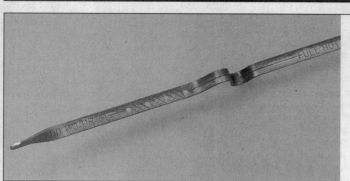

6.6 The automatic transaxle fluid level must be maintained within the cross-hatched area on the dipstick, between the upper and lower holes

7.2 The power steering fluid reservoir is located near the front (drivebelt end) of the engine; turn the cap clockwise or removal

7.6 The marks on the dipstick indicate the safe fluid level range

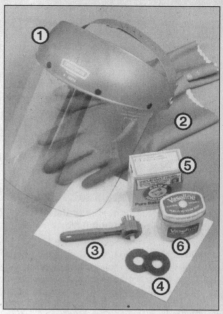

8.1 Tools and materials required for battery maintenance

1 **Face shield/safety goggles** - When removing corrosion with a brush, the acidic particles can easily fly up into your eyes
2 **Rubber gloves** - Another safety item to consider when servicing the battery - remember that's acid inside the battery!
3 **Battery terminal/cable cleaner** - This wire brush cleaning tool will remove all traces of corrosion from the battery and cable
4 **Treated felt washers** - Placing one of these on each terminal, directly under the cable end, will help prevent corrosion (be sure to get the correct type for side-terminal batteries)
5 **Baking soda** - A solution of baking soda and water can be used to neutralize corrosion
6 **Petroleum jelly** - A layer of this on the battery terminal bolts will help prevent corrosion

end of the dipstick is a dark reddish-brown color, or if the fluid has a burned smell, the fluid should be changed. If you're in doubt about the condition of the fluid, purchase some new fluid and compare the two for color and smell.

7 Power steering fluid level check (every 3000 miles or 3 months)

Refer to illustrations 7.2 and 7.6
1 . Unlike manual steering, the power steering system relies on fluid which may, over a period of time, require replenishing.
2 The fluid reservoir for the power steering pump is located behind the radiator near the front (drivebelt end) of the engine **(see illustration)**.
3 For the check, the front wheels should be pointed straight ahead and the engine should be off.
4 Use a clean rag to wipe off the reservoir cap and the area around the cap. This will help prevent any foreign matter from entering the reservoir during the check.
5 Twist off the cap and check the temperature of the fluid at the end of the dipstick with your finger.
6 Wipe off the fluid with a clean rag, reinsert it, then withdraw it and read the fluid level. The level should be at the HOT mark if the fluid was hot to the touch **(see illustration)**. It should be at the COLD mark if the fluid was cool to the touch. Note that on some models the marks (FULL HOT and COLD) are on opposite sides of the dipstick. At no time should the fluid level drop below the ADD mark.
7 If additional fluid is required, pour the specified type directly into the reservoir, using a funnel to prevent spills.
8 If the reservoir requires frequent fluid additions, all power steering hoses, hose connections, the power steering pump and the rack and pinion assembly should be carefully checked for leaks.

8 Battery check, maintenance and charging (every 3000 miles or 3 months)

Refer to illustrations 8.1 and 8.6
Warning: Certain precautions must be followed when checking and servicing the battery. Hydrogen gas, which is highly flammable, is always present in the battery cells, so keep lighted tobacco and all other open flames and sparks away from the battery. The electrolyte inside the battery is actually dilute sulfuric acid, which will cause injury if splashed on your skin or in your eyes. It will also ruin clothes and painted surfaces. When removing the battery cables, always detach the negative cable first and hook it up last!
1 Battery maintenance is an important procedure which will help ensure you aren't stranded because of a dead battery. Several tools are required for this procedure **(see illustration)**.
2 A sealed battery is standard equipment on all vehicles covered by this manual. Although this type of battery has many advantages over the older, capped cell type, it should still be routinely maintained according to the procedures which follow.

Check
3 The battery is located on the left side of the engine compartment. The exterior of the

8.6 Check the tightness of the battery hold-down bolt (arrow)

battery should be inspected periodically for damage such as a cracked case or cover.

4   Check the tightness of the battery cable terminals and connections to ensure good electrical connections and check the entire length of each cable for cracks and frayed conductors.

5   If corrosion (visible as white, fluffy deposits) is evident, remove the cables from the terminals, clean them with a battery brush and reinstall the cables. Corrosion can be kept to a minimum by using special treated fiber washers available at auto parts stores or by applying a layer of petroleum jelly to the terminals and cables after they are assembled.

6   Make sure that the battery tray is in good condition and the hold-down clamp bolt is tight **(see illustration)**. If the battery is removed from the tray, make sure no parts remain in the bottom of the tray when the battery is reinstalled. When reinstalling the hold-down clamp bolt, do not overtighten it.

7   Information on removing and installing the battery can be found in Chapter 5. Information on jump starting can be found at the front of this manual. For more detailed battery checking procedures, refer to the Haynes *Automotive Electrical Manual*.

Cleaning

8   Corrosion on the hold-down components, battery case and surrounding areas can be removed with a solution of water and baking soda. Thoroughly rinse all cleaned areas with plain water.

9   Any metal parts of the vehicle damaged by corrosion should be covered with a zinc-based primer, then painted.

Charging

Warning: *When batteries are being charged, hydrogen gas, which is very explosive and flammable, is produced. Do not smoke or allow open flames near a charging or a recently charged battery. Wear eye protection when near the battery during charging. Also, make sure the charger is unplugged before connecting or disconnecting the battery from the charger.*

10   Slow-rate charging is the best way to restore a battery that's discharged to the point where it will not start the engine. It's also a good way to maintain the battery charge in a vehicle that's only driven a few miles between starts. Maintaining the battery charge is particularly important in the winter when the battery must work harder to start the engine and electrical accessories that drain the battery are in greater use.

11   It's best to use a one or two-amp battery charger (sometimes called a "trickle" charger). They are the safest and put the least strain on the battery. They are also the least expensive. For a faster charge, you can use a higher amperage charger, but don't use one rated more than 1/10th the amp/hour rating of the battery. Rapid boost charges that claim to restore the power of the battery in one to two hours are hardest on the battery and can damage batteries that aren't in good condition. This type of charging should only be used in emergency situations.

12   The average time necessary to charge a battery should be listed in the instructions that come with the charger. As a general rule, a trickle charger will charge a battery in 12 to 16 hours.

13   Remove all of the cell caps (if equipped) and cover the holes with a clean cloth to prevent spattering electrolyte. Disconnect the negative battery cable and hook the battery charger leads to the battery posts (positive to positive, negative to negative), then plug in the charger. Make sure it is set at 12-volts if it has a selector switch.

14   If you're using a charger with a rate higher than two amps, check the battery regularly during charging to make sure it doesn't overheat. If you're using a trickle charger, you can safely let the battery charge overnight after you've checked it regularly for the first couple of hours.

15   If the battery has removable cell caps, measure the specific gravity with a hydrometer every hour during the last few hours of the charging cycle. Hydrometers are available inexpensively from auto parts stores - follow the instructions that come with the hydrometer. Consider the battery charged when there's no change in the specific gravity reading for two hours and the electrolyte in the cells is gassing (bubbling) freely. The specific gravity reading from each cell should be very close to the others. If not, the battery probably has a bad cell(s).

16   Some batteries with sealed tops have built-in hydrometers on the top that indicate the state of charge by the color displayed in the hydrometer window. Normally, a bright-colored hydrometer indicates a full charge and a dark hydrometer indicates the battery still needs charging. Check the battery manufacturer's instructions to be sure you know what the colors mean.

17   If the battery has a sealed top and no built-in hydrometer, you can hook up a digital voltmeter across the battery terminals to check the charge. A fully charged battery should read 12.5-volts or higher.

Check for a chafed area that could fail prematurely.

Check for a soft area indicating the hose has deteriorated inside.

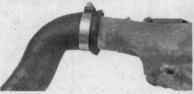

Overtightening the clamp on a hardened hose will damage the hose and cause a leak.

Check each hose for swelling and oil-soaked ends. Cracks and breaks can be located by squeezing the hose.

9.4 Hoses, like drivebelts, have a habit of falling at the worst possible time – to prevent the inconvenience of a blown radiator or heater, inspect them carefully as shown here

9   Cooling system check (every 3000 miles or 3 months)

Refer to illustration 9.4

1   Many major engine failures can be attributed to a faulty cooling system. If the vehicle is equipped with an automatic transaxle, the cooling system also cools the transaxle fluid and plays an important role in prolonging transaxle life.

2   The cooling system should be checked with the engine cold. Do this before the vehicle is driven for the day or after the engine has been shut off for at least three hours.

3   Remove the radiator cap by turning it to the left until it reaches a stop. If you hear any hissing sounds (indicating there is still pressure in the system), wait until it stops. Now press down on the cap with the palm of your

Chapter 1 Tune-up and routine maintenance

hand and continue turning to the left until the cap can be removed. Thoroughly clean the cap, inside and out, with clean water. Also clean the filler neck on the radiator. All traces of corrosion should be removed. The coolant inside the radiator should be relatively transparent. If it is rust colored, the system should be drained and refilled (see Section 29). If the coolant level is not up to the top, add additional antifreeze/coolant mixture (see Section 4).

4 Carefully check the large upper and lower radiator hoses along with any smaller diameter heater hoses which run from the engine to the firewall. Inspect each hose along its entire length, replacing any hose which is cracked, swollen or shows signs of deterioration. Cracks may become more apparent if the hose is squeezed **(see illustration)**.

5 Make sure all hose connections are tight. A leak in the cooling system will usually show up as white or rust colored deposits on the areas adjoining the leak. If wire-type clamps are used at the ends of the hoses, it may be wise to replace them with more secure screw-type clamps.

6 Use compressed air or a soft brush to remove bugs, leaves, etc. from the front of the radiator or air conditioning condenser. Be careful not to damage the delicate cooling fins or cut yourself on them.

7 Every other inspection, or at the first indication of cooling system problems, have the cap and system pressure tested. If you don't have a pressure tester, most gas stations and repair shops will do this for a minimal charge.

10 Underhood hose check and replacement (every 3000 miles or 3 months)

General

Refer to illustration 10.1

1 **Warning:** *Replacement of air conditioning hoses must be left to a dealer service department or air conditioning shop that has the equipment to depressurize the system safely. Never remove air conditioning components or hoses* **(see illustration)** *until the system has been depressurized.*

2 High temperatures under the hood can cause the deterioration of the rubber and plastic hoses used for engine, accessory and emission systems operation. Periodic inspection should be made for cracks, loose clamps, material hardening and leaks. Information specific to the cooling system hoses can be found in Section 9.

3 Some, but not all, hoses are secured to the fittings with clamps. Where clamps are used, check to be sure they haven't lost their tension, allowing the hose to leak. If clamps aren't used, make sure the hose hasn't expanded and/or hardened where it slips over the fitting, allowing it to leak.

Vacuum hoses

4 It's quite common for vacuum hoses, especially those in the emissions system, to be color coded or identified by colored stripes molded into each hose. Various systems require hoses with different wall thicknesses, collapse resistance and temperature resistance. When replacing hoses, be sure the new ones are made of the same material.

5 Often the only effective way to check a hose is to remove it completely from the vehicle. If more than one hose is removed, be sure to label the hoses and fittings to ensure correct installation.

6 When checking vacuum hoses, be sure to include any plastic T-fittings in the check. Inspect the fittings for cracks and the hose where it fits over the fitting for distortion, which could cause leakage.

7 A small piece of vacuum hose (1/4-inch inside diameter) can be used as a stethoscope to detect vacuum leaks. Hold one end of the hose to your ear and probe around vacuum hoses and fittings, listening for the "hissing" sound characteristic of a vacuum leak. **Warning:** *When probing with the vacuum hose stethoscope, be careful not to allow your body or the hose to come into contact with moving engine components such as the drivebelt, cooling fan, etc.*

Fuel hose

Warning: *Gasoline is extremely flammable, so take extra precautions when you work on any part of the fuel system. Don't smoke or allow open flames or bare light bulbs near the work area, and don't work in a garage where a natural gas-type appliance (such as a water heater or clothes dryer) with a pilot light is present. If you spill any fuel on your skin, rinse it off immediately with soap and water. When you perform any kind of work on the fuel system, wear safety glasses and have a Class B type fire extinguisher on hand. The fuel system is under pressure, so if any lines must be disconnected, the pressure in the system must be relieved first (see Chapter 4 for more information).*

8 Check all rubber fuel lines for deterioration and chafing. Check especially for cracks in areas where the hose bends and just before fittings, such as where a hose attaches to the fuel filter and fuel injection unit.

9 High quality fuel line, usually identified by the word *Fluroelastomer* printed on the hose, should be used for fuel line replacement. Never, under any circumstances, use unreinforced vacuum line, clear plastic tubing or water hose for fuel lines.

10 Spring-type clamps are commonly used on fuel lines. These clamps often lose their tension over a period of time, and can be "sprung" during the removal process. As a result spring-type clamps be replaced with screw-type clamps whenever a hose is replaced.

Metal lines

11 Sections of steel tubing often used for fuel line between the fuel pump and fuel injection unit. Check carefully for cracks, kinks and flat spots in the line.

12 If a section of metal fuel line must be replaced, only seamless steel tubing should be used, since copper and aluminum tubing do not have the strength necessary to withstand normal engine vibration.

13 Check the metal brake lines where they enter the master cylinder and brake proportioning unit (if used) for cracks in the lines and loose fittings. Any sign of brake fluid leakage calls for an immediate thorough inspection of the brake system.

11 Wiper blade inspection and replacement (every 3000 miles or 3 months)

Refer to illustrations 11.5, 11.7 and 11.8

1 The windshield and rear wiper and blade assemblies should be inspected periodically for damage, loose components and cracked or worn blade elements.

2 Road film can build up on the wiper blades and affect their efficiency, so they should be washed regularly with a mild detergent solution.

3 The action of the wiping mechanism can loosen the bolts, nuts and fasteners, so they should be checked and tightened, as necessary, at the same time the wiper blades are checked.

4 If the wiper blade elements (sometimes called inserts) are cracked, worn or warped, they should be replaced with new ones.

Windshield wiper

5 Remove the wiper blade assembly from the wiper arm by using a small screwdriver to lift the release lever while pulling on the blade

10.1 Air conditioning hoses are easily identified by the metal tubes used at all bends (arrow) – DO NOT disconnect or accidentally damage the air conditioning hoses (the system is under high pressure)

Chapter 1 Tune-up and routine maintenance

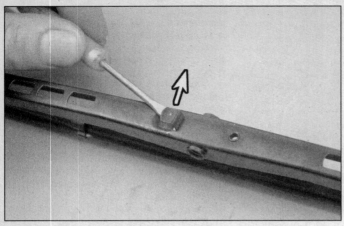

11.5 Lift the release lever, then slide the blade assembly away from the arm

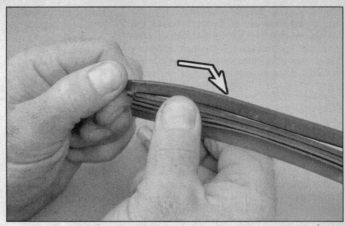

11.7 Hold the end of the wiper bridge tightly and pull the element out of the bridge end, then slide the element out

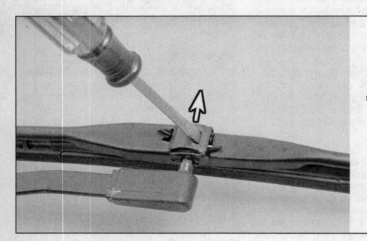

11.8 Insert a small screwdriver into the release slot, then slide the rear wiper off the arm pin

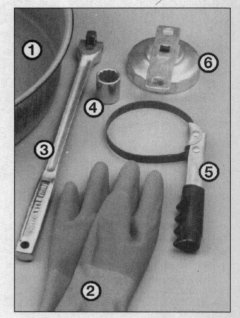

12.3 These tools are required when changing the engine oil and filter

1 **Drain pan** - It should be fairly shallow in depth, but wide to prevent spills
2 **Rubber gloves** - When removing the drain plug and filter, you will get oil on your hands (the gloves will prevent burns)
3 **Breaker bar** - Sometimes the oil drain plug is tight, and a long breaker bar is needed to loosen it
4 **Socket** - To be used with the breaker bar or a ratchet (must be the correct size to fit the drain plug - six-point preferred)
5 **Filter wrench** - This is a metal band-type wrench, which requires clearance around the filter to be effective
6 **Filter wrench** - This type fits on the bottom of the filter and can be turned with a ratchet or breaker bar (different-size wrenches are available for different types of filters)

to release it **(see illustration)**.
6 With the blade removed from the vehicle, you can remove the rubber element from the blade.
7 Grasp the end of the wiper bridge securely with one hand and the element with the other. Pull one end of the element out of the bridge to free it, then slide the element out **(see illustration)**.

Rear wiper

8 Insert a small screwdriver into the opening on top of the blade bridge to release it, then slide the assembly off the wiper arm pin **(see illustration)**.
9 Using pliers, pinch the clips on the end of the element together, then slide the element out of the blade assembly.

All models

10 Compare the new element with the old for length, design, etc.
11 Reverse the removal procedure and slide the new element into place. It will automatically lock at the correct location.
12 Reinstall the blade assembly on the arm, wet the windshield or rear window glass and test for proper operation.

12 Engine oil and filter change (every 3000 miles or 3 months)

Refer to illustrations 12.3, 12.9, 12.14 and 12.18

1 Frequent oil changes are the most important preventive maintenance procedures that can be done by the home mechanic. As engine oil ages, it becomes diluted and contaminated, which leads to premature engine wear.
2 Although some sources recommend oil filter changes every other oil change, we feel that the minimal cost of an oil filter and the relative ease with which it is installed dictate that a new filter be used every time the oil is changed.
3 Gather together all necessary tools and materials before beginning the procedure **(see illustration)**.
4 In addition, you should have plenty of clean rags and newspapers handy to mop up any spills. Access to the underside of the vehicle is greatly improved if the vehicle can be lifted on a hoist, driven onto ramps or supported by jackstands. **Warning:** Do not work under a vehicle which is supported only by a jack.

Chapter 1 Tune-up and routine maintenance

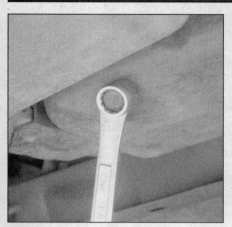

12.9 The engine oil drain plug is located at the rear of the oil pan – it is usually very tight, so use a box-end wrench to avoid rounding off the hex

12.14 The oil filter is usually on very tight as well and will a require special wrench for removal – DO NOT use the wrench to tighten the new filter!

12.18 Lubricate the oil filter gasket with clean engine oil before installing the filter on the engine

5   If this is your first oil change, get under the vehicle and familiarize yourself with the locations of the oil drain plug and the oil filter. The engine and exhaust components will be warm during the actual work, so note how they are situated to avoid touching them when working under the vehicle.

6   Warm the engine to normal operating temperature. If the new oil or any tools are needed, use this warm-up time to gather everything necessary for the job. The correct type of oil for your application can be found in *Recommended lubricants and fluids* at the beginning of this Chapter.

7   With the engine oil warm (warm engine oil will drain better and more built-up sludge will be removed with the oil), raise and support the vehicle. Make sure it's safely supported.

8   Move all necessary tools, rags and newspapers under the vehicle. Position the drain pan under the drain plug. Keep in mind that the oil will initially flow from the pan with some force, so place the pan accordingly.

9   Being careful not to touch any of the hot exhaust components, remove the drain plug at the bottom of the oil pan **(see illustration)**. Depending on how hot the oil is, you may want to wear gloves while unscrewing the plug the final few turns.

10   Allow the old oil to drain into the pan. It may be necessary to move the pan farther under the engine as the oil flow slows to a trickle.

11   After all the oil has drained, wipe off the drain plug with a clean rag. Small metal particles may cling to the plug which would immediately contaminate the new oil.

12   Clean the area around the drain plug opening and reinstall the plug. Tighten the plug securely with the wrench. If a torque wrench is available, use it to tighten the plug.

13   Move the drain pan into position under the oil filter.

14   Use the filter wrench to loosen the oil filter **(see illustration)**. Chain or metal band filter wrenches may distort the filter canister, but this is of no concern as the filter will be discarded anyway.

15   Completely unscrew the old filter. Be careful - it's full of oil. Empty the oil inside the filter into the drain pan.

16   Compare the old filter with the new one to make sure they are the same type.

17   Use a clean rag to remove all oil, dirt and sludge from the area where the oil filter mounts on the engine. Check the old filter to make sure the rubber gasket isn't stuck to the engine. If the gasket is stuck to the engine (use a flashlight if needed), remove it.

18   Apply a light coat of oil to the rubber gasket on the new oil filter **(see illustration)**. Open a can of oil and partially fill the oil filter with fresh oil. Oil pressure will not build in the engine until the oil pump has filled the filter with oil, so partially filling it at this time will reduce the amount of time the engine runs with no oil pressure.

19   Attach the new filter to the engine, following the tightening directions printed on the filter canister or box. Most filter manufacturers recommend against using a filter wrench due to the possibility of overtightening and damage to the seal.

20   Remove all tools, rags, etc. from under the vehicle, being careful not to spill the oil in the drain pan, then lower the vehicle.

21   Move to the engine compartment and locate the oil filler cap.

22   Pour the fresh oil through the filler opening. A funnel can be used.

23   Pour three quarts of fresh oil into the engine. Wait a few minutes to allow the oil to drain into the pan, then check the level on the dipstick (see Section 4 if necessary). If the oil level is above the ADD mark, start the engine and allow the new oil to circulate.

24   Run the engine for only about a minute and then shut it off. Immediately look under the vehicle and check for leaks at the oil pan drain plug and around the oil filter. If either is leaking, tighten with a bit more force.

25   With the new oil circulated and the filter now completely full, recheck the level on the dipstick and add more oil as necessary.

26   During the first few trips after an oil change, make it a point to check frequently for leaks and proper oil level.

27   The old oil drained from the engine cannot be reused in its present state and should be disposed of. Oil reclamation centers, auto repair shops and gas stations will normally accept the oil, which can be refined and used again. After the oil has cooled it can be drained into a container (capped plastic jugs, topped bottles, milk cartons, etc.) for transport to a disposal site.

13 Driveaxle boot check (every 3000 miles or 3 months)

Refer to illustration 13.2

1   The driveaxle boots are very important because they prevent dirt, water and foreign material from entering and damaging the constant velocity (CV) joints.

2   Inspect the boots for tears and cracks as well as loose clamps **(see illustration)**. If

13.2 Push on the driveaxle boots to check for cracks

14.5 Inspect the rack and pinion steering boots (there's one on each side) for tears or leaking grease

15.2 Check the rubber exhaust system hangers for cracks and damage

there is any evidence of cracks or leaking lubricant, they must be replaced as described in Chapter 8.

14 Suspension and steering check (every 3000 miles or 3 months)

Refer to illustration 14.5

1 Raise the front of the vehicle periodically and visually check the suspension and steering components for wear.
2 Be alert for excessive play in the steering wheel before the front wheels react, excessive sway around corners, body movement over rough roads and binding at some point as the steering wheel is turned. If you notice any of the above symptoms, the steering and suspension systems should be checked.
3 Support the vehicle on jackstands placed under the frame rails. Because of the work to be done, make sure the vehicle cannot fall off the stands.
4 Check the front wheel hub nuts and make sure they are securely locked in place.
5 Working under the vehicle, check for loose bolts, broken or disconnected parts and deteriorated rubber bushings on all suspension and steering components. Look for grease or fluid leaking from the steering assembly **(see illustration)**. Check the power steering hoses and connections for leaks.
6 Have an assistant turn the steering wheel from side-to-side and check the steering components for free movement, chafing and binding. If the steering doesn't react with the movement of the steering wheel, try to determine where the slack is located.

15 Exhaust system check (every 3000 miles or 3 months)

Refer to illustration 15.2

1 With the engine cold (at least three hours after the vehicle has been driven), check the complete exhaust system from the engine to the end of the tailpipe. Ideally, the inspection should be done with the vehicle on a hoist to permit unrestricted access. If a hoist is not available, raise the vehicle and support it securely on jackstands.
2 Check the exhaust pipes and connections for evidence of leaks, severe corrosion and damage. Make sure that all brackets and hangers are in good condition and tight **(see illustration)**.
3 At the same time, inspect the underside of the body for holes, corrosion, open seams, etc. which may allow exhaust gases to enter the interior. Seal all body openings with silicone or body putty.
4 Rattles and other noises can often be traced to the exhaust system, especially the mounts and hangers. Try to move the pipes, muffler and catalytic converter. If the components can come in contact with the body or suspension parts, secure the exhaust system with new mounts.
5 Check the running condition of the engine by inspecting inside the end of the tailpipe. The exhaust deposits here are an indication of engine state-of-tune. If the pipe is black and sooty or coated with white deposits, the engine is in need of a tune-up, including a thorough fuel system inspection and adjustment.

16 Tire rotation (every 7500 miles or 12 months)

Refer to illustration 16.2

1 The tires should be rotated at the specified intervals and whenever uneven wear is noticed.
2 Front wheel drive vehicles require a special tire rotation pattern **(see illustration)**.
3 Refer to the information in *Jacking and towing* at the front of this manual for the proper procedures to follow when raising the vehicle and changing a tire. If the brakes are going to be checked, don't apply the parking brake as stated. Make sure the tires are

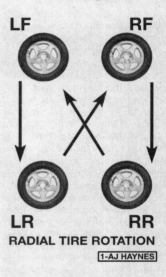

16.2 Tire rotation diagram

blocked to prevent the vehicle from rolling as it's raised.
4 The entire vehicle should be raised at the same time. This can be done on a hoist or by jacking up each corner and then lowering the vehicle onto jackstands placed under the frame rails. Always use four jackstands and make sure the vehicle is safely supported.
5 After rotation, check and adjust the tire pressures as necessary and be sure to check the lug nut tightness.

17 Brake check (every 7500 miles or 12 months)

Warning: *Brake system dust may contain asbestos, which is hazardous to your health. DO NOT blow it out with compressed air or inhale it. DO NOT use gasoline or solvents to remove the dust. Use brake system cleaner or denatured alcohol only.*

Chapter 1 Tune-up and routine maintenance

17.3 The brake wear indicated (arrow) will contact the disc and make a squealing noise when the pad is worn

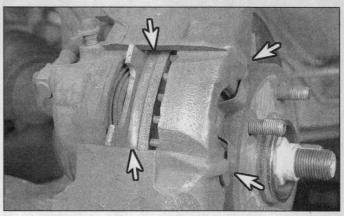

17.5 Look through the openings in the front of the caliper to check the brake pads (arrows) – the pad lining, which rubs against the disc, can also be inspected by looking through each end of the caliper

Note: *For detailed photographs of the brake system, refer to Chapter 9.*

1 In addition to the specified intervals, the brakes should be inspected every time the wheels are removed or whenever a defect is suspected. Raise the vehicle and place it securely on jackstands. Remove the wheels (see *Jacking and towing* at the front of this manual, if necessary).

Disc brakes (front)

Refer to illustrations 17.3 and 17.5

2 Disc brakes can be checked without removing any parts except the wheels. Extensive rotor damage can occur if the pads are not replaced when needed.

3 The disc brake pads have built-in wear indicators which make a high-pitched squealing or cricket-like warning sound when the pads are worn **(see illustration)**. **Caution:** *Expensive damage to the disc can result if the pads are not replaced soon after the wear indicators start squealing.*

4 The disc brake calipers, which contain the pads, are now visible. There is an outer pad and an inner pad in each caliper. All pads should be inspected.

5 Each caliper has a "window" to inspect the pads **(see illustration)**. If the pad material has worn to about 1/8-inch thick or less, the pads should be replaced.

6 If you're unsure about the exact thickness of the remaining lining material, remove the pads for further inspection or replacement (refer to Chapter 9).

7 Before installing the wheels, check for leakage and/or damage at the brake hoses and connections. Replace the hose or fittings as necessary, referring to Chapter 9.

8 Check the condition of the brake rotor. Look for score marks, deep scratches and overheated areas (they will appear blue or discolored). If damage or wear is noted, the disc can be removed and resurfaced by an automotive machine shop or replaced with a new one. Refer to Chapter 9 for more detailed inspection and repair procedures.

Drum brakes (rear)

Refer to illustration 17.14

9 Raise the vehicle and support it securely on jackstands. Block the front tires to prevent the vehicle from rolling; however, don't apply the parking brake or it will lock the drums in place.

10 Remove the wheels, referring to *Jacking and towing* at the front of this manual if necessary.

11 Mark the hub so it can be reinstalled in the same position. Use a scribe, chalk, etc. on the drum, hub and backing plate.

12 Remove the brake drum.

13 With the drum removed, carefully clean the brake assembly with brake system cleaner. **Warning:** *Don't blow the dust out with compressed air and don't inhale any of it (it may contain asbestos, which is harmful to your health).*

14 Note the thickness of the lining material on both front and rear brake shoes. If the material has worn away to within 1/8-inch of the recessed rivets or metal backing, the shoes should be replaced **(see illustration)**. The shoes should also be replaced if they're cracked, glazed (shiny areas), or covered with brake fluid.

15 Make sure all the brake assembly springs are connected and in good condition.

16 Check the brake components for signs of fluid leakage. With your finger or a small screwdriver, carefully pry back the rubber cups on the wheel cylinder located at the top of the brake shoes. Any leakage here is an indication that the wheel cylinders should be overhauled immediately (see Chapter 9). Also, check all hoses and connections for signs of leakage.

17 Wipe the inside of the drum with a clean rag and denatured alcohol or brake cleaner. Again, be careful not to breathe the dangerous asbestos dust.

18 Check the inside of the drum for cracks, score marks, deep scratches and "hard spots" which will appear as small discolored areas. If imperfections cannot be removed with fine emery cloth, the drum must be taken to an automotive machine shop for resurfacing.

19 Repeat the procedure for the remaining wheel. If the inspection reveals that all parts are in good condition, reinstall the brake drums, install the wheels and lower the vehicle to the ground.

Parking brake

20 The parking brake is operated by a foot pedal and locks the rear brake system. The easiest, and perhaps most obvious, method of periodically checking the operation of the parking brake assembly is to park the vehicle on a steep hill with the parking brake set and the transaxle in Neutral. If the parking brake cannot prevent the vehicle from rolling, it needs service (see Chapter 9).

18 Fuel system check (every 7500 miles or 12 months)

Refer to illustrations 18.3 and 18.4
Warning: *Gasoline is extremely flammable, so take extra precautions when you work on any part of the fuel system. Don't smoke or allow open flames or bare light bulbs near the work area, and don't work in a garage where a natural-gas type appliance (such as a water heater or clothes dryer) with a pilot light is present. If you spill any fuel on your skin, rinse it off immediately with soap and water. When you perform any kind of work on the fuel system, wear safety glasses and have a Class B type fire extinguisher on hand. The fuel system is under pressure, so if any lines must be disconnected, the pressure in the system must be relieved first (see Chapter 4 for more information).*

1 The fuel system is most easily checked with the vehicle raised on a hoist so the components underneath the vehicle are readily visible and accessible.

2 If the smell of gasoline is noticed while driving or after the vehicle has been in the

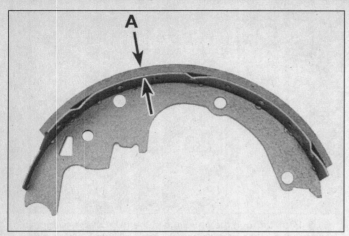

17.14 If the lining is bonded to the brake shoe, measure the lining thickness from the outer surface to the metal shoe, as shown here; if the lining is riveted to the shoe, measure from the lining outer surface to the rivet head

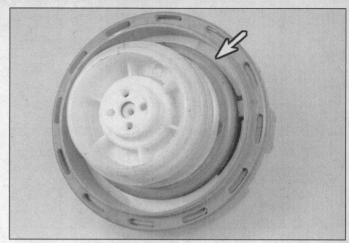

18.3 Make sure the gas cap gasket (arrow) is not damaged and shows signs that it's making good contact all the way around, otherwise the fuel system won't seal properly

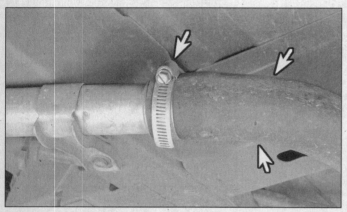

18.4 Inspect the fuel clamp to make sure it is tight and check the hose for cracks or damage (arrows)

19.2a Remove the wingnut (arrow) and lift off the cover

sun, the system should be thoroughly inspected immediately.

3   Remove the gas tank cap and check for damage, corrosion and an unbroken sealing imprint on the gasket **(see illustration)**. Replace the cap with a new one if necessary.

4   With the vehicle raised, inspect the gas tank and filler neck for punctures, cracks and other damage. The connection between the filler neck and tank is especially critical. Sometimes a rubber filler neck will leak due to loose clamps or deteriorated rubber, problems a home mechanic can usually rectify **(see illustration)**. *Warning: Do not, under any circumstances, try to repair a fuel tank yourself (except rubber components). A welding torch or any open flame can easily cause the fuel vapors to explode if the proper precautions are not taken.*

5   Carefully check all rubber hoses and metal lines leading away from the fuel tank. Check for loose connections, deteriorated hoses, crimped lines and other damage. Follow the lines to the front of the vehicle, carefully inspecting them all the way. Repair or replace damaged sections as necessary.

19 Air filter replacement (every 7500 miles or 12 months)

1   At the specified intervals, the air filter should be replaced with a new one. The filter should be inspected between changes.

Air filter replacement

3.1L engine

Refer to illustrations 19.2a and 19.2b

2   The air filter is located inside the air cleaner housing located at the front of the engine, near the battery. Remove the wing nut, lift off the cover, then lift the filter out **(see illustrations)**.

3800 engine

Refer to illustrations 19.3a and 19.3b

3   Unscrew the four plastic wingnuts, pull the housing cover back, then lift the filter from the housing **(see illustrations)**.

All models

4   While the filter housing cover is off, be careful not to drop anything down into the

19.2b With the cover off, lift the air filter element from the housing

throttle body or air cleaner assembly.

5   Wipe out the inside of the air cleaner housing with a clean rag.

6   Place the new filter in the air cleaner housing. Make sure it seats properly in the bottom of the housing. Install the top plate or cover.

Chapter 1 Tune-up and routine maintenance

19.3a Unscrew the four plastic wingnuts at the corners of the air cleaner housing, then . . .

19.3b . . . pull the cover toward you and lift the filter element

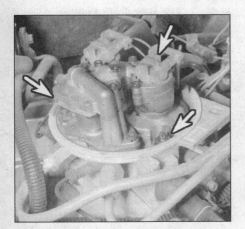

20.4 Nuts on the three Torx-head studs (arrows) hold the throttle body in place (rear nuts not visible in this photo)

21.5 Insert a breaker bar into the tensioner and rotate it counterclockwise to remove or install the belt

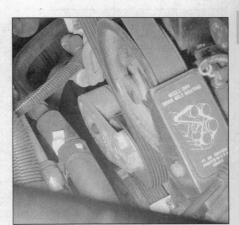

21.7 The serpentine drivebelt routing diagram is located on the belt guard on most models

20 Throttle Body Injection (TBI) mounting bolt torque check (3.1L engine only) (every 7500 miles or 12 months)

Refer to illustration 20.4

1 The TBI throttle body is attached to the top of the intake manifold by three nuts. They can sometimes work loose from vibration and temperature changes during normal engine operation and cause a vacuum leak.
2 If you suspect a vacuum leak exists at the bottom of the throttle body, use a rubber hose as a stethoscope. Start the engine and place one end of the hose next to your ear as you probe around the base of the throttle body with the other end. You will hear a hissing sound if a leak exists (be careful of hot and moving engine components).
3 Remove the air cleaner assembly (see Chapter 4).
4 Locate the throttle body mounting studs (see illustration).
5 Tighten the nut securely and evenly. Do not overtighten them, as the stud threads could strip.
6 If, after the nuts are properly tightened, a vacuum leak still exists, the throttle body must be removed and a new gasket installed. See Chapter 4 for more information.
7 After tightening the fasteners, reinstall the air cleaner and return all hoses to their original positions.

21 Drivebelt check and replacement (every 7500 miles or 12 months)

Refer to illustrations 21.5 and 21.7

1 A single serpentine drivebelt is located at the front of the engine and plays an important role in the overall operation of the engine and its components. Due to its function and material make up, the belt is prone to wear and should be periodically inspected. The serpentine belt drives the alternator, power steering pump, water pump and air conditioning compressor.
2 With the engine off, open the hood and use your fingers (and a flashlight, if necessary), to move along the belt checking for cracks and separation of the belt plies. Also check for fraying and glazing, which gives the belt a shiny appearance. Both sides of the belt should be inspected, which means you will have to twist the belt to check the underside.
3 Check the ribs on the underside of the belt. They should all be the same depth, with none of the surface uneven.
4 The tension of the belt is maintained by the tensioner assembly and isn't adjustable. The belt should be replaced at the mileage specified in the maintenance schedule at the front of this chapter, or if it is damaged or worn.
5 To replace the belt, rotate the tensioner counterclockwise to release belt tension **(see illustration)**. The tensioner will swing down once the tension of the belt is released.
6 Remove the belt from the auxiliary components and carefully release the tensioner.
7 Route the new belt over the various pulleys, again rotating the tensioner to allow the belt to be installed, then release the belt tensioner. **Note:** *These models have a drivebelt routing decal on the belt guard to help during drivebelt installation* **(see illustration)**.

Chapter 1 Tune-up and routine maintenance

24.2 Open the liftgate and pry off the cover for access to the jack

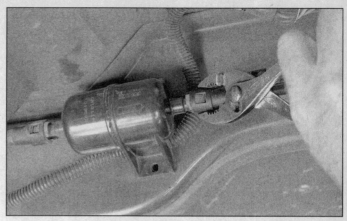

25.3a Grasp each of the fuel line-to-fuel filter fittings and turn them - turn to dislodge any dirt

22 Seat belt check (every 7500 miles or 12 months)

1 Check the seat belts, buckles, latch plates and guide loops for obvious damage and signs of wear.
2 See if the seat belt reminder light comes on when the key is turned to the Run or Start position. A chime should also sound.
3 The seat belts are designed to lock up during a sudden stop or impact, yet allow free movement during normal driving. Make sure the retractors return the belt against your chest while driving and rewind the belt fully when the buckle is unlatched.
4 If any of the above checks reveal problems with the seat belt system, replace parts as necessary.

23 Starter safety switch check (every 7500 miles or 12 months)

Warning: *During the following checks there's a chance the vehicle could lunge forward, possibly causing damage or injuries. Allow plenty of room around the vehicle, apply the parking brake and hold down the regular brake pedal during the checks.*

1 Try to start the engine in each gear. The engine should crank only in Park or Neutral.
2 Make sure the steering column lock allows the key to go into the Lock position only when the shift lever is in Park.
3 The ignition key should come out only in the Lock position.

24 Spare tire and jack check (every 7500 miles or 12 months)

Refer to illustration 24.2

1 Periodically checking the security and condition of the spare tire and jack will help to familiarize you with the procedures necessary for emergency tire replacement and also ensure that no components work loose during normal vehicle operation.
2 Following the instructions in your owner's manual or under *Jacking and towing* near the front of this manual, check the spare tire and jack. The jack is located in the right rear corner or the passenger compartment, under a cover **(see illustration)**. The spare tire is located under the rear of the vehicle.
3 Using a reliable air pressure gauge, check the pressure in the spare tire. It should be kept at the pressure marked on the tire sidewall.
4 Make sure the jack operates freely and all components are undamaged.
5 When finished, make sure the jack and tire are securely.

25 Fuel filter replacement (every 30,000 miles or 24 months)

Refer to illustrations 25.3a through 25.3f
Warning: *Gasoline is extremely flammable, so extra precautions must be taken when working on any part of the fuel system. Don't smoke or allow open flames or unshielded light bulbs in or near the work area. Also, don't work in a garage where a natural gas appliance (clothes dryer or water heater) with a pilot light is present. Have a fire extinguisher rated for gasoline fires handy and know how to use it! Wear eye protection during this procedure.*

1 Relieve the fuel system pressure (see

25.3b Use compressed air or aerosol carburetor cleaner to blow or wash the dirt from the fitting

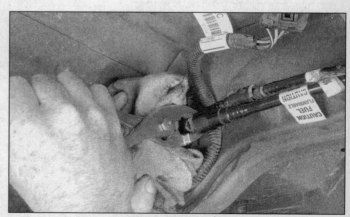

25.3c Wrap a rag around the fuel line (otherwise the residual pressure in the line will spray out), use pliers to gently depress the white plastic quick-disconnect tabs and detach the fuel lines from the filter

Chapter 1 Tune-up and routine maintenance

25.3d Remove the filter on earlier models by using a large screwdriver to break off the plastic rivet (later models are held in place by bolts)

25.3e Push the quick-disconnect clips into place on the new filter

Chapter 4, Section 2).
2   Raise the vehicle and support it securely on jackstands.
3   Refer to the accompanying illustrations and replace the fuel filter **(see illustrations)**.

26 Chassis lubrication (every 30,000 miles or 24 months)

Refer to illustrations 26.1, 26.5 and 26.6

1   Refer to *Recommended lubricants and fluids* at the front of this Chapter to obtain the necessary grease, etc. You'll also need a grease gun **(see illustration)**.
2   For easier access under the vehicle, raise it with a jack and place jackstands under the frame. Make sure it's safely supported by the stands. If the wheels are to be

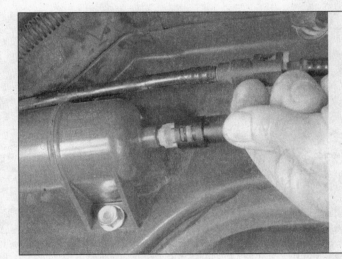

25.3f Install the new filter (with the arrow pointing toward the engine), then push the fuel lines onto the filter until the quick-disconnect fittings click into place

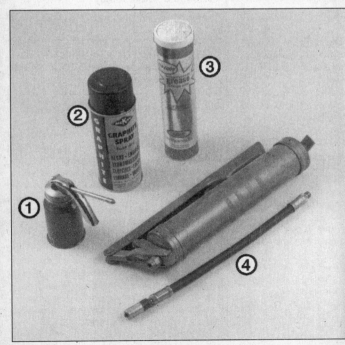

26.1 Materials required for chassis and body lubrication

1   **Engine oil** - Light engine oil in a can like this can be used for door and hood hinges
2   **Graphite spray** - Used to lubricate lock cylinders
3   **Grease** - Grease, in a variety of types and weights, is available for use in a grease gun. Check the Specifications for your requirements
4   **Grease gun** - A common grease gun, shown here with a detachable hose and nozzle, is needed for chassis lubrication. After use, clean it thoroughly

26.5 After wiping the grease fitting clean, push the nozzle firmly into place and pump the grease into the balljoint – usually two pumps of the gun will be sufficient

26.6 Pump the grease into the steering rod end fitting until the rubber seal is firm to the touch

27.7 After removing the front and side pan bolts, loosen the rear bolts and allow the fluid to drain, then remove the bolts and lower the pan from the vehicle

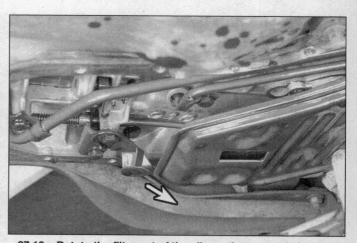

27.10a Rotate the filter out of the clip on the transaxle (arrow), then pull it straight down to remove it

removed at this interval for tire rotation or brake inspection, loosen the lug nuts slightly while the vehicle is still on the ground.

3   Before beginning, force a little grease out of the nozzle to remove any dirt from the end of the gun. Wipe the nozzle clean with a rag.

4   With the grease gun and plenty of clean rags, crawl under the vehicle and begin lubricating the components.

5   Wipe the balljoint grease fitting clean and push the nozzle firmly over it **(see illustration)**. Squeeze the trigger on the grease gun to force grease into the component. Continue pumping grease into the fitting until it oozes out of the joint between the two components. If it escapes around the grease gun nozzle, the fitting is clogged or the nozzle is not completely seated on the fitting. Resecure the gun nozzle to the fitting and try again. If necessary, replace the fitting with a new one.

6   Wipe the excess grease from the components and the grease fitting. Repeat the procedure for the steering rod end fittings.

These should be lubricated until the rubber seal is firm to the touch. Do not pump too much grease into the fittings as it could rupture the seal **(see illustration)**.

7   While you are under the vehicle, clean and lubricate the parking brake cable, along with the cable guides and levers. This can be done by smearing some of the chassis grease onto the cable and its related parts with your fingers.

8   Open the hood and smear a little chassis grease on the hood latch mechanism. Have an assistant pull the hood release lever from inside the vehicle as you lubricate the cable at the latch.

9   Lubricate all the hinges (door, hood, etc.) with engine oil to keep them in proper working order.

10   The key lock cylinders can be lubricated with spray graphite or silicone lubricant, which is available at auto parts stores.

11   Lubricate the door weatherstripping with silicone spray. This will reduce chafing and retard wear.

27   Automatic transaxle fluid and filter change (every 30,000 miles or 24 months)

Refer to illustrations 27.7, 27.10a, 27.10b, 27.11a, 27.11b and 27.12

1   At the specified time intervals, the transaxle fluid should be drained and replaced. Since the fluid will remain hot long after driving, perform this procedure only after everything has cooled down completely.

2   Before beginning work, purchase the specified transaxle fluid (see *Recommended lubricants and fluids* at the front of this Chapter) and a new filter.

3   Other tools necessary for this job include jackstands to support the vehicle in a raised position, a drain pan capable of holding several quarts, newspapers and clean rags.

4   Raise and support the vehicle on jackstands.

5   With a drain pan in place, remove the

Chapter 1 Tune-up and routine maintenance 1-23

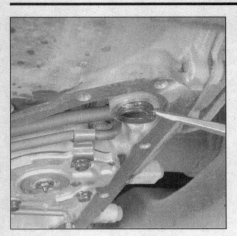

27.10b Use a screwdriver to pry out the old seal

27.11a Press the new seal fully into the transaxle openings

27.11b Press the filter into the seal and rotate it into place in the clip

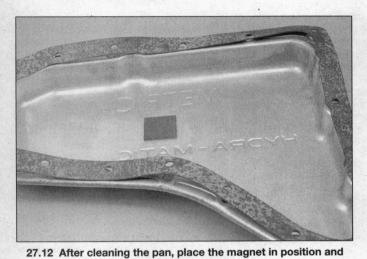

27.12 After cleaning the pan, place the magnet in position and install the gasket

28.5 The vacuum air control valve (arrow) is located in the air cleaner duct fitting (in this case it is in the down or heat off position)

front and side transaxle pan mounting bolts.
6 Loosen the rear pan bolts one turn.
7 Carefully pry the transaxle pan loose with a screwdriver, allowing the fluid to drain **(see illustration)**.
8 Remove the remaining bolts, pan and gasket. Carefully clean the gasket surface of the transaxle to remove all traces of the old gasket and sealant.
9 Drain the fluid from the transaxle pan, clean the pan with solvent and dry it with compressed air. Be careful not to lose the magnet.
10 Remove the filter and pry out the seal **(see illustrations)**.
11 Install a new seal and filter **(see illustrations)**.
12 Make sure the gasket surface on the transaxle pan is clean, then install the magnet and a new gasket **(see illustration)**. Put the pan in place against the transaxle and install the bolts. Working around the pan, tighten each bolt a little at a time until the final torque figure is reached.
13 Lower the vehicle and add the specified amount of automatic transmission fluid through the filler tube (see Section 6).
14 With the shift lever in Park and the parking brake set, run the engine at a fast idle, but don't race it.
15 Move the shift lever through each gear and back to Park. Check the fluid level.
16 Check under the vehicle for leaks during the first few trips.

28 Thermostatic air cleaner check (3.1L engine only) (every 30,000 miles or 24 months)

Refer to illustration 28.5

1 The thermostatic air cleaner draws heated air to the fuel injection throttle body from different locations, depending on engine temperature.
2 This is a simple visual check; however, the outside air duct must be removed.
3 Locate the vacuum air valve in the air cleaner assembly. It's located inside the air cleaner duct fitting, where you removed the outside air duct from. Make sure the flexible heat duct under the air cleaner is securely attached and undamaged.
4 The check should be done when the engine and outside air are cold (less than 65-degrees F). Start the engine and look through duct fitting at the valve (which should move up to the "heat on" position). With the valve up, air can't enter through the end of the duct fitting, but instead enters the air cleaner through the heat duct attached to the exhaust manifold. If you can't see into the duct fitting, a mirror may be helpful.
5 As the engine warms up to operating temperature, the valve should move down to the "heat off" position to allow air through duct fitting **(see illustration)**. Depending on outside temperature, it may take 10 to 15 minutes. To speed up the check you can reconnect the outside air duct, drive the vehicle and then check to see if the valve has moved down.
6 If the air cleaner isn't operating properly, see Chapter 6 for more information.

29 Cooling system servicing (draining, flushing and refilling) (every 30,000 or 24 months)

Refer to illustration 29.6

1   Periodically, the cooling system should be drained, flushed and refilled to replenish the antifreeze mixture and prevent formation of rust and corrosion, which can impair the performance of the cooling system and cause engine damage.

2   At the same time the cooling system is serviced, all hoses and the radiator cap should be inspected and replaced if defective (see Section 9).

3   Since antifreeze is a corrosive and poisonous solution, be careful not to spill any of the coolant mixture on the vehicle's paint or your skin. If this happens, rinse it off immediately with plenty of clean water. Consult local authorities about the dumping of antifreeze before draining the cooling system. In many areas, reclamation centers have been set up to collect automobile oil and drained antifreeze/water mixtures, rather than allowing them to be added to the sewage system.

4   With the engine cold, remove the radiator cap.

5   Move a large container under the radiator to catch the coolant as it's drained.

6   Drain the radiator by opening the drain plug at the bottom on the left side **(see illustration)**. If the drain plug is corroded and can't be turned easily, or if the radiator isn't equipped with a plug, disconnect the lower radiator hose to allow the coolant to drain. Be careful not to get antifreeze on your skin or in your eyes.

7   After the coolant stops flowing out of the radiator, move the container under the engine block drain plug(s), then remove the plugs. The 3.1L engine has plugs on both sides of the block, the 3800 has only one. On models equipped with an engine oil cooler, disconnect the inlet hose located on the left side of the cooler.

8   Disconnect the hose from the coolant reservoir and remove the reservoir. Flush it out with clean water.

9   Place a garden hose in the radiator filler neck and flush the system until the water runs clear at all drain points.

10   In severe cases of contamination or clogging of the radiator, remove it (see Chapter 3) and reverse flush it. This involves inserting the hose in the bottom radiator outlet to allow the water to run against the normal flow, draining through the top. A radiator repair shop should be consulted if further cleaning or repair is necessary.

11   When the coolant is regularly drained and the system refilled with the correct antifreeze/water mixture, there should be no need to use chemical cleaners or descalers.

12   To refill the system, install the block plugs, reconnect any radiator or oil cooler hoses and install the reservoir and the overflow hose.

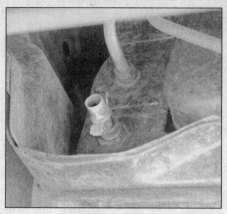

29.6   The drain plug is located at the lower left corner of the radiator

13   Slowly fill the radiator with the recommended mixture of antifreeze and water (see Section 4) to the base of the filler neck. Wait two minutes and recheck the coolant level, adding if necessary, then install the radiator cap. Add more coolant to the reservoir until it reaches the lower mark.

14   On models with rear compartment heaters, avoid using the rear fan for the first three times the engine reaches operating temperature. This will ensure full coolant flow to the rear heater core and bleed air out of the core and the lines.

15   Keep a close watch on the coolant level and the cooling system hoses during the first few miles of driving. Tighten the hose clamps and/or add more coolant as necessary. The coolant level should be a little above the HOT mark on the reservoir with the engine at normal operating temperature.

30 Positive Crankcase Ventilation (PCV) valve check and replacement (every 30,000 miles or 24 months)

Refer to illustrations 30.1 and 30.10

3.1L engine

1   With the engine idling at normal operating temperature, pull the valve (with hose attached) out of the rubber grommet in the valve cover **(see illustration)**.

2   Place your finger over the end of the valve. If there is no vacuum at the valve, check for a plugged hose, manifold port, or the valve itself. Replace any plugged or deteriorated hoses.

3   Turn off the engine and shake the PCV valve, listening for a rattle. If the valve doesn't rattle, replace it with a new one.

4   To replace the valve, pull it out of the end of the hose, noting its installed position and direction.

5   When purchasing a replacement PCV valve, make sure it's for your particular vehicle, model year and engine size. Compare the old valve with the new one to make sure they

30.1   On 3.1L engines the PCV valve is located in the valve cover – to check it, pull it out and feel for suction and shake it to make sure it rattles

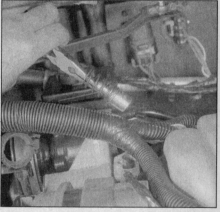

30.10   Use needle nose pliers to remove the PCV valve and spring from the cavity in the intake manifold

are the same.

6   Push the valve into the end of the hose until it's seated.

7   Inspect the rubber grommet for damage and replace it with a new one if necessary.

8   Push the PCV valve and hose securely into position in the valve cover.

3800 engine

9   The PCV valve on these models is located in the intake manifold under a cover. Remove the three bolts and detach the cover.

10   Use needle nose pliers to withdraw the PCV valve and spring from the manifold **(see illustration)**. Be careful not to lose the O-ring at the end of the valve.

11   Shake the PCV valve, listening for a rattle. If the valve doesn't rattle, replace it with a new one. Make sure the new PCV valve is for your particular vehicle, model year and engine size. Compare the old valve with the new one to make sure they are the same.

12   Insert the new PCV valve, O-ring and spring into position and install the cover and bolts.

Chapter 1 Tune-up and routine maintenance 1-25

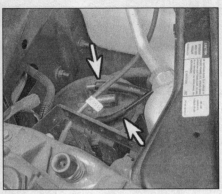

31.2 Check the evaporative canister for damage and the hose connections (arrows) for cracks and damage

31 Evaporative emissions control system check (every 30,000 miles or 12 months)

Refer to illustration 31.2

1   The function of the evaporative emissions control system is to draw fuel vapors from the gas tank and fuel system, store them in a charcoal canister and then burn them during normal engine operation.
2   The most common symptom of a fault in the evaporative emissions system is a strong fuel odor in the engine compartment. If a fuel odor is detected, inspect the charcoal canister, located at the front of the engine compartment. Check the canister and all hoses for damage and deterioration **(see illustration)**.
3   The evaporative emissions control system is explained in more detail in Chapter 6.

32 Spark plug replacement (every 30,000 miles or 12 months)

Refer to illustrations 32.2, 32.5a, 32.5b, 32.6, 32.8 and 32.10

1   The spark plugs are located at the front (radiator) and rear (firewall) sides of the engine. The spark plugs can be reached from the engine compartment, although on 3.1L engines it will be necessary to remove the air cleaner assembly for access to the rear plugs. **Note:** *On models equipped with the 3800 engine, it may be easier to replace the spark plugs on the rear cylinder bank from underneath. If you find this to be the case, raise the vehicle and support it securely on jackstands.*
2   In most cases, the tools necessary for spark plug replacement include a spark plug socket which fits onto a ratchet (spark plug sockets are padded inside to prevent damage to the porcelain insulators on the new plugs), various extensions and a gap gauge to check and adjust the gaps on the new plugs **(see illustration)**. A special plug wire removal tool is available for separating the wire boots from the spark plugs, and is a good idea on these models because the boots fit very tightly. A torque wrench should be used to tighten the new plugs. It is a good idea to allow the engine to cool before removing or installing the spark plugs.
3   The best approach when replacing the spark plugs is to purchase the new ones in advance, adjust them to the proper gap and replace the plugs one at a time. When buying the new spark plugs, be sure to obtain the correct plug type for your particular engine. The plug type can be found in the Specifications at the front of this Chapter and on the Emission Control Information label located under the hood. If these two sources list different plug types, consider the emission control label correct.
4   Allow the engine to cool completely before attempting to remove any of the plugs. While you are waiting for the engine to cool, check the new plugs for defects and adjust the gaps.
5   Check the gap by inserting the proper thickness gauge between the electrodes at the tip of the plug **(see illustration)**. The gap between the electrodes should be the same

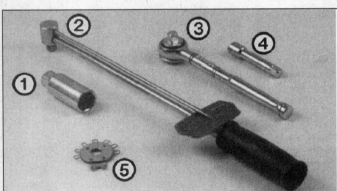

32.2 Tools required for changing spark plugs

1 **Spark plug socket** - This will have special padding inside to protect the spark plug's porcelain insulator
2 **Torque wrench** - Although not mandatory, using this tool is the best way to ensure the plugs are tightened properly
3 **Ratchet** - Standard hand tool to fit the spark plug socket
4 **Extension** - Depending on model and accessories, you may need special extensions and universal joints to reach one or more of the plugs
5 **Spark plug gap gauge** - This gauge for checking the gap comes in a variety of styles. Make sure the gap for your engine is included

32.5a Spark plug manufacturers recommend using a wire type gauge when checking the gap – if the wire does not slide between the electrodes with a slight drag, adjustment is required

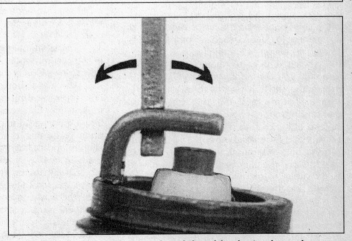

32.5b To change the gap, bend the *side* electrodes only, as indicated by the arrows, and be very *careful* not to crack or chip the porcelain insulator surrounding the center electrode

1-26 Chapter 1 Tune-up and routine maintenance

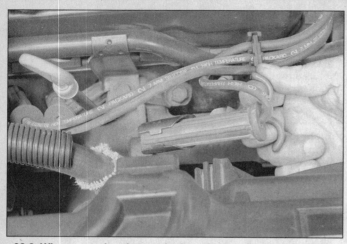

32.6 When removing the spark plug wires, pull only on the boot and use a twisting, pulling motion

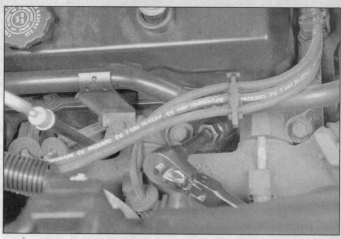

32.8 A socket and extension will be required to remove the spark plug on these models

32.10 A length of 3/16-inch ID rubber hose will save time and prevent damaged threads when installing the spark plugs

as the one specified on the Emissions Control Information label. The wire should slide between the electrodes with a slight amount of drag. If the gap is incorrect, use the adjuster on the gauge body to bend the curved side electrode slightly until the proper gap is obtained (see illustration). If the side electrode is not exactly over the center electrode, bend it with the adjuster until it is. Check for cracks in the porcelain insulator (if any are found, the plug should not be used).

6 With the engine cool, remove the spark plug wire from one spark plug. Pull only on the boot at the end of the wire - do not pull on the wire. A plug wire removal tool should be used if available (see illustration).

7 If compressed air is available, use it to blow any dirt or foreign material away from the spark plug hole. A common bicycle pump will also work. The idea here is to eliminate the possibility of debris falling into the cylinder as the spark plug is removed.

8 Place the spark plug socket over the plug and remove it from the engine by turning it in a counterclockwise direction (see illustration).

9 Compare the spark plug to those shown in the color photos on the inside rear cover of this manual to get an indication of the general running condition of the engine.

10 Thread one of the new plugs into the hole until you can no longer turn it with your fingers, then tighten it with a torque wrench (if available) or the ratchet. It might be a good idea to slip a short length of rubber hose over the end of the plug to use as a tool to thread it into place (see illustration). The hose will grip the plug well enough to turn it, but will start to slip if the plug begins to cross-thread in the hole - this will prevent damaged threads and the accompanying repair costs.

11 Before pushing the spark plug wire onto the end of the plug, inspect it following the procedures outlined in Section 33.

12 Attach the plug wire to the new spark plug, again using a twisting motion on the boot until it's seated on the spark plug.

13 Repeat the procedure for the remaining spark plugs, replacing them one at a time to prevent mixing up the spark plug wires.

33 Spark plug wire, distributor cap and rotor check and replacement (every 30,000 miles or 24 months)

Refer to illustration 33.12

Note: *3800 engines are equipped with distributorless ignition systems. The spark plug wires are connected directly to the ignition coils.*

1 The spark plug wires should be checked at the recommended intervals and whenever new spark plugs are installed in the engine.

2 The wires should be inspected one at a time to prevent mixing up the order, which is essential for proper engine operation.

3 Disconnect the plug wire from the spark plug. To do this, grab the rubber boot, twist slightly and pull the wire off. Do not pull on the wire itself, only on the rubber boot.

4 Check inside the boot for corrosion, which will look like a white crusty powder. Push the wire and boot back onto the end of the spark plug. It should be a tight fit on the plug. If it isn't, remove the wire and use pliers to carefully crimp the metal connector inside the boot until it fits securely on the end of the spark plug.

5 Using a clean rag, wipe the entire length of the wire to remove any built-up dirt and grease. Once the wire is clean, check for burns, cracks and other damage. Do not bend the wire excessively or pull the wire lengthwise - the conductor inside might break.

6 Disconnect the wire from the coil. Again, pull only on the rubber boot. Check for corrosion and a tight fit in the same manner as the spark plug end. Replace the wire at the coil.

7 Check the remaining spark plug wires one at a time, making sure they are securely fastened at the ignition coil and the spark plug when the check is complete.

8 If new spark plug wires are required, purchase a set for your specific engine model. Wire sets are available pre-cut, with the rubber boots already installed. Remove and replace the wires one at a time to avoid mix-ups in the firing order.

9 Loosen the distributor cap mounting screws (note that the screws have a shoulder so they don't come completely out). Pull up on the cap, with the wires attached, to separate it from the distributor, then position it out of the way.

10 The rotor is now visible at the end of the distributor shaft. Check it carefully for cracks and carbon tracks. Make sure the center terminal spring tension is adequate and look for corrosion and wear on the rotor tip. If in doubt about its condition, replace it with a new one.

Chapter 1 Tune-up and routine maintenance

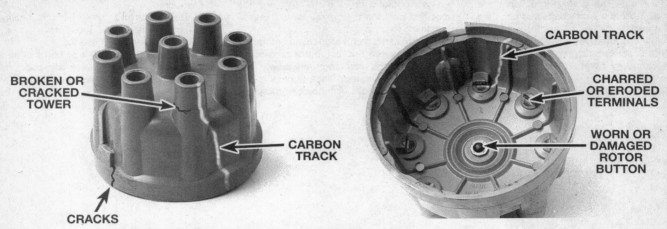

33.12 Shown here are some of the common defects to look for when inspecting the distributor cap (if in doubt about its condition, install a new one)

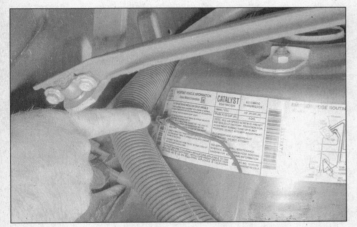

34.4 Unplug the EST wire before checking the ignition timing

34.5 The ignition timing marks (arrow) are located at the front of the engine

11   If replacement is required, remove the screws, detach it from the shaft and install a new one.
12   Check the distributor cap for carbon tracks, cracks and other damage. Closely examine the terminal on the inside of the cap for excessive corrosion and damage (see illustration). Again, if in doubt about the condition of the cap, replace it with a new one.
13   If it is necessary to replace the distributor cap, separate it from the distributor and transfer the spark plug wires, one at a time, to the new cap. Be careful not to mix up the wires!
14   Reattach the cap to the distributor, then tighten the screws.

34 Ignition timing check and adjustment (3.1L engine only) (every 30,000 miles or 12 months)

Refer to illustrations 34.4 and 34.5
Note: *If the information in this Section differs from the Vehicle Emission Control Information label (located on the right front shock tower in the engine compartment of your vehicle), the label should be considered correct.*

1   The engine must be at normal operating temperature and the air conditioner must be Off.
2   Apply the parking brake and block the wheels to prevent movement of the vehicle. The transmission must be in Park.
3   If the SERVICE ENGINE SOON light is on, don't proceed with the ignition timing check (see Chapter 6 for more information).
4   The Electronic Spark Timing (EST) system must be bypassed prior to checking the ignition timing. Locate the single tan wire with a black stripe attached to the engine wiring harness at the right front shock tower, near the VECI label, and unplug the connector (see illustration). Don't unplug the 4-wire harness connector at the distributor.
5   Locate the timing marks at the front of the engine (they should be visible from above after the hood is opened) (see illustration). The crankshaft pulley or vibration damper has a notch or groove in it and a small metal plate with notches and numbers is attached to the timing cover. Clean the plate with solvent so the numbers are visible.
6   Use chalk or white paint to mark the notch or groove in the pulley/vibration damper.
7   Highlight the notch or point on the timing plate that corresponds to the ignition timing specification on the Emission Control Information label (0-degree or TDC will most likely be specified).
8   Hook up the timing light by following the manufacturer's instructions (an inductive pick-up timing light is preferred). Generally, the power leads are attached to the battery terminals and the pick-up lead is attached to the number one spark plug wire. See the Specifications Section in Chapter 2, Part A for the location of cylinder number one. **Caution:** *If an inductive pick-up timing light isn't available, don't puncture the spark plug wire to attach the timing light pick-up lead. Instead, use an adapter between the spark plug and plug wire. If the insulation on the plug wire is damaged, the secondary voltage will jump to ground at the damaged point and the engine will misfire.*

9 Make sure the timing light wires are routed away from the drivebelts and fan, then start the engine.

10 Allow the idle speed to stabilize, then point the flashing timing light at the timing marks - be very careful of moving engine components!

11 The mark on the pulley/vibration damper will appear stationary. If it's aligned with the specified point on the timing plate, the ignition timing is correct.

12 If the marks aren't aligned, adjustment is required. Loosen the distributor hold-down bolt and turn the distributor very slowly until the marks are aligned. Since access to the bolt is tight, a special distributor wrench may be needed.

13 Tighten the bolt and recheck the timing.

14 Turn off the engine and remove the timing light (and adapter, if used).

15 Reconnect the EST wire harness connector, then clear any ECM trouble codes set during the ignition timing procedure (see Chapter 6).

Chapter 2 Part A
3.1L V6 engine

Contents

Section		Section
Camshaft and bearings - removal, inspection and installation	See Chapter 2C	Oil pan - removal and installation ... 15
Crankshaft front oil seal - replacement	12	Oil pump - removal and installation ... 16
Cylinder compression check	See Chapter 2C	Rear main oil seal - replacement ... See Chapter 2C
Cylinder head(s) - removal and installation	9	Repair operations possible with the engine in the vehicle ... 2
Drivebelt check, adjustment and replacement	See Chapter 1	Rocker arms and pushrods - removal, inspection and installation ... 6
Driveplate - removal and installation	See Chapter 2C	Spark plug replacement ... See Chapter 1
Engine mounts - check and replacement	17	Timing chain and sprockets - inspection, removal and installation ... 14
Engine oil and filter change	See Chapter 1	Timing chain cover - removal and installation ... 13
Engine overhaul - general information	See Chapter 2C	Top Dead Center (TDC) for number one piston - locating ... 3
Engine - removal and installation	See Chapter 2C	Valve covers - removal and installation ... 4
Exhaust manifolds - removal and installation	8	Valve springs, retainers and seals - replacement ... 7
General information	1	Vibration damper - removal and installation ... 11
Hydraulic lifters - removal, inspection and installation	10	Water pump - removal and installation ... See Chapter 3
Intake manifold - removal and installation	5	

Specifications

General

Cylinder numbers (drivebelt end-to-transaxle end)
- Front bank (radiator side) 2-4-6
- Rear bank 1-3-5
- Firing order 1-2-3-4-5-6

Cylinder location and distributor rotation

The blackened terminal shown on the distributor cap indicates the Number One spark plug wire position

Torque specifications

Ft-lbs (unless otherwise indicated)

Camshaft sprocket bolts	18
Cylinder head bolts	
First step	41
Second step	Rotate an additional 90-degrees (1/4-turn)
Exhaust manifold-to-cylinder head bolts	24
Intake manifold-to-cylinder head bolts/nuts	
Step one	13
Step two	19
Oil pan bolts	89 in-lbs
Oil pan nuts	89 in-lbs
Oil pan bolts (2 at rear main)	18
Oil pump mounting bolt	30
Rocker arm nuts	See Section 6, Step 10
Timing chain cover bolts	
6 mm bolts	20
8 mm bolts	28
Timing chain dampener	15
Valve cover-to-cylinder head fasteners	89 in-lbs
Vibration damper bolt	75

1 General information

Note: *On models equipped with the Delco Loc II audio system, be sure the lockout feature is turned off before performing any procedure which requires disconnecting the battery.*

This Part of Chapter 2 is devoted to in-vehicle repair procedures for the 3.1L V6 engine, which utilizes a cast-iron block with six cylinders arranged in a "V" shape at a 60-degree angle between the two banks. The overhead valve cylinder head is equipped with replaceable valve guides and seats. Hydraulic lifters actuate the valves through tubular pushrods.

All information concerning engine removal and installation and engine block and cylinder head overhaul can be found in Part C of this Chapter. The following repair procedures are based on the assumption the engine is installed in the vehicle. If the engine has been removed from the vehicle and mounted on a stand, many of the steps outlined in this Part of Chapter 2 will not apply.

The Specifications included in this Part of Chapter 2 apply only to the procedures contained in this Part. Part C of Chapter 2 contains the Specifications necessary for cylinder head and engine block rebuilding.

2 Repair operations possible with the engine in the vehicle

Many major repair operations can be accomplished without removing the engine from the vehicle.

Clean the engine compartment and the exterior of the engine with some type of degreaser before any work is done. It'll make the job easier and help keep dirt out of the internal areas of the engine.

Depending on the components involved, it may be helpful to remove the hood to improve access to the engine as repairs are performed (refer to Chapter 11 if necessary). Cover the fenders to prevent damage to the paint. Special pads are available, but an old bedspread or blanket will also work.

If vacuum, exhaust, oil or coolant leaks develop, indicating a need for gasket or seal replacement, the repairs can generally be done with the engine in the vehicle. The intake and exhaust manifold gaskets, timing chain cover gasket, oil pan gasket, crankshaft front oil seal and cylinder head gaskets are all accessible with the engine in place.

Exterior engine components, such as the intake and exhaust manifolds, the oil pan (and the oil pump), the water pump, the starter motor, the alternator and the fuel system components can be removed for repair with the engine in place.

Since the cylinder heads can be removed without pulling the engine, valve component servicing can also be accomplished with the engine in the vehicle. Replacement of the timing chain and sprockets is also possible with the engine in the vehicle.

In extreme cases caused by a lack of necessary equipment, repair or replacement of piston rings, pistons, connecting rods and rod bearings is possible with the engine in the vehicle. However, this practice is not recommended because of the cleaning and preparation work that must be done to the components involved.

3 Top Dead Center (TDC) for number one piston - locating

Note: *The following procedure is based on the assumption that the distributor is correctly installed. If you are trying to locate TDC to install the distributor correctly, piston position must be determined by feeling for compression at the number one spark plug hole, then aligning the ignition timing marks as described in Step 8.*

1 Top Dead Center (TDC) is the highest point in the cylinder that each piston reaches as it travels up-and-down when the crankshaft turns. Each piston reaches TDC on the compression stroke and again on the exhaust stroke, but TDC generally refers to piston position on the compression stroke.

2 Positioning the piston(s) at TDC is an essential part of many procedures such as rocker arm removal, camshaft and timing chain/sprocket removal and distributor removal.

3 Before beginning this procedure, be sure to place the transaxle in Neutral and apply the parking brake or block the rear wheels. Also, disable the ignition system by unplugging the primary (low-voltage) electrical connector(s) from the distributor. Remove the spark plugs (see Chapter 1, Section 32).

4 In order to bring any piston to TDC, the crankshaft must be turned using one of the methods outlined below. When looking at the front of the engine, normal crankshaft rotation is clockwise.

   a) *The preferred method is to turn the crankshaft with a socket and ratchet attached to the bolt threaded into the front of the crankshaft.*

   b) *A remote starter switch, which may save some time, can also be used. Follow the instructions included with the switch. Once the piston is close to TDC, use a socket and ratchet as described in the previous paragraph.*

   c) *If an assistant is available to turn the ignition switch to the Start position in short bursts, you can get the piston close to TDC without a remote starter switch. Make sure your assistant is out of the vehicle, away from the ignition switch, then use a socket and ratchet as described in Paragraph a) to complete the procedure.*

5 Note the position of the terminal for the number one spark plug wire on the distributor cap. If the terminal isn't marked, follow the plug wire from the number one cylinder spark

Chapter 2 Part A 3.1L V6 engine

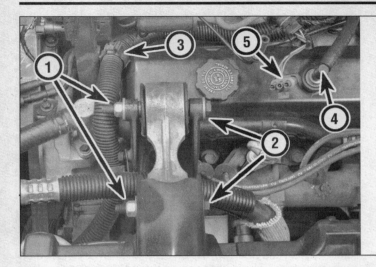

4.4 Remove or disconnect the following items to get at the valve cover:

1. Remove the torque strut nuts
2. Remove the torque strut bolts and remove the torque strut
3. Disconnect the wiring harness from its retaining clip
4. Detach the PCV valve from the valve cover
5. Unplug the IAC valve electrical connector (already unplugged)

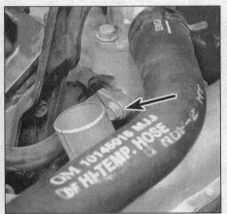

4.9a To remove the front valve cover, remove this coolant tube bolt (arrow) from the thermostat housing . . .

4.9b . . . remove the coolant tube bracket bolt (1), pinch the coolant tube hose clamps (2), detach the hose from the water pump, pull the coolant tube out of the thermostat housing and remove the valve cover nuts or bolts (3) (some models may require a Torx bit)

plug to the cap.

6 Use a felt-tip pen or chalk to make a mark on the distributor body directly under the terminal.

7 Detach the cap from the distributor and set it aside (see Chapter 1, Section 33 if necessary).

8 Turn the crankshaft (see Step 3 above) until the notch in the crankshaft pulley is aligned with the "0" mark on the timing scale.

9 Look at the distributor rotor - it should be pointing directly at the mark you made on the distributor body.

10 If the rotor is 180-degrees off, the number one piston is at TDC on the exhaust stroke. To get the piston to TDC on the compression stroke, turn the crankshaft one complete turn (360-degrees) clockwise. The rotor should now be pointing at the mark on the distributor. When the rotor is pointing at the number one spark plug wire terminal in the distributor cap and the ignition timing marks are aligned, the number one piston is at TDC on the compression stroke.

11 After the number one piston has been positioned at TDC on the compression stroke, TDC for any of the remaining pistons can be located by turning the crankshaft 120-degrees in the direction of normal rotation and following the firing order. Mark the remaining spark plug wire terminal locations on the distributor body just like you did for the number one terminal, then number the marks to correspond with the cylinder numbers. As you turn the crankshaft, the rotor will also turn. When it's pointing directly at one of the marks on the distributor, the piston for that particular cylinder is at TDC on the compression stroke.

4 Valve covers - removal and installation

Removal

1 Disconnect the negative battery cable from the battery.

2 Remove the air cleaner assembly (see Chapter 4, Section 8).

Front valve cover

Refer to illustrations 4.4, 4.9a and 4.9b

3 Unplug the electrical connector for the Idle Air Control (IAC) motor (see Chapter 4, Section 12).

4 Remove the engine torque strut (see illustration).

5 Detach the wiring harness from the clip at the timing cover end of the valve cover.

6 Detach the spark plug wires from the spark plugs (see Chapter 1, Section 33). Be sure each wire is labeled before removal to ensure correct reinstallation.

7 Remove the PCV tube from the valve cover (see Chapter 1, Section 30).

8 Drain the coolant (see Chapter 1, Section 29).

9 Unbolt the coolant tube from the cylinder head and the thermostat housing, disconnect the coolant tube hose from the water pump and remove the tube (see illustrations).

10 Remove the valve cover mounting nuts/bolts. **Note:** *Some models are equipped with special bolts. Remove them with a T-30 Torx driver.*

11 Detach the valve cover. **Note:** *If the cover sticks to the cylinder head, use a block of wood and a hammer to dislodge it. If the cover still won't come loose, pry on it carefully, but don't distort the sealing flange.*

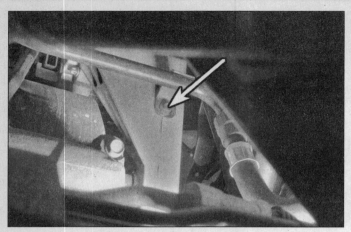

4.16a Remove this bolt (arrow) and detach the air conditioning line from the backside of the alternator mounting bracket

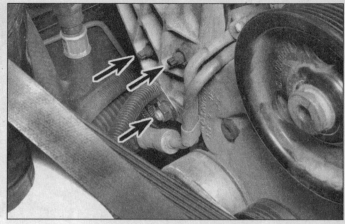

4.16b To remove the alternator bracket from the engine, remove the lower nut (lower arrow), detach the wiring harness clip and the power steering line clip, then remove the nuts from all three studs

Rear valve cover

Refer to illustrations 4.16a and 4.16b

12 Unplug the coolant temperature sensor and the oxygen sensor (see Chapter 6, Section 4).
13 Detach the spark plug wires from the spark plugs (see Chapter 1, Section 33). Be sure each wire is labeled before removal to ensure correct reinstallation.
14 Remove the serpentine drivebelt (see Chapter 1, Section 21).
15 Remove the alternator (see Chapter 5, Section 13).
16 Detach the air conditioning line bracket from the alternator bracket **(see illustration)**. Remove the alternator mounting bracket **(see illustration)**.
17 Remove the valve cover mounting nuts/bolts. **Note:** *Some models are equipped with special bolts. Remove them with a T-30 Torx driver.*
18 Detach the valve cover. **Note:** *If the cover sticks to the cylinder head, use a block of wood and a hammer to dislodge it. If the cover still won't come loose, pry on it carefully, but don't distort the sealing flange.*

Installation

19 The mating surfaces of each cylinder head and valve cover must be perfectly clean when the covers are installed. Use a gasket scraper to remove all traces of sealant or old gasket material, then clean the mating surfaces with lacquer thinner or acetone (if there's sealant or oil on the mating surfaces when the cover is installed, oil leaks may develop).
20 Place the valve cover and new gasket in position, then install the nuts or bolts. Tighten the fasteners in several steps to the torque listed in this Chapter's Specifications.
21 Complete the installation by reversing the removal procedure. Start the engine and check carefully for oil leaks at the valve cover-to-head joints.

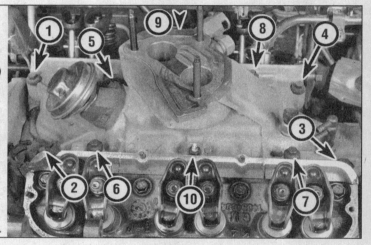

5.9 the intake manifold has nine bolts and two nuts (arrows) - the numerical sequence shown is for TIGHTENING the intake manifold bolts/nuts during reassembly - loosen them in the opposite order shown

5 Intake manifold - removal and installation

Removal

Refer to illustrations 5.9 and 5.10

1 Relieve the fuel system pressure (see Chapter 4, Section 2).
2 Disconnect the negative battery cable from the battery.
3 Drain the coolant (see Chapter 1, Section 29).
4 Disconnect the upper radiator hose from the thermostat housing. If you're replacing the intake manifold, remove the thermostat (see Chapter 3, Section 3).
5 Unplug the electrical connectors for the coolant temperature sensor and the EGR temperature switch (see Chapter 6, Sections 4 and 7).
6 If you're planning to replace the intake manifold, remove the throttle body (see Chapter 4, Section 12); if you're planning to reuse the same intake manifold, leave the throttle body attached.
7 If you're planning to replace the intake manifold, remove the EGR valve (see Chapter 6, Section 7).
8 Remove the valve covers (see Section 4).
9 Remove the intake manifold bolts **(see illustration)** and remove the manifold. Don't pry between the manifold and heads, as damage to the soft aluminum gasket sealing surfaces may result. If you're installing a new manifold, transfer all fittings and sensors to the new manifold.
10 Cut the old intake manifold gaskets as shown **(see illustration)** and remove and discard them.

Installation

Refer to illustration 5.13

Note: *The mating surfaces of the cylinder heads, block and manifold must be perfectly clean when the manifold is installed. Gasket removal solvents are available at most auto parts stores and may be helpful when removing old gasket material that's stuck to the heads and manifold (since the manifold is made of aluminum, aggressive scraping can cause damage). Be sure to follow the directions printed on the container.*

11 Use a gasket scraper to remove all traces of sealant and old gasket material,

Chapter 2 Part A 3.1L V6 engine

5.10 To remove the old intake manifold gaskets (or install new ones), cut the old intake manifold gaskets at the points (arrows) where the thin strips run behind the pushrods

5.13 Apply a bead of sealant to, or position an end seal on, the ridges between the heads (arrows)

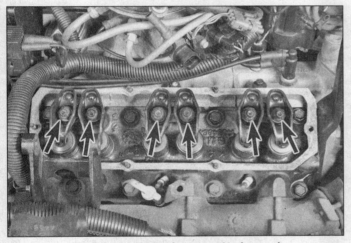

6.2 Remove the rocker arm nuts (arrows)

6.3 A perforated cardboard box can be used to store the pushrods to ensure they are reinstalled in their original locations - note the label indicating the transaxle end of the engine

then clean the mating surfaces with lacquer thinner or acetone. If there's old sealant or oil on the mating surfaces when the manifold is installed, oil or vacuum leaks may develop. Use a vacuum cleaner to remove any gasket material that falls into the intake ports or the lifter valley.

12 Use a tap of the correct size to chase the threads in the bolt holes, if necessary, then use compressed air (if available) to remove the debris from the holes. **Warning:** *Wear safety glasses or a face shield to protect your eyes when using compressed air!*

13 Apply a 3/16-inch (5 mm) bead of RTV sealant to the front and rear ridges of the engine block between the heads **(see illustration).** Note: *Some gasket sets may include end seals which you can use instead.*

14 Cut the new intake manifold gaskets in the same places you cut the old ones and install them.

15 Carefully lower the manifold into place and install the mounting bolts/nuts finger tight.

16 Tighten the mounting bolts/nuts in two steps, using the tightening sequence shown in **illustration 5.9**, until they're all at the torque listed in this Chapter's Specifications.

17 Install the remaining components in the reverse order of removal.

18 Change the oil and filter and refill the cooling system (see Chapter 1, Sections 12 and 29). Start the engine and check for leaks.

6 Rocker arms and pushrods - removal, inspection and installation

Removal

Refer to illustrations 6.2 and 6.3

1 Remove the valve covers (see Section 4) and the intake manifold (see Section 5).

2 Beginning at the drivebelt end of one cylinder head, remove the rocker arm mounting nuts one at a time and detach the rocker arms, nuts, and pivot balls **(see illustration).**

Store each set of rocker arm components separately in a marked plastic bag to ensure they're reinstalled in their original locations. **Note:** *If you only need to remove the pushrods, loosen the rocker arm nuts and turn the rocker arms to allow room for pushrod removal.*

3 Remove the pushrods and store them separately to make sure they don't get mixed up during installation **(see illustration).** Note: *Intake and exhaust pushrods are different lengths - make sure you don't mix them up.*

Inspection

4 Inspect each rocker arm for wear, cracks and other damage, especially where the pushrods and valve stems make contact.

5 Check the pivot seat in each rocker arm and the pivot ball faces. Look for galling, stress cracks and unusual wear patterns. If the rocker arms are worn or damaged, replace them with new ones and install new pivot balls as well.

6 Make sure the hole at the pushrod end

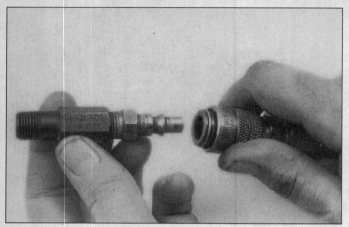

7.4 This is what the air hose adapter that threads into the spark plug hole looks like - they're commonly available from auto parts stores

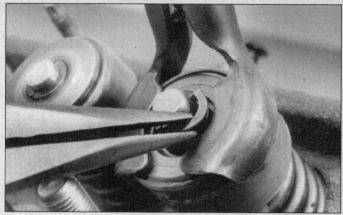

7.9 You'll need a valve spring compressor to compress the spring and remove the keepers to replace the valve seals or springs with the head installed - the type shown here is available at most auto parts stores and is fairly inexpensive

of each rocker arm is open.

7 Inspect the pushrods for cracks and excessive wear at the ends. Roll each pushrod across a piece of plate glass to see if it's bent (if it wobbles, it's bent).

Installation

8 Lubricate the lower end of each pushrod with clean engine oil or moly-base grease and install them in their original locations. Make sure each pushrod seats completely in the lifter socket.
9 Apply moly-base grease to the ends of the valve stems and the upper ends of the pushrods.
10 Apply moly-base grease to the pivot balls to prevent damage to the mating surfaces before engine oil pressure builds up. Install the rocker arms, pivot balls and nuts and tighten the nuts as follows: With the No. 1 cylinder at TDC, tighten the intake valves 3/4 turns past zero lash for cylinders 1, 5 and 6. Do the same thing to the exhaust valves for cylinders 1, 2 and 3. As the nuts are tightened, make sure the pushrods engage properly in the rocker arms. Then bring the No. 4 cylinder to TDC and tighten the intake valves 3/4 turns past zero lash for cylinders 2, 3 and 4, and do the same thing to the exhaust valves for cylinders 4, 5 and 6. **Note:** *The point of zero lash is reached when the pushrod can no longer be easily rotated as the rocker arm nut is tightened.*
11 Install the valve covers and intake manifold.

7 Valve springs, retainers and seals - replacement

Refer to illustrations 7.4, 7.9 and 7.17
Note: *Broken valve springs and defective valve stem seals can be replaced without removing the cylinder heads. Two special tools and a compressed air source are normally required to perform this operation, so read through this Section carefully and rent or buy the tools before beginning the job. If compressed air isn't available, a length of nylon rope can be used to keep the valves from falling into the cylinder during this procedure.*

1 Remove the valve cover(s) from the affected cylinder head(s) (see Section 4).
2 Remove the spark plug from the cylinder which has the defective component. If all of the valve stem seals are being replaced, all of the spark plugs should be removed.
3 Turn the crankshaft until the piston in the affected cylinder is at Top Dead Center on the compression stroke (see Section 3). If you're replacing all of the valve stem seals, begin with cylinder number one and work on the valves for one cylinder at a time. Move from cylinder-to-cylinder following the firing order sequence (see this Chapter's Specifications).
4 Thread an adapter into the spark plug hole **(see illustration)** and connect an air hose from a compressed air source to it. Most auto parts stores can supply the air hose adapter. **Note:** *Many cylinder compression gauges utilize a screw-in fitting that may work with your air hose quick-disconnect fitting.*
5 Remove the nut, pivot ball and rocker arm for the valve with the defective part and pull out the pushrod. If all of the valve stem seals are being replaced, all of the rocker arms and pushrods should be removed (refer to Section 6).
6 Apply compressed air to the cylinder. **Warning:** *The piston may be forced down by compressed air, causing the crankshaft to turn suddenly. If the wrench used when positioning the number one piston at TDC is still attached to the bolt in the crankshaft nose, it could cause damage or injury when the crankshaft moves.*
7 The valves should be held in place by the air pressure. If the valve faces or seats are in poor condition, leaks may prevent air pressure from retaining the valves - try the next method instead.
8 If you don't have access to compressed air, position the piston at a point approximately 45-degrees before TDC on the compression stroke, then feed a long piece of nylon rope through the spark plug hole until it fills the combustion chamber. Be sure to leave the end of the rope hanging out of the engine so it can be removed easily. Use a large ratchet and socket to rotate the crankshaft in the normal direction of rotation until slight resistance is felt.
9 Stuff shop rags into the cylinder head holes above and below the valves to prevent parts and tools from falling into the engine, then use a valve spring compressor to compress the spring. Remove the keepers with small needle-nose pliers or a magnet. **Note:** *A couple of different types of tools are available for compressing the valve springs with the head in place. One type grips the lower spring coils and presses on the retainer as the knob is turned, while the other type utilizes the rocker arm stud and nut for leverage* **(see illustration)**. *Both types work very well, although the lever type is usually less expensive.*
10 Remove the spring retainer, shield and valve spring, then remove the umbrella-type guide seal. **Note:** *If air pressure fails to hold the valve in the closed position during this operation, the valve face or seat is probably damaged. If so, the cylinder head will have to be removed for additional repair operations.*
11 Wrap a rubber band or tape around the top of the valve stem so the valve won't fall into the combustion chamber, then release the air pressure. **Note:** *If a rope was used instead of air pressure, turn the crankshaft slightly in the direction opposite normal rotation.*
12 Inspect the valve stem for damage. Rotate the valve in the guide and check the end for eccentric movement, which would indicate that the valve is bent.
13 Move the valve up-and-down in the guide and make sure it doesn't bind. If the

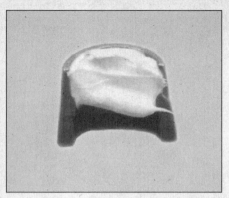

7.17 Keepers don't always stay in place, so apply a small dab of grease to each one as shown here before installation - it'll hold them in place on the valve stem as the spring is released

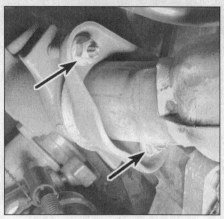

8.5 To disconnect the exhaust crossover pipe from either exhaust manifold, simply remove these two nuts (arrows) from the flange

valve stem binds, either the valve is bent or the guide is damaged. In either case, the head will have to be removed for repair.
14 Reapply air pressure to the cylinder to retain the valve in the closed position, then remove the tape or rubber band from the valve stem. If a rope was used instead of air pressure, rotate the crankshaft in the normal direction of rotation until slight resistance is felt.
15 Lubricate the valve stem with engine oil and install a new guide seal.
16 Install the spring and shield in position over the valve.
17 Install the valve spring retainer. Compress the valve spring and carefully position the keepers in the groove. Apply a small dab of grease to the inside of each keeper to hold it in place **(see illustration)**.
18 Remove the pressure from the spring tool and make sure the keepers are seated.
19 Disconnect the air hose and remove the adapter from the spark plug hole. If a rope was used in place of air pressure, pull it out of the cylinder.
20 Install the rocker arm(s) and pushrod(s) (see Section 6).
21 Install the spark plug(s) and hook up the wire(s).
22 Install the valve cover(s) (see Section 4).
23 Start and run the engine, then check for oil leaks and unusual sounds coming from the valve cover area.

8 Exhaust manifolds - removal and installation

1 Disconnect the negative battery cable from the battery.
2 Remove the air cleaner assembly (see Chapter 4, Section 8).

Front manifold

Refer to *illustrations 8.5, 8.8 and 8.9*
3 Allow the engine to cool completely, then drain the coolant (see Chapter 1, Section 29).
4 Disconnect the coolant tube (see Section 4).
5 Remove the retaining nuts from the crossover flange **(see illustration)** and disconnect the crossover pipe from the front exhaust manifold.
6 Remove the air conditioning compressor, leaving the hoses connected to it (see Chapter 3, Section 15).
7 Remove the torque strut (see Section 4).
8 Remove the air conditioning compressor/torque strut bracket **(see illustration)**.
9 Remove the exhaust manifold retaining bolts **(see illustration)** and detach the manifold from the cylinder head.
10 Clean the mating surfaces to remove all traces of old gasket material, then inspect the manifold for distortion and cracks. Warpage can be checked with a precision straightedge held against the mating flange. If a feeler gauge thicker than 0.030-inch can be inserted between the straightedge and flange surface, take the manifold to an automotive machine shop for resurfacing.
11 Place the manifold in position with a new gasket and install the mounting bolts finger tight.
12 Starting in the middle and working out toward the ends, tighten the mounting bolts a little at a time until all of them are at the torque listed in this Chapter's Specifications.
13 Install the remaining components in the reverse order of removal.
14 Start the engine and check for exhaust leaks between the manifold and cylinder head and between the manifold and exhaust pipe.

Rear manifold

Refer to *illustration 8.19*
15 You'll have to work from underneath the vehicle to get at the rear exhaust manifold. Set the parking brake, block the rear wheels, raise the front of the vehicle and support it securely on jackstands.
16 Unplug the oxygen sensor wire (see Chapter 6, Section 4).
17 Remove the retaining nuts from the crossover flange and disconnect the cross-

8.8 To remove the air conditioning compressor/torque strut bracket, remove these five bolts (the upper left bolt is a recessed Torx head type)

8.9 Front exhaust manifold retaining bolts (arrows)

8.19 Rear exhaust manifold retaining bolts (arrows)

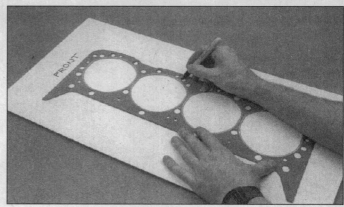

9.10a To avoid mixing up the head bolts, use a new gasket to transfer the bolt hole pattern to a piece of cardboard, then punch holes to accept the bolts

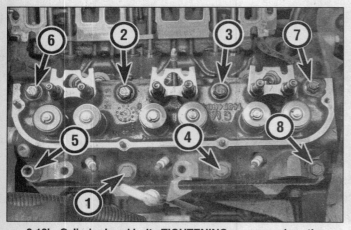

9.10b Cylinder head bolts TIGHTENING sequence (use the reverse of this order for removing the bolts)

9.14 Remove the old gasket and carefully scrape off all old gasket material and sealant

over pipe from the rear exhaust manifold (see illustration 8.5).

18 Remove the two exhaust pipe-to-manifold bolts. You may have to apply penetrating oil to the fastener threads - they're usually corroded. Disconnect the exhaust pipe from the exhaust manifold (see Chapter 4, Section 14).

19 Remove the exhaust manifold bolts (see illustration).

20 Clean the mating surfaces to remove all traces of old sealant, then check for warpage and cracks. Warpage can be checked with a precision straightedge held against the mating flange. If a feeler gauge thicker than 0.030-inch can be inserted between the straightedge and flange surface, take the manifold to an automotive machine shop for resurfacing.

21 Place the manifold in position with a new gasket and install the bolts finger tight.

22 Starting in the middle and working out toward the ends, tighten the mounting bolts a little at a time until all of them are at the torque listed in this Chapter's Specifications.

23 Install the remaining components in the reverse order of removal.

24 Start the engine and check for exhaust leaks between the manifold and cylinder head and between the manifold and exhaust pipe.

9 Cylinder head(s) - removal and installation

Caution: *Allow the engine to cool completely before loosening the cylinder head bolts.*
Note: *On engines with high mileage and during an overhaul, camshaft lobe height should be checked prior to cylinder head removal (see Chapter 2, Part C, Section 12 for instructions).*

Removal

Refer to illustrations 9.10a and 9.10b

1 Disconnect the negative battery cable from the battery.
2 Remove the air cleaner assembly (see Chapter 4, Section 8).
3 Remove the intake manifold (see Section 5).
4 If you're removing the front cylinder head, remove the oil dipstick tube mounting bolt.
5 Disconnect all wires and vacuum hoses from the cylinder head(s). Be sure to label them to simplify reinstallation.
6 Disconnect the spark plug wires and remove the spark plugs (see Chapter 1, Section 32). Be sure the plug wires are labeled to simplify reinstallation.
7 Remove the exhaust manifold from the cylinder head being removed (see Section 8).
8 Remove the valve cover(s) (see Section 4).
9 Remove the rocker arms and pushrods (see Section 6).
10 Using the new head gasket, outline the cylinders and bolt pattern on a piece of cardboard **(see illustration)**. Be sure to indicate the front (drivebelt end) of the engine for reference. Punch holes at the bolt locations. Loosen each of the cylinder head mounting bolts 1/4-turn at a time until they can be removed by hand - work from bolt-to-bolt in a pattern that's the reverse of the tightening sequence **(see illustration)**. Store the bolts in the cardboard holder as they're removed - this will ensure they are reinstalled in their original locations, which is absolutely essential.
11 Lift the head(s) off the engine. If resistance is felt, don't pry between the head and

Chapter 2 Part A 3.1L V6 engine

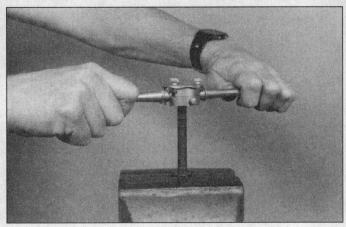

9.16 Use the correct die to clean the head bolt threads prior to installation

9.17a Position the new gasket over the dowel pins (arrows) . . .

9.17b . . . with the correct side facing up

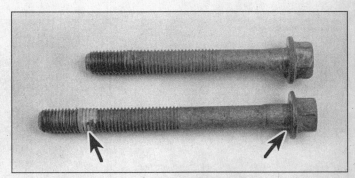

9.19 Two different length cylinder head bolts are used - apply sealant to the threads and the undersides of the bolt heads (arrows)

block as damage to the mating surfaces will result. Recheck for head bolts that may have been overlooked, then use a hammer and block of wood to tap up on the head and break the gasket seal. Be careful because there are locating dowels in the block which position each head. As a last resort, pry each head up at the rear corner only and be careful not to damage anything. After removal, place the head on blocks of wood to prevent damage to the gasket surfaces.

12 Refer to Chapter 2, Part C, for cylinder head disassembly, inspection and valve service procedures.

Installation

Refer to illustrations 9.14, 9.16, 9.17a, 9.17b and 9.19

13 The mating surfaces of each cylinder head and block must be perfectly clean when the head is installed.
14 Use a gasket scraper to remove all traces of carbon and old gasket material **(see illustration)**, then clean the mating surfaces with lacquer thinner or acetone. If there's oil on the mating surfaces when the head is installed, the gasket may not seal correctly and leaks may develop. When working on the block, it's a good idea to cover the lifter valley with shop rags to keep debris out of the engine. Use a shop rag or vacuum cleaner to remove any debris that falls into the cylinders.
15 Check the block and head mating surfaces for nicks, deep scratches and other damage. If damage is slight, it can be removed with a file; if it's excessive, machining may be the only alternative.
16 Use a tap of the correct size to chase the threads in the head bolt holes. Run a die down each one of the cylinder head bolts **(see illustration)**. Dirt, corrosion, sealant and damaged threads will affect torque readings.
17 Position the new gasket over the dowel pins in the block. Some gaskets are marked TOP or THIS SIDE UP to ensure correct installation **(see illustrations)**.
18 Carefully position the head on the block without disturbing the gasket.
19 Apply sealant such as Permatex no. 2 or equivalent to the threads and the undersides of the bolt heads. Install the bolts in the correct locations - two different lengths are used **(see illustration)**. Here's where the cardboard holder comes in handy.
20 Tighten the bolts, using the recommended sequence **(see illustration 9.10b)**, to the torque listed in this Chapter's Specifications. Then, using the same sequence, turn each bolt the amount of angle listed in this Chapter's Specifications.
21 The remaining installation steps are the reverse of removal.
22 Change the engine oil and filter (see Chapter 1, Section 12).

10 Hydraulic lifters - removal, inspection and installation

1 A noisy valve lifter can be isolated when the engine is idling. Hold a mechanic's stethoscope or a length of hose near the location of each valve while listening at the other end. Another method is to remove the valve cover and, with the engine idling, touch each of the valve spring retainers, one at a time. If a valve lifter is defective, it'll be evident from the shock felt at the retainer each time the valve seats.
2 The most likely causes of noisy valve lifters are dirt trapped inside the lifter and lack of oil flow, viscosity or pressure. Before condemning the lifters, check the oil for fuel contamination, correct level, cleanliness and correct viscosity.

Removal

Refer to illustrations 10.5a, 10.5b and 10.6

3 Remove the valve cover(s) and intake manifold (see Sections 4 and 5).
4 Remove the rocker arms and pushrods (see Section 6).

10.5a A magnetic pick-up tool . . .

10.5b . . . or a scribe can be used to remove the lifters

10.6 Store the lifters in a box like this to ensure installation in their original locations

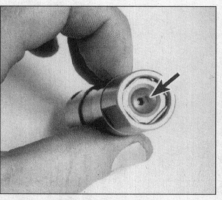
10.9a Check the pushrod seat (arrow) in the top of each lifter for wear

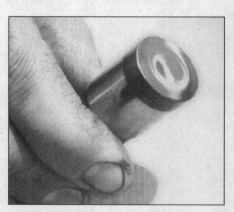
10.9b If the bottom of any lifter is worn concave, scratched or galled, replace the entire set with new lifters

10.9c The foot of each lifter should be slightly convex - the side of another lifter can be used as a straightedge to check it; if it appears flat, it is worn and must not be reused

10.9d If the foot of any lifter is scratched or galled, both the entire lifter set and the camshaft will have to be replaced

5   There are several ways to extract the lifters from the bores. A special tool designed to grip and remove lifters is manufactured by many tool companies and is widely available, but it may not be required in every case. On newer engines without a lot of varnish buildup, the lifters can often be removed with a small magnet or even with your fingers. A machinist's scribe with a bent end can be used to pull the lifters out by positioning the point under the retainer ring in the top of each lifter **(see illustrations)**. **Caution:** *Don't use pliers to remove the lifters unless you intend to replace them with new ones (along with the camshaft). The pliers may damage the precision machined and hardened lifters, rendering them useless.*

6   Before removing the lifters, arrange to store them in a clearly labeled box to ensure they're reinstalled in their original locations. Remove the lifters and store them where they won't get dirty **(see illustration)**. *Note: Some engines may have both standard and 0.010-inch oversize lifters installed at the factory. They are marked on the block.*

Inspection and installation

Refer to illustrations 10.9a, 10.9b, 10.9c and 10.9d

7   Parts for valve lifters are not available separately. The work required to remove them from the engine again if cleaning is unsuccessful outweighs any potential savings from repairing them.

8   Clean the lifters with solvent and dry them thoroughly without mixing them up.

9   Check each lifter wall, foot and pushrod seat for scuffing, score marks and uneven wear **(see illustrations)**. If the lifter walls are damaged or worn (which isn't very likely), inspect the lifter bores in the engine block as

Chapter 2 Part A 3.1L V6 engine

11.5 To get at the vibration damper, you'll need to remove the splash shield bolts (arrows) and splash shield

11.7 To remove the vibration damper-to-crankshaft bolt, fabricate a damper holder from a section of angle stock with a couple of holes drilled in its end - the bore centers of the holes must match the bore centers of two of the puller holes in the damper - then bolt the bar to the damper and use it to hold the damper while you break the bolt loose

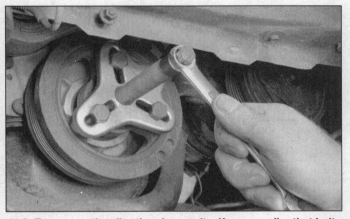

11.8 To remove the vibration damper itself, use a puller that bolts to the vibration damper hub - jaw-type pullers will damage the vibration damper

12.2 Carefully pry the old seal out of the timing chain cover - don't damage the crankshaft in the process

well. If the pushrod seats are worn, check the pushrod ends.
10 When reinstalling used lifters, make sure they're replaced in their original bores. Soak new lifters in oil to remove trapped air. Coat all lifters with moly-base grease or engine assembly lube prior to installation.
11 The remaining installation steps are the reverse of removal.
12 Run the engine and check for oil leaks.

11 Vibration damper - removal and installation Removal

Refer to illustrations 11.5, 11.7 and 11.8

1 Disconnect the negative battery cable from the battery.
2 Loosen the lug nuts on the right front wheel.
3 Raise the vehicle and support it securely on jackstands.
4 Remove the right front wheel.
5 Remove the right front inner fender splash shield **(see illustration)**.
6 Remove the serpentine drivebelt (see Chapter 1, Section 21).
7 Fabricate a damper holder from a three-foot length of angle stock by drilling two holes in the one end the same distance apart as two of the puller holes in the damper. Attach it to the damper with two bolts of the correct size and use it to hold the damper while you break the vibration damper-to-crankshaft bolt loose **(see illustration)**. Or, remove the driveplate access cover (see Chapter 2, Part C, Section 5) and position a large screwdriver in the ring gear teeth to keep the crankshaft from turning while an assistant removes the damper bolt.
8 Pull the damper off the crankshaft with a bolt-type puller **(see illustration)**. Leave the Woodruff key in place in the end of the crankshaft. **Caution:** *Don't use a jaw-type puller, as it will damage the damper.*

Installation

9 Installation is the reverse of removal. Be sure to apply a thin film of multi-purpose grease to the seal contact surface of the damper hub (if it isn't lubricated, the seal lip could be damaged and oil leakage would result). Align the damper hub keyway with the Woodruff key.
10 Tighten the vibration damper-to-crankshaft bolt to the torque listed in this Chapter's Specifications.
11 Reinstall the remaining parts in the reverse order of removal.

12 Crankshaft front oil seal - replacement

Refer to illustrations 12.2 and 12.3

1 Remove the vibration damper (see Section 11).
2 Note how the seal is installed - the new one must be installed to the same depth and facing the same way. Carefully pry the oil seal out of the cover with a seal puller or a screwdriver **(see illustration)**. Be very careful not to distort the cover or scratch the crankshaft!

12.3 A large socket and a hammer can be used to drive the new seal into place

13.5 The drivebelt tensioner is secured to the timing chain cover by a bolt (arrow)

13.12 Loosen this hose clamp (arrow) and detach the radiator outlet hose from the water pump inlet

13.13 Timing chain cover bolt locations (arrows)

3   Apply clean engine oil or multi-purpose grease to the outer edge of the new seal, then install it in the cover with the lip (spring side) facing IN. Drive the seal into place **(see illustration)** with a seal installation tool or a large socket and a hammer (if a large socket isn't available, a piece of pipe of the proper diameter will also work). Make sure the seal enters the bore squarely and stop when the front face is at the proper depth.
4   Reinstall the vibration damper.

13 Timing chain cover - removal and installation

Removal

Refer to illustrations 13.5, 13.12 and 13.13

1   Disconnect the negative battery cable from the battery.
2   Loosen, but do not remove, the water pump pulley bolts, then remove the serpentine drivebelt (see Chapter 1, Section 21).
3   Remove the water pump (see Chapter 3, Section 8);
4   Remove the vibration damper (see Section 11).

5   Unbolt and remove the drivebelt tensioner **(see illustration)**.
6   Drain the coolant and engine oil (see Chapter 1, Sections 29 and 12).
7   Remove the alternator (see Chapter 5, Section 13) and the mounting bracket **(see illustrations 4.16a and 4.16b)**.
8   Unbolt the power steering pump and tie it aside (see Chapter 10, Section 19). Leave the hoses connected.
9   Unbolt the driveplate cover.
10 Remove the starter (see Chapter 5, Section 16).
11 Remove the oil pan (see Section 15).
12 Disconnect the radiator hose from the water pump **(see illustration)**.
13 Remove the timing chain cover-to-engine block bolts **(see illustration)**. Note that some of the bolts require T-40 and T-50 Torx bits for removal.
14 Separate the cover from the engine. If it's stuck, tap it with a soft-face hammer, but don't try to pry it off.
15 Use a gasket scraper to remove all traces of old gasket material and sealant from the cover and engine block. The cover is made of aluminum, so be careful not to nick or gouge it. Clean the gasket sealing surfaces with lacquer thinner or acetone.

Installation

16 Apply a thin layer of RTV sealant to both sides of the new gasket, then position the gasket on the engine block (the dowel pins should keep it in place). Attach the cover to the engine and install the bolts.
17 Apply sealant to the bottom of the gasket **(see illustration 13.16)**. Follow a criss-cross pattern when tightening the fasteners and work up to the torque listed in this Chapter's Specifications in three steps.
18 The remainder of installation is the reverse of removal.
19 Add oil and coolant, start the engine and check for leaks.

14 Timing chain and sprockets - inspection, removal and installation

Inspection

1   The timing chain should be replaced with a new one if the engine has high mileage, the chain has visible damage, or total freeplay midway between the sprockets exceeds one inch. Failure to replace a worn

Chapter 2 Part A 3.1L V6 engine

14.3 The timing marks (arrows) on the sprockets should be aligned as shown - a straight line should pass through the center of the camshaft, the camshaft sprocket timing mark, the crankshaft sprocket timing mark and the center of the crankshaft (there are also two notches in the dampener for alignment purposes)

14.4 A screwdriver will keep the camshaft sprockets from turning while loosening the mounting bolts

timing chain may result in erratic engine performance, loss of power and decreased fuel mileage. Loose chains can "jump" timing. In the worst case, chain "jumping" or breakage will result in severe engine damage.

Removal

Refer to illustrations 14.3 and 14.4

2 Remove the timing chain cover (see Section 13).
3 Temporarily install the vibration damper bolt and turn the crankshaft with the bolt to align the timing marks on the crankshaft and camshaft sprockets with the marks on the dampener **(see illustration)**. At this point the number one and four pistons will be at Top Dead Center, with the number four piston on the compression stroke (this can be verified by checking the distributor rotor, which should point to the number four spark plug wire terminal on the distributor cap). **Note:** *Do not attempt to remove either sprocket or the timing chain until this is done and do not turn the crankshaft or camshaft after the sprockets/chain are removed.*
4 Remove the camshaft sprocket bolts **(see illustration)**. Do not turn the camshaft in the process (if you do, realign the timing marks before the bolts are removed).
5 Remove the camshaft sprocket. If it sticks to the camshaft, gently tap one side of the sprocket with a plastic mallet. Slip the timing chain and camshaft sprocket off the engine.
6 Timing chains and sprockets should be replaced in sets. If you intend to install a new timing chain, remove the crankshaft sprocket with a puller and install a new one. Be sure to align the key in the crankshaft with the keyway in the sprocket during installation.
7 Inspect the timing chain dampener (guide) for cracks and wear and replace it if necessary. The dampener is held to the engine block by two bolts.
8 Clean the timing chains and sprockets

with solvent and dry them with compressed air (if available). **Warning:** *Wear eye protection when using compressed air.*
9 Inspect the components for wear and damage. Look for teeth that are deformed, chipped, pitted and cracked.

Installation

Refer to illustration 14.10

10 Lubricate the thrust (rear) surface of the camshaft sprocket with moly-base grease or engine assembly lube **(see illustration)**. Install the timing chain over the camshaft sprocket with the slack hanging down.
11 With the timing marks aligned **(see illustration 14.3a)**, slip the chain over the crankshaft sprocket and install the camshaft sprocket onto the camshaft. If necessary, draw the sprocket onto the camshaft with the three bolts, but don't attempt to drive it on with a hammer, as this could dislodge the welch plug at the other end of the camshaft. Tighten the bolts to the torque listed in this Chapter's Specifications.

14.10 Lubricate the thrust (rear) surface of the camshaft sprocket before installing it

12 Check again to make sure the timing marks on the sprocket are properly aligned. If they aren't, remove the camshaft sprocket and move the chain until the marks align.
13 If you desire to position the engine with the number one piston at Top Dead Center on the compression stroke (to simplify distributor installation after installing a new camshaft, for example), turn the crankshaft one complete revolution (360-degrees) - the marks on the crankshaft and camshaft sprockets will now both be at the 12 o'clock position. However, if the distributor has not been removed, it isn't necessary to perform this Step.
14 Lubricate the chain and sprocket with clean engine oil. Install the timing chain cover (see Section 13).
15 The remaining installation steps are the reverse of removal.

15 Oil pan - removal and installation

Removal

Refer to illustrations 15.8, 15.9 and 15.11

1 Disconnect the cable from the negative battery terminal.
2 Remove the serpentine drivebelt (see Chapter 1, Section 21).
3 Raise the front of the vehicle and place it securely on jackstands. Apply the parking brake and block the rear wheels to keep it from rolling off the stands. Drain the engine oil (see Chapter 1, Section 12).
4 Remove the vibration damper (see Section 11).
5 Drain the oil (see Chapter 1, Section 12).
6 Remove the driveplate access cover (see Chapter 2, Part C, **illustrations 5.13a, 5.13b and 5.13c**).
7 Remove the starter (see Chapter 5, Section 16).
8 Support the engine from underneath

15.8 To remove the oil pan, you'll have to raise the engine with a floor jack - place the jack pad directly underneath the rear transaxle bracket

15.9 Remove these three bolts (arrows) from the engine mounting bracket and raise the engine high enough to get at the oil pan bolts

with a floor jack. Put the pad of the jack directly underneath the rear transaxle mounting bracket **(see illustration)**.

9 Remove the bolts that attach the engine to the engine mounting bracket **(see illustration)**.

10 Raise the engine slightly with the floor jack.

11 Remove the oil pan bolts **(see illustration)**, then carefully separate the oil pan from the block. Don't pry between the block and the pan - you'll damage the sealing surface of the pan flange and oil leaks may develop. Instead, tap the pan with a soft-face hammer to break the gasket seal.

Installation

12 Clean the pan with solvent and remove all old sealant and gasket material from the block and pan mating surfaces. Clean the mating surfaces with lacquer thinner or acetone and make sure the bolt holes in the block are clear. Check the oil pan flange for distortion, particularly around the bolt holes. If necessary, place the pan on a block of wood and use a hammer to flatten and restore the gasket surface.

13 Always use a new gasket whenever the oil pan is installed. Apply a bead of RTV sealant to the gasket tabs at the rear main bearing caps.

14 Place the oil pan in position on the block and install the nuts/bolts.

15 After the fasteners are installed, tighten them to the torque listed in this Chapter's Specifications. Starting at the center, follow a criss-cross pattern and work up to the final torque in three steps.

16 The remaining steps are the reverse of the removal procedure.

17 Refill the engine with oil, run it until normal operating temperature is reached and check for leaks.

16 Oil pump - removal and installation

Refer to illustration 16.2

1 Remove the oil pan (see Section 15).
2 Unbolt the oil pump and lower it from the engine **(see illustration)**.
3 If the pump is defective, replace it with a new one - don't reuse the original or attempt to rebuild it.
4 Prime the pump by pouring motor oil into the pick-up screen while turning the pump driveshaft.
5 To install the pump, turn the hexagonal driveshaft so it mates with the oil pump drive.
6 Install the pump mounting bolt and tighten it to the torque listed in this Chapter's Specifications.

17 Engine mounts - check and replacement

Warning: *We strongly urge you to use proper lifting methods and devices. DO NOT place any part of your body under the engine/transaxle assembly when it's supported only by a jack. Failure of the jack could result in severe injury, even death.*

1 Engine mounts seldom require attention, but broken or deteriorated mounts should be replaced immediately or the added strain placed on the driveline components may cause damage or wear.

Check

2 During the check, the engine must be raised slightly to remove the weight from the mounts.
3 Raise the vehicle and place it securely on jackstands, then position a jack under the engine oil pan. Place a block of wood between the jack head and the oil pan, then carefully raise the engine just enough to take the weight off the mounts.
4 Check the mounts to see if the rubber is cracked, hardened or separated from the metal plates. Sometimes the rubber will split right down the center.
5 Check for relative movement between the mount plates and the engine or frame (use a large screwdriver or pry bar to attempt to move the mounts). If movement is noted, lower the engine and tighten the mount fasteners.
6 Rubber preservative should be applied to the mounts to slow deterioration.

Replacement

Refer to illustration 17.9

7 Disconnect the negative battery cable from the battery, then raise the vehicle and place it securely on jackstands, if you haven't already done so.
8 Remove the engine torque strut **(see illustration 4.4)**.
9 Raise the engine slightly with a jack or hoist. Remove the fasteners and detach the mount from the frame **(see illustration)**.
10 Move the mount-to-block bracket bolts/nuts and detach the mount.
11 Installation is the reverse of removal. Use thread locking compound on the threads and be sure to tighten all fasteners securely.

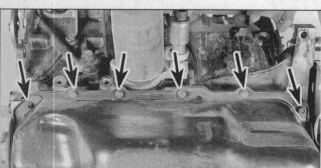

15.11 Oil pan bolts (arrows) on the front (radiator side) flange of the oil pan (bolts on rear flange identical)

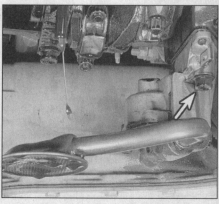

16.2 Oil pump mounting bolt location (arrow)

Chapter 2 Part B
3800 V6 engine

Contents

	Section
Camshaft and bearings - removal, inspection and installation	See Chapter 2C
Crankshaft front oil seal - replacement	12
Cylinder compression check	See Chapter 2C
Cylinder heads - removal and installation	10
Drivebelt check, adjustment and replacement	See Chapter 1
Driveplate - removal and installation	See Chapter 2C
Engine mount - check and replacement	19
Engine oil and filter change	See Chapter 1
Engine overhaul - general information	See Chapter 2C
Engine - removal and installation	See Chapter 2C
Exhaust manifolds - removal and installation	9
General information	1
Intake manifold - removal and installation	7
Oil filter adapter and pressure regulator valve – removal and installation	14
Oil pan - removal and installation	17
Oil pump pick-up tube and screen assembly – removal and installation	18
Oil pump - removal, inspection and installation	15
Rear main oil seal - replacement	See Chapter 2C
Repair operations possible with the engine in the vehicle	2
Rocker arms and pushrods - removal, inspection and installation	5
Spark plug replacement	See Chapter 1
Timing chain and sprockets - removal and installation	16
Timing chain cover - removal and installation	13
Top Dead Center (TDC) for number one piston - locating	3
Valve covers - removal and installation	4
Valve lifters - removal, inspection and installation	8
Valve springs, retainers and seals - replacement	6
Vibration damper - removal and installation	11
Water pump - removal and installation	See Chapter 3

Specifications

General

Cylinder numbers (drivebelt end-to-transaxle end)	
Front bank (radiator side)	1-3-5
Rear bank	2-4-6
Firing order	1-6-5-4-3-2

Valve lifters

Diameter	0.8420 to 0.8427 inch
Lifter-to-bore clearance	0.0008 to 0.0025 inch

Oil pump

Outer gear-to-housing clearance	0.008 to 0.015 inch
Inner gear-to-outer gear (tip) clearance	0.006 inch
Gear end clearance	0.001 to 0.0035 inch
Pump cover warpage limit	0.002 inch

Cylinder location and distributor rotation

Torque specifications

Ft-lbs (unless otherwise indicated)

Camshaft sprocket bolt	
Step one	74
Step two	Turn an additional 105-degrees
Cylinder head bolts	
Step one	35
Step two	Turn all bolts an additional 130-degrees
Step three	Turn the center four bolts an additional 30-degrees
Exhaust manifold-to-cylinder head bolts	38
Intake manifold-to-cylinder head bolts/nuts	88 in-lbs
Oil pan bolts	124 in-lbs
Oil filter adapter-to-timing chain cover bolts	22
Oil pump	
Cover-to-timing chain cover bolts	97 in-lbs
Pick-up tube and screen assembly bolts	132 in-lbs
Valve cover nuts/bolts	88 in-lbs
Rocker arm pivot bolts	28
Timing chain cover bolts	22
Vibration damper center bolt	
Step one	110
Step two	Turn an additional 76-degrees

1 General information

Note: *On models equipped with the Delco Loc II audio system, be sure the lockout feature is turned off before performing any procedure which requires disconnecting the battery.*

This Part of Chapter 2 is devoted to in-vehicle repair procedures for the 3800 V6 engine. As on other GM V6s, the crankshaft is counterbalanced by the flywheel, the crankshaft pulley and weights cast into the crankshaft, but it's also equipped with a balance shaft - driven by the camshaft - which rides in the block above the camshaft.

Information concerning the camshaft, balance shaft, driveplate, rear main oil seal and engine removal and installation, as well as engine block and cylinder head overhaul, is in Part C of this Chapter.

The following repair procedures are based on the assumption that the engine is installed in the vehicle. If the engine has been removed from the vehicle and mounted on a stand, many of the steps included in this Part of Chapter 2 will not apply.

The Specifications included in this Part of Chapter 2 apply only to the engine and procedures in this Part. The Specifications necessary for rebuilding the block and cylinder heads are found in Part C.

2 Repair operations possible with the engine in the vehicle

Many major repair operations can be accomplished without removing the engine from the vehicle.

Clean the engine compartment and the exterior of the engine with some type of pressure washer before any work is done. A clean engine will make the job easier and will help keep dirt out of the internal areas of the engine.

Depending on the components involved, it may be a good idea to remove the hood to improve access to the engine as repairs are performed (refer to Chapter 11, Section 9 if necessary).

If vacuum, exhaust, oil or coolant leaks develop, indicating a need for gasket or seal replacement, the repairs can generally be made with the engine in the vehicle. The intake and exhaust manifold gaskets, oil pan gasket and cylinder head gaskets are all accessible with the engine in place.

Exterior engine components such as the intake and exhaust manifolds, the oil pan, the oil pump, the water pump, the starter motor, the alternator and the fuel injection system can be removed for repair with the engine in place. The timing chain and sprockets can also be replaced with the engine in the vehicle, but the camshaft (and balance shaft, if equipped) cannot be removed with the engine in place.

Since the cylinder heads can be removed without pulling the engine, valve component servicing can also be accomplished with the engine in the vehicle.

In extreme cases caused by a lack of necessary equipment, repair or replacement of piston rings, pistons, connecting rods and rod bearings is possible with the engine in the vehicle. However, this practice is not recommended because of the cleaning and preparation work that must be done to the components involved.

3 Top Dead Center (TDC) for number one piston - locating

Refer to illustrations 3.6 and 3.8

1 Top Dead Center (TDC) is the highest point in the cylinder each piston reaches as it travels up-and-down when the crankshaft turns. Each piston reaches TDC on the compression stroke and again on the exhaust stroke, but TDC generally refers to piston position on the compression stroke.

2 Positioning the piston(s) at TDC is an essential part of certain procedures such as camshaft removal and timing chain/sprocket removal.

3 Remove the spark plugs (see Chapter 1, Section 32).

4 Place the transaxle in Park, apply the parking brake and block the rear wheels. Raise the front of the vehicle and support it securely on jackstands.

5 When looking at the drivebelt end of the engine, normal crankshaft rotation is clockwise. In order to bring any piston to TDC, the crankshaft must be turned with a socket and ratchet attached to the bolt threaded into the center of the vibration damper on the crankshaft.

6 Have an assistant turn the crankshaft with a socket and ratchet as described above while you hold a finger over the number one

3.6 Hold a finger over the spark plug hole until you feel air escaping

Chapter 2 Part B 3800 V6 engine

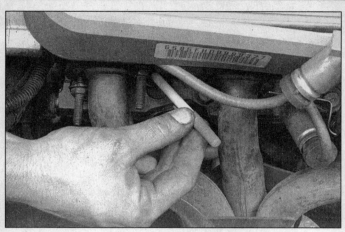

3.8 Insert a soft plastic pen into the hole to detect piston movement

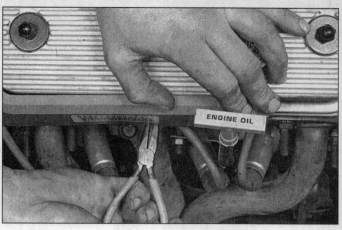

4.2 Unclip the harness cover

spark plug hole **(see illustration)**. Note: *See the cylinder numbering diagram in this Chapter's Specifications for the location of the number one cylinder.*

7   When the piston approaches TDC, air pressure will be felt at the spark plug hole. Instruct your assistant to turn the crankshaft slowly.

8   Insert a plastic pen into the spark plug hole **(see illustration)**. As the piston rises the pen will be pushed out. Note the point where the pen stops moving out - this is TDC.

9   After the number one piston has been positioned at TDC on the compression stroke, TDC for any of the remaining pistons can be located by turning the crankshaft clockwise, 120-degrees at a time, following the firing order. Or, you can simply repeat Steps 6 through 8 to find TDC for any particular cylinder.

4   Valve covers - removal and installation

Front (radiator side) cover

Refer to illustration 4.2

1   Disconnect the negative battery cable from the battery.

2   Remove the spark plug wires from the spark plugs and remove the harness cover **(see illustration)**. Number each wire before removal to ensure correct reinstallation.

3   Remove the serpentine drivebelt (see Chapter 1, Section 21).

4   Remove the alternator brace.

5   Remove the valve cover mounting bolts/nuts.

6   Detach the valve cover. Note: *If the cover sticks to the cylinder head, use a soft-face hammer to dislodge it.*

Rear (firewall side) cover

7   Disconnect the negative battery cable from the battery.

8   Remove the spark plug wires from the spark plugs and remove the wire holder. Be sure each wire is labeled before removal to ensure correct reinstallation.

9   Remove the serpentine drivebelt (see Chapter 1, Section 21).

10   Detach the power steering pump and braces (see Chapter 10, Section 19).

11   Remove the valve cover mounting bolts/nuts.

12   Detach the valve cover. Note: *If the cover sticks to the cylinder head, use a soft-face hammer to dislodge it.*

Installation

13   The mating surfaces of the cylinder head and valve cover must be perfectly clean when the covers are installed. Use a gasket scraper to remove all traces of sealant or old gasket, then clean the mating surfaces with lacquer thinner or acetone (if there's sealant or oil on the mating surfaces when the cover is installed, oil leaks may develop). The valve covers are made of aluminum, so be extra careful not to nick or gouge the mating surfaces with the scraper.

14   Apply a non-hardening thread locking compound to the mounting bolt/nut threads. Place the valve cover and new gasket in position, then install the bolts/nuts.

15   Tighten the bolts/nuts in several steps to the torque listed in this Chapter's Specifications.

16   Complete the installation by reversing the removal procedure. Be sure to add coolant if it was drained.

17   Start the engine and check for oil leaks at the valve cover-to-head joints.

5   Rocker arms and pushrods - removal, inspection and installation

Removal

Refer to illustration 5.3

1   Detach the valve covers from the cylinder heads (see Section 4).

2   Loosen the rocker arm pivot bolts one at

5.3 If more than one pushrod is being removed, store them in a perforated cardboard box to prevent mix-ups during installation - note the label indicating the front of the engine

a time and detach the rocker arms, bolts, pivots and pivot retainers. Store each set of rocker components separately in a marked plastic bag to ensure that they're reinstalled in their original locations.

3   Remove the pushrods and store them separately to make sure they don't get mixed up during installation **(see illustration)**.

Inspection

4   Check each rocker arm for wear, cracks and other damage, especially where the pushrods and valve stems contact the rocker arm.

5   Check the pivot seat in each rocker arm and the pivot faces. Look for galling, stress cracks and unusual wear patterns. If the rocker arms are worn or damaged, replace them with new ones and install new pivots as well.

6   Make sure the hole at the pushrod end of each rocker arm is open.

7   Inspect the pushrods for cracks and

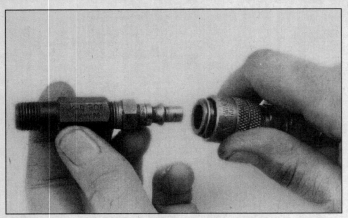

6.4 This is what the air hose adapter that threads into the spark plug hole looks like - they're commonly available from auto parts stores

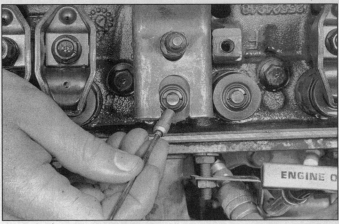

6.8 After compressing the valve spring, remove the keepers with a magnet (as shown here) or small needle-nose pliers

excessive wear at the ends. Roll each pushrod across a piece of plate glass to see if it's bent (if it wobbles, it's bent).

Installation

8   Lubricate the lower end of each pushrod with clean engine oil or moly-base grease and install them in their original locations. Make sure each pushrod seats completely in the lifter socket.

9   Apply moly-base grease to the ends of the valve stems, the upper ends of the pushrods and to the pivot faces to prevent damage to the mating surfaces before engine oil pressure builds up.

10   Coat the rocker arm pivot bolts with a non-hardening thread locking compound. Install the rocker arms, pivots, pivot retainers and bolts and tighten the bolts to the torque listed in this Chapter's Specifications. As the bolts are tightened, make sure the pushrods seat properly in the rocker arms.

11   Install the valve covers (see Section 4).

6   Valve springs, retainers and seals - replacement

Refer to illustrations 6.4, 6.8 and 6.17

Note: *Broken valve springs and defective valve stem seals can be replaced without removing the cylinder heads. Two special tools and a compressed air source are normally required to perform this operation, so read through this Section carefully and rent or buy the tools before beginning the job. If compressed air is not available, a length of nylon rope can be used to keep the valves from falling into the cylinder during this procedure.*

1   Remove the valve cover from the affected cylinder head. If all of the valve stem seals are being replaced, remove both valve covers (see Section 4).

2   Remove the spark plug from the cylinder which has the defective component. If all of the valve stem seals are being replaced, all of the spark plugs should be removed.

3   Turn the crankshaft until the piston in the affected cylinder is at Top Dead Center on the compression stroke (see Section 3). If you're replacing all of the valve stem seals, begin with cylinder number one and work on the valves for one cylinder at a time. Move from cylinder-to-cylinder following the firing order sequence (see this Chapter's Specifications).

4   Thread an adapter into the spark plug hole **(see illustration)** and connect an air hose from a compressed air source to it. Most auto parts stores can supply the air hose adapter. **Note:** *Many cylinder compression gauges utilize a screw-in fitting that may work with your air hose quick-disconnect fitting.*

5   Remove the bolt, pivot and rocker arm for the valve with the defective part and pull out the pushrod (see Section 5). If all of the valve stem seals are being replaced, all of the rocker arms and pushrods should be removed.

6   Apply compressed air to the cylinder. The valves should be held in place by the air pressure. If the valve faces or seats are in poor condition, leaks may prevent the air pressure from retaining the valves - refer to the alternative procedure below.

7   If you don't have access to compressed air, an alternative method can be used. Position the piston at a point approximately 45-degrees before TDC on the compression stroke, then feed a long piece of nylon rope through the spark plug hole until it fills the combustion chamber. Be sure to leave the end of the rope hanging out of the engine so it can be removed easily. Use a large breaker bar and socket to rotate the crankshaft in the normal direction of rotation until slight resistance is felt.

8   Stuff shop rags into the cylinder head holes above and below the valves to prevent parts and tools from falling into the engine, then use a valve spring compressor to compress the spring. Remove the keepers with small needle-nose pliers or a magnet **(see illustration)**. **Note:** *A couple of different types of tools are available for compressing the valve springs with the head in place. One type grips the lower spring coils and presses on the retainer as the knob is turned, while the other type utilizes the rocker arm bolt for leverage* **(see illustration)**. *Both types work very well, but the lever type is usually less expensive.*

9   Remove the spring retainer and valve spring, then remove the guide seal. **Note:** *If air pressure fails to hold the valve in the closed position during this operation, the valve face or seat is probably damaged. If so, the cylinder head will have to be removed for additional repair operations.*

10   Wrap a rubber band or tape around the top of the valve stem so the valve won't fall into the combustion chamber, then release the air pressure. **Note:** *If a rope was used instead of air pressure, turn the crankshaft slightly in the direction opposite normal rotation.*

11   Inspect the valve stem for damage. Rotate the valve in the guide and check the end for eccentric movement, which would indicate that the valve is bent.

12   Move the valve up-and-down in the guide and make sure it doesn't bind. If the valve stem binds, either the valve is bent or the guide is damaged. In either case, the head will have to be removed for repair.

13   Reapply air pressure to the cylinder to retain the valve in the closed position, then remove the tape or rubber band from the valve stem. If a rope was used instead of air pressure, rotate the crankshaft in the normal direction of rotation until slight resistance is felt.

14   Lubricate the valve stem with clean engine oil. Install a new guide seal.

15   Install the spring in position over the valve.

16   Install the valve spring retainer and compress the valve spring.

17   Position the keepers in the upper groove. Apply a small dab of grease to the inside of each keeper to hold it in place if

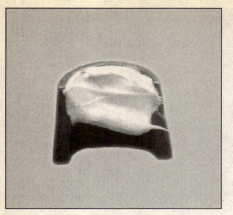

6.17 Apply a small dab of grease to each keeper as shown here before installation - it'll hold them in place on the valve stem as the spring is released

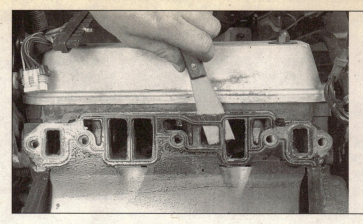

7.15 Remove all traces of gasket material, but be careful not to gouge the mating surfaces of the manifold and cylinder heads

necessary **(see illustration)**. Remove the pressure from the spring tool and make sure the keepers are seated.
18 Disconnect the air hose and remove the adapter from the spark plug hole. If a rope was used in place of air pressure, pull it out of the cylinder.
19 Install the rocker arm(s) and pushrod(s).
20 Install the spark plug(s) and connect the wire(s).
21 Refer to Section 4 and install the valve cover(s).
22 Start and run the engine, then check for oil leaks and unusual sounds coming from the valve cover area.

7 Intake manifold - removal and installation

Removal

1 Relieve the fuel system pressure (see Chapter 4, Section 2).
2 Disconnect the negative battery cable from the battery.
3 Remove the air intake duct and the throttle body (see Chapter 4, Section 12 or 13).
4 Remove the fuel rail and injectors (see Chapter 4, Section 13).
5 Remove the heat shield from the exhaust manifold crossover tube (see Section 9).
6 Detach the control cables from their brackets (see Chapter 4, Section 9).
7 Remove the power steering pump-to-manifold brace (see Chapter 10, Section 19). Note the location of all fasteners for reassembly.
8 Loosen the alternator and swing it out of the way. Detach the alternator mounting bracket (see Chapter 5, Section 13). Note the location of all fasteners for reassembly.
9 Drain the cooling system (see Chapter 1, Section 29).
10 Disconnect the heater coolant tubes and hoses at the manifold.
11 Disconnect the bypass hose.
12 Label and disconnect the fuel and vacuum lines and all wire harness connectors at the manifold. When disconnecting the fuel line fittings, wrap shop rags around them and be prepared for some residual fuel leakage. Cap the fittings to prevent contamination.
13 Remove the intake manifold bolts and remove the manifold. Do not pry between the manifold and heads, as damage to the gasket sealing surfaces may result.
14 If you're installing a new manifold, transfer all fittings and sensors to the new manifold.

Installation

Refer to illustrations 7.15 and 7.19
Note: *The mating surfaces of the cylinder heads, block and manifold must be perfectly clean when the manifold is installed. Gasket removal solvents in aerosol cans are available at most auto parts stores and may be helpful when removing old gasket material that's stuck to the heads and manifold (since the manifold is made of aluminum, aggressive scraping can cause damage). Be sure to follow the directions printed on the container.*
15 Use a gasket scraper to remove all traces of sealant and old gasket material **(see illustration)**, then clean the mating surfaces with lacquer thinner or acetone. If there's old sealant or oil on the mating surfaces when the manifold is installed, oil or vacuum leaks may develop. Use a vacuum cleaner to remove any gasket material that falls into the intake ports or the lifter valley.
16 Use a tap of the correct size to chase the threads in the bolt holes, then use compressed air (if available) to remove the debris from the holes. **Warning:** *Wear safety glasses or a face shield to protect your eyes when using compressed air.*
17 If steel manifold gaskets are used, apply a thin, even coat of sealant such as K&W Copper Coat (or equivalent) to both sides of the gaskets before positioning them on the heads. Apply RTV sealant to the ends of the new manifold-to-block seals, then install them. Make sure the pointed end of the seal fits snugly against the head and block.
18 Carefully lower the manifold into place.

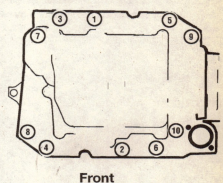

7.19 Intake manifold bolt tightening sequence

Apply a non-hardening thread locking compound to the mounting bolt threads and install the bolts finger tight.
19 Tighten the mounting bolts in two stages, following the recommended sequence **(see illustration)**, until they're all at the torque listed in this Chapter's specifications.
20 Install the remaining components in the reverse order of removal.
21 Change the oil and filter (see Chapter 1, Section 12) and fill the cooling system (see Chapter 1, Section 29). Start the engine and check for oil and vacuum leaks.

8 Valve lifters - removal, inspection and installation

1 A noisy valve lifter can be isolated when the engine is idling. Hold a mechanic's stethoscope or a length of hose near the position of each valve while listening at the other end.
2 The most likely causes of noisy valve lifters are dirt trapped between the plunger and the lifter body or lack of oil flow, viscosity or pressure. Before condemning the lifters, we recommend checking the oil for fuel contamination, proper level, cleanliness and correct viscosity.

8.6a The guide is held in place by two bolts (arrows)

8.6b The lifter guides slip over the lifters

8.7 You can usually remove the lifters by hand

8.8 You may have to remove the lifters on a high-mileage engine with a special tool – be sure to store the lifters in an organized manner to make sure they're reinstalled in their original locations

Removal

Refer to illustrations 8.6a, 8.6b, 8.7 and 8.8

3   Remove the valve cover(s) (see Section 4).
4   Remove the rocker arms and pushrods (see Section 5).
5   Remove the intake manifold (see Section 7).
6   Remove the lifter guide retainer and lifter guides **(see illustrations)**.
7   There are several ways to extract the lifters from the bores. A special tool designed to grip and remove lifters is manufactured by many tool companies and is widely available, but it may not be required in every case. On newer engines without a lot of varnish buildup, the lifters can often be removed with a small magnet or even with your fingers **(see illustration)**. A machinist's scribe with a bent end can be used to pull the lifters out by positioning the point under the retainer ring in the top of each lifter. **Caution:** *Don't use pliers to remove the lifters unless you intend to replace them with new ones. The pliers will damage the precision machined and hardened lifters, rendering them useless.*
8   Before removing the lifters, arrange to store them in a clearly labeled box to ensure that they're reinstalled in their original locations. Remove the lifters and store them where they won't get dirty **(see illustration)**.

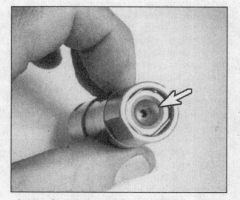

8.11a Check the pushrod seat (arrow) in the top of each lifter for wear

Inspection

Refer to illustrations 8.11a and 8.11b

9   Parts for valve lifters are not available separately. The work required to remove them from the engine again if repair is unsuccessful outweighs any potential savings from repairing them.
10 Clean the lifters with solvent and dry them thoroughly without mixing them up.
11 Check the bearing surface of each lifter body and the pushrod seat for scuffing, score marks and uneven wear. If the lifter walls are damaged or worn, inspect the lifter bores in

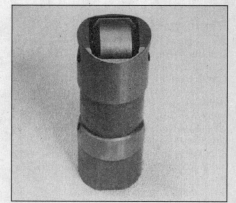

8.11b The roller on roller lifters must turn freely - check for wear and excessive play as well

9.3 The exhaust manifolds are bolted to the crossover pipe (arrows)

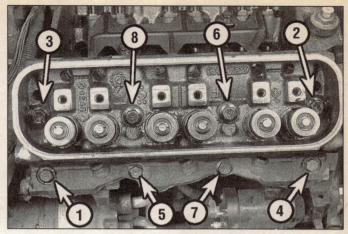

10.9 Cylinder head bolt LOOSENING sequence

the engine block. If the pushrod seats are worn **(see illustration)**, check the pushrod ends. Check the rollers carefully for wear and damage and make sure they turn freely without excessive play **(see illustration)**.

Installation

12 Used roller lifters can be reinstalled with a new camshaft (provided they are in good shape) and the original camshaft can be used if new lifters are installed (as long as it isn't worn out).
13 When reinstalling used lifters, make sure they're replaced in their original bores. Soak new lifters in oil to remove trapped air. Coat all lifters with moly-base grease or engine assembly lube prior to installation.
14 The remaining installation steps are the reverse of removal.
15 Run the engine and check for oil leaks.

9 Exhaust manifolds - removal and installation

Warning: *Allow the engine to cool completely before beginning this procedure.*
Note: *Exhaust system fasteners are frequently difficult to remove - they get frozen in place because of the heating/cooling cycle to which they're constantly exposed. To ease removal, apply penetrating oil to the threads of all exhaust manifold and exhaust pipe fasteners and allow it to soak in.*

Front manifold

Refer to illustration 9.3

1 Disconnect the negative battery cable.
2 Remove the cooling fan for access (see Chapter 3, Section 4).
3 Unbolt the manifold from the crossover pipe **(see illustration)**.
4 Remove the dipstick tube hold-down nut and wiggle the dipstick tube out of the block.
5 Detach the spark plug wires from the front spark plugs (see Chapter 1, Section 33).
6 Remove the heat shield, if equipped.
7 Unbolt and remove the exhaust manifold.

Rear manifold

8 Disconnect the negative battery cable.
9 Disconnect the spark plug wires from the rear spark plugs (see Chapter 1, Section 33).
10 Detach the throttle cable bracket.
11 Remove the crossover pipe heat shield.
12 Remove the two bolts attaching the crossover pipe to the rear exhaust manifold.
13 Remove the plastic vacuum tank mounted on the cowl.
14 Raise the vehicle and place it on safety stands.
15 Remove the manifold heat shield, if equipped.
16 Remove the catalytic converter heat shield and hanger.
17 Detach the transaxle dipstick tube bracket and remove the tube.
18 Unplug the connector for the oxygen sensor lead.
19 Remove the three retaining nuts from the underside of the exhaust manifold.
20 Lower the vehicle.
21 Remove the engine lifting bracket.
22 Remove the three upper retaining nuts from the exhaust manifold and detach the manifold from the head.
23 If you're replacing the manifold itself, remove the oxygen sensor (see Chapter 6, Section 4) and install it on the new manifold.

Installation

24 Clean the mating surfaces of the manifold and cylinder head to remove all traces of old gasket material, then check the manifold for warpage and cracks. If the manifold gasket was blown, take the manifold to an automotive machine shop for resurfacing.
25 Place the manifold in position with a new gasket and install the bolts finger tight.
26 Starting in the middle and working out toward the ends, tighten the mounting bolts a little at a time until all of them are at the torque listed in this Chapter's Specifications.
27 Install the remaining components in the reverse order of removal.
28 Start the engine and check for exhaust leaks between the manifold and cylinder head and between the manifold and exhaust pipe.

10 Cylinder heads - removal and installation

Removal

Refer to illustrations 10.9 and 10.10

1 Disconnect the negative battery cable at the battery.
2 Disconnect the spark plug wires and remove the spark plugs (see Chapter 1, Section 32). Be sure to label the plug wires to simplify reinstallation.
3 Remove all accessories and accessory mounting brackets. Note the size and location of each fastener to facilitate reassembly.
4 Disconnect all wires and hoses from the cylinder head(s). Be sure to label them to simplify reinstallation.
5 Remove the intake manifold (see Section 7).
6 Detach the exhaust manifold(s) from the cylinder head(s) being removed (see Section 9).
7 Remove the valve cover(s) (see Section 4).
8 Remove the rocker arms and pushrods (see Section 5).
9 Loosen the head bolts in 1/4-turn increments until they can be removed by hand. Work from bolt-to-bolt as shown **(see illustration)**.
10 Lift the head off the engine. If resistance is felt, don't pry between the head and block as damage to the mating surfaces will result. Recheck for head bolts that may have been overlooked, then use a hammer and block of

10.10 Pry carefully - don't force a tool between the gasket surfaces

10.13 Carefully remove all traces of old gasket material

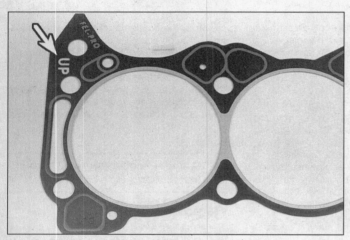

10.16 Look for gasket markings (arrow) to ensure correct installation

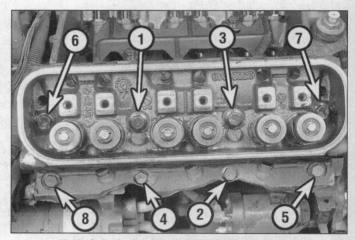

10.19 Cylinder head bolt TIGHTENING sequence

wood to tap the head and break the gasket seal. Be careful because there are locating dowels in the block which position each head. As a last resort, pry each head up at the rear corner only and be careful not to damage anything **(see illustration)**. After removal, place the head on blocks of wood to prevent damage to the gasket surfaces.

11 Refer to Chapter 2, Part C, for cylinder head disassembly, inspection and valve service procedures.

Installation

Refer to illustrations 10.13, 10.16 and 10.19

12 The mating surfaces of the cylinder heads and block must be perfectly clean when the heads are installed.

13 Use a gasket scraper to remove all traces of carbon and old gasket material **(see illustration)**, then clean the mating surfaces with lacquer thinner or acetone. If there's oil on the mating surfaces when the heads are installed, the gaskets may not seal correctly and leaks may develop. When working on the block, it's a good idea to cover the lifter valley with shop rags to keep debris out of the engine. Use a shop rag or vacuum cleaner to remove any debris that falls into the cylinders.

14 Check the block and head mating surfaces for nicks, deep scratches and other damage. If damage is slight, it can be removed with a file; if it's excessive, machining may be the only alternative.

15 Use a tap of the correct size to chase the threads in the head bolt holes. Dirt, corrosion, sealant and damaged threads will affect torque readings.

16 Position the new gaskets over the dowel pins in the block. If steel gaskets are used, apply a thin, even coat of sealant such as K&W Copper Coat (or equivalent) to both sides of the gaskets before positioning them on the block. Install "non-retorquing" type gaskets dry (no sealant), unless the manufacturer states otherwise. Most gaskets are marked UP or TOP **(see illustration)** because they must be installed a certain way.

17 Carefully position the heads on the block without disturbing the gaskets.

18 Apply a non-hardening sealant, such as Permatex no. 2 (or equivalent) to the threads and the undersides of the bolt heads.

19 Tighten the cylinder head bolts as directed in this Chapter's Specifications in the sequence shown **(see illustration)**. This must be done in three steps, following the sequence each time.

20 The remaining installation steps are the reverse of removal.

21 Change the oil and filter (see Chapter 1, Section 12).

11 Vibration damper - removal and installation

Refer to illustrations 11.7a, 11.7b and 11.9

Caution: *The vibration damper is serviced as an assembly. Do not attempt to separate the pulley from the balancer hub.*

1 Disconnect the negative cable from the battery.

2 Loosen the lug nuts on the right front wheel.

3 Raise the vehicle and support it securely on jackstands.

Chapter 2 Part B 3800 V6 engine

11.7a Keep the ring gear from turning by wedging a large screwdriver in the ring gear teeth

11.7b Remove the bolt in the center of the damper hub (arrow)

11.9 Be sure to align the key way (arrow) with the key in the crankshaft

4   Remove the right front wheel.
5   Remove the right front fender inner splash shield.
6   Remove the drivebelt (see Chapter 1, Section 21).
7   Remove the lower bellhousing cover plate and position a large screwdriver in the ring gear teeth **(see illustration)** to keep the crankshaft from turning while an assistant removes the crankshaft balancer bolt **(see illustration)**. The bolt is normally quite tight, so use a large breaker bar and a six-point socket.
8   The damper should pull off the crankshaft by hand. Leave the Woodruff key in place in the end of the crankshaft.
9   Installation is the reverse of removal. Align the keyway with the key **(see illustration)** and avoid bending the metal tabs. Be sure to apply multi-purpose grease to the seal contact surface on the back side of the damper (if it isn't lubricated, the seal lip could be damaged and oil leakage would result).
10 Apply sealant to the threads and tighten the crankshaft bolt to the torque listed in this Chapter's Specifications.

11 Reinstall the remaining components in the reverse order of removal.

12 Crankshaft front oil seal - replacement

Refer to illustrations 12.2 and 12.5

1   Remove the vibration damper (see Section 11). If there's a groove worn into the seal contact surface on the vibration damper, sleeves are available that fit over the groove, restoring the contact surface to like-new condition. These sleeves are sometimes included with the seal kit. Check with your parts supplier for details.
2   Pry the old oil seal out with a seal removal tool or a screwdriver **(see illustration)**. Be very careful not to nick or otherwise damage the crankshaft in the process.
3   Apply a thin coat of RTV sealant to the outer edge of the new seal. Lubricate the seal lip with multi-purpose grease or clean engine oil.
4   Place the seal squarely in position in the bore and press it into place with special tool, available at most auto parts stores. Make sure the seal enters the bore squarely and seats completely.
5   If the special tool is unavailable, carefully guide the seal into place with a large socket or piece of pipe and a hammer **(see illustration)**. The outer diameter of the socket or pipe should be the same size as the seal outer diameter.
6   Install the vibration damper (see Section 11).
7   Reinstall the remaining parts in the reverse order of removal.
8   Start the engine and check for oil leaks at the seal.

13 Timing chain cover - removal and installation

Refer to illustrations 13.9 and 13.11

Note: *A special tool is required to complete this procedure - read through all steps before beginning.*

1   Disconnect the negative battery cable from the battery.
2   Set the parking brake and put the transmission in Park. Raise the front of the vehicle and support it securely on jackstands. Remove the splash shield from the right inner fender.
3   Drain the oil and coolant (see Chapter 1, Sections 12 and 29). Remove the coolant hoses from the timing chain cover.
4   Remove the right (passenger's side) engine mount (see Section 19).
5   Remove the serpentine drivebelt (see Chapter 1, Section 21).
6   Remove the vibration damper (see Section 11) and the sensor shield.
7   Unplug the connectors from the oil pressure, camshaft and crankshaft sensors.
8   Remove the oil pan-to-timing chain cover bolts (see Section 17).
9   Remove the timing chain cover bolts

12.2 Pry out the old seal with a seal removal tool (shown here) or a screwdriver

12.5 Gently drive the new seal into place with a hammer and large socket

13.9 Typical timing chain cover bolt locations (arrows)

13.11 Remove all traces of old gasket material

(see illustration). Note that two of the bolts also secure the crankshaft sensor. Lift the sensor off when removing these bolts.
10   Separate the cover from the front of the engine.
11   Use a gasket scraper to remove all traces of old gasket material and sealant from the cover and engine block **(see illustration)**. The cover is made of aluminum, so be careful not to nick or gouge it. Clean the gasket sealing surfaces with lacquer thinner or acetone.
12   The oil pump cover must be removed and the cavity packed with petroleum jelly (see Section 15) before the cover is installed.
13   Apply a thin layer of RTV sealant to both sides of the new gasket, then position the gasket on the engine block (the dowel pins should keep it in place). Attach the cover to the engine. The oil pump drive must engage with the crankshaft.
14   Apply a non-hardening thread sealant (Permatex no. 2 or equivalent) to the bolt threads, then install them finger tight. Install the crankshaft sensor, but leave the bolts finger tight until Step 16. Follow a criss-cross pattern when tightening the other bolts and work up to the torque listed in this Chapter's specifications in three steps to avoid warping the cover.
15   Use a special tool, available at most auto parts stores, to position the crankshaft sensor on the timing cover (see Chapter 6, Section 4). Tighten the bolts to the torque listed in this Chapter's Specifications for the timing chain cover.
16   The remainder of installation is the reverse of removal.
17   Add oil and coolant, start the engine and check for leaks.

14   Oil filter adapter and pressure regulator valve - removal and installation

Refer to illustration 14.4
1   Remove the oil filter (see Chapter 1, Section 12).

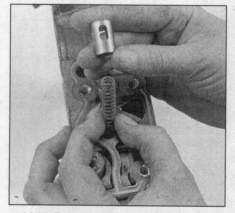

14.14 Remove the pressure regulator valve and spring, then check the valve for wear and damage

2   Remove the timing chain cover (see Section 13).
3   Remove the four bolts holding the oil filter adapter to the timing chain cover. The cover is spring loaded, so remove the bolts while keeping pressure on the cover, then release the spring pressure carefully.
4   Remove the pressure regulator valve and spring **(see illustration)**. Use a gasket scraper to remove all traces of the old gasket.
5   Clean all parts with solvent and dry them with compressed air. **Warning:** *Wear eye protection.* Check for wear, score marks and valve binding.
6   Installation is the reverse of removal. Be sure to use a new gasket.
7   Tighten the bolts to the torque listed in this Chapter's Specifications.
8   Run the engine and check for oil leaks.

15   Oil pump - removal, inspection and installation

Removal

Refer to illustration 15.4
1   Remove the oil filter (see Chapter 1,

15.4 The oil pump cover is attached to the inside of the timing chain cover - a T-30 Torx driver is required for removal of the screws

Section 12).
2   Remove the oil filter adapter, pressure regulator valve and spring (see Section 14).
3   Remove the timing chain cover (see Section 13).
4   Remove the oil pump cover-to-timing chain cover bolts **(see illustration)**.
5   Lift out the cover and oil pump gears as an assembly. Note how the gears are installed; they should be reinstalled facing the same way.

Inspection

Refer to illustrations 15.10 and 15.11
6   Clean the parts with solvent and dry them with compressed air (if available). **Warning:** *Wear eye protection!*
7   Inspect all components for wear and score marks. Replace any worn out or damaged parts.
8   Inspect the bypass valve (see Section 14).
9   Reinstall the gears in the timing chain cover.
10   Measure the outer gear-to-housing clearance with a feeler gauge **(see illustration)**.

15.10 Measuring the outer gear-to-housing clearance with a feeler gauge

15.11 Measuring the inner gear-to-housing clearance with a feeler gauge

16.1 The shim has a notch in it to fit over the key in the crankshaft

11 Measure the inner gear-to-outer gear clearance at several points **(see illustration)**.
12 Use a dial indicator or straightedge and feeler gauges to measure the gear end clearance (distance from the gear to the gasket surface of the cover).
13 Check for pump cover warpage by laying a precision straightedge across the cover and trying to slip a feeler gauge between the cover and straightedge.
14 Compare the measurements to this Chapter's Specifications. Replace all worn or damaged components with new ones.

Installation

15 Remove the gears and pack the pump cavity with petroleum jelly.
16 Install the gears - make sure petroleum jelly is forced into every cavity. Failure to do so could cause the pump to lose its prime when the engine is started, causing damage from lack of oil pressure.
17 Install the pump cover, using a new GM gasket only - its thickness is critical for maintaining the correct clearances.
18 Install the pressure regulator spring and valve.
19 Install the timing chain cover.
20 Install the oil filter and check the oil level (see Chapter 1, Section 12). Start and run the engine and check for correct oil pressure, then look carefully for oil leaks at the timing chain cover.

16 Timing chain and sprockets - removal and installation

Removal

Refer to illustrations 16.1, 16.3 and 16.6

1 Remove the timing chain cover (see Section 13), then slide the shim off the nose of the crankshaft **(see illustration)**.
2 The timing chain should be replaced with a new one if the total freeplay midway between the sprockets exceeds one inch. Failure to replace the timing chain may result

16.3 The marks on the crankshaft and camshaft sprockets (arrows) must be aligned adjacent to each other as shown here (the mark on the camshaft sprocket is a dimple - the one on the crankshaft sprocket is a raised dot)

in erratic engine performance, loss of power and lowered fuel mileage.
3 Temporarily install the vibration damper bolt and turn the crankshaft clockwise to align the timing marks on the crankshaft and camshaft sprockets directly opposite each other **(see illustration)**.
4 Remove the timing chain damper.
5 Remove the camshaft sprocket bolt. Try not to turn the camshaft in the process (if you do, realign the timing marks after the bolts are loosened).
6 Alternately pull the camshaft sprocket and then the crankshaft sprocket forward and remove the sprockets and timing chain as an assembly **(see illustration)**.
7 Clean the timing chain components with solvent and dry them with compressed air (if available). **Warning:** *Wear eye protection.*
8 Remove the camshaft gear.
9 Inspect the components for wear and damage. Look for teeth that are deformed, chipped, pitted, polished or discolored.

16.6 Guide the chain and sprockets off as an assembly

Installation

Note: *If the crankshaft has been disturbed, install the sprocket temporarily and turn the crankshaft until the mark on the crankshaft sprocket is exactly at the top. If the camshaft was disturbed, install the sprocket temporarily and turn the camshaft until the timing mark is at the bottom, opposite the mark on the crankshaft sprocket (see illustration 16.3).*

10 Assemble the timing chain on the sprockets, then slide the sprocket and chain assembly onto the shafts with the timing marks aligned **(see illustration 16.3)**. Be sure to install the camshaft gear. Aligning the gear with the balance shaft gear is covered in Chapter 2, Part C, Section 24).
11 Install the camshaft sprocket bolt and tighten it to the torque listed in this Chapter's Specifications.
12 Attach the timing chain damper assembly to the block.
13 Lubricate the chain and sprocket with clean engine oil and install the timing chain cover (see Section 13).

17 Oil pan - removal and installation

Refer to illustration 17.8

1 Disconnect the cable from the negative battery terminal.
2 Raise the vehicle and place it securely on jackstands.
3 Drain the engine oil and replace the oil filter (see Chapter 1, Section 12).
4 Remove the driveplate access cover.
5 Remove the starter (see Chapter 5).
6 Support the engine/transaxle from underneath with a floor jack. Place the jack directly underneath the rear transaxle mounting bracket.
7 Remove the engine mount-to-frame bolts and raise the engine slightly to gain clearance for oil pan removal.
8 Remove the oil pan retaining bolts **(see illustration)** and carefully separate the oil pan from the block. Don't pry between the block and the pan or damage to the sealing surfaces may result and oil leaks may develop. Instead, tap the pan with a soft-face hammer to break the gasket seal.
9 Clean the pan with solvent and remove all old sealant and gasket material from the block and pan mating surfaces. Clean the mating surfaces with lacquer thinner or acetone and make sure the bolt holes in the block are clear. Check the oil pan flange for distortion, particularly around the bolt holes. If necessary, place the pan on a block of wood and use a hammer to flatten and restore the gasket surface.
10 Always use a new gasket whenever the oil pan is installed.
11 Place the oil pan in position on the block and install the bolts.
12 After the bolts are installed, tighten them to the torque listed in this Chapter's specifications. Starting at the center, follow a criss-cross pattern and work up to the final torque in three steps.
13 The remaining steps are the reverse of the removal procedure.
14 Refill the engine with oil (see Chapter 1, Section 12). Run the engine until normal operating temperature is reached and check for leaks.

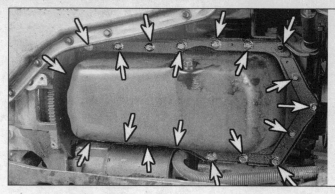

17.8 The oil pan bolts (arrows) are located around the perimeter of the oil pan (viewed from below)

18 Oil pump Pick-up tube and screen assembly - removal and installation

Refer to illustration 18.2

1 Remove the oil pan (see Section 17).
2 Unbolt the oil pump pick-up tube and screen assembly and detach it from the engine **(see illustration)**.
3 Clean the screen and housing assembly with solvent and dry it with compressed air, if available. **Warning:** *Wear eye protection.*
4 If the oil screen is damaged or has metal chips in it, replace it. An abundance of metal chips indicates a major engine problem which must be corrected.
5 Make sure the mating surfaces of the pipe flange and the engine block are clean and free of nicks and install the screen assembly with a new gasket.
6 Install the oil pan (see Section 17).
7 Be sure to refill the engine with oil before starting it (see Chapter 1, Section 12).

19 Engine mount - check and replacement

Warning: *A special tool is available to support the engine during repair operations. Similar fixtures are available from rental yards. DO NOT place any part of your body under the engine/transaxle when it's supported only by a jack. Improper lifting methods or failure of the lifting device could result in serious injury or death.*

Note: *See Chapter 7, Section 8 for transaxle mount information.*

1 Engine mounts seldom require attention, but broken or deteriorated mounts should be replaced immediately or the added strain placed on the driveline components may cause damage or wear.

Check

Refer to illustration 19.4

2 During the check, the engine must be raised slightly to remove the weight from the mounts.
3 Raise the vehicle and support it securely on jackstands, then position a jack under the engine oil pan. Place a large block of wood between the jack head and the oil pan, then carefully raise the engine just enough to take the weight off the mounts. **Warning:** *DO NOT place any part of your body under the engine when it's supported only by a jack!*
4 Check the mounts to see if the rubber is cracked, hardened or separated from the metal plates **(see illustration)**. Sometimes the rubber will split right down the center.
5 Check for relative movement between the mount plates and the engine or frame (use a large screwdriver or prybar to attempt to move the mounts). If movement is noted, lower the engine and tighten the mount fasteners.
6 Rubber preservative may be applied to the mounts to slow deterioration.

Replacement

7 Disconnect the negative battery cable from the battery, then raise the vehicle and support it securely on jackstands (if not already done).
8 Raise the engine slightly with a jack or hoist. Install an engine support as described in the Warning above. Remove the fasteners and detach the mount from the frame bracket.
9 Remove the mount-to-block bracket bolts/nuts and detach the mount.
10 Installation is the reverse of removal. Use thread locking compound on the threads and be sure to tighten everything securely.

18.2 The oil pump pick-up tube is retained by two bolts (arrows)

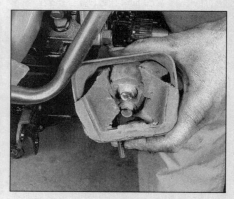

19.4 Here's a typical broken engine mount - the rubber has separated from the metal plate (removed from vehicle for clarity)

Chapter 2 Part C
General engine overhaul procedures

Contents

Section		Section
Camshaft, balance shaft and bearings – removal and inspection	12	Engine overhaul - general information ... 2
Camshaft and balance shaft - installation	24	Engine overhaul - reassembly sequence ... 21
Crankshaft - inspection	19	Engine rebuilding alternatives ... 6
Crankshaft - installation and main bearing oil clearance check	23	Engine - removal and installation ... 5
		Engine removal - methods and precautions ... 4
Crankshaft - removal	14	General information ... 1
Cylinder compression check	3	Initial start-up and break-in after overhaul ... 28
Cylinder head - cleaning and inspection	9	Main and connecting rod bearings - inspection ... 20
Cylinder head - disassembly	8	Pistons and connecting rods - inspection ... 18
Cylinder head - reassembly	11	Pistons and connecting rods - installation and rod bearing oil clearance check ... 26
Cylinder honing	17	Pistons and connecting rods - removal ... 13
Driveplate - removal and installation	27	Piston rings - installation ... 22
Engine block - cleaning	15	Rear main oil seal - replacement ... 25
Engine block - inspection	16	Valves - servicing ... 10
Engine overhaul - disassembly sequence	7	

Specifications

3.1L engine

General

RPO Code	LG6
VIN code	D
Displacement	189 cubic inches
Cylinder compression pressure	100 psi minimum
Maximum variation between cylinders	30-percent
Firing order	1-2-3-4-5-6
Oil pressure	15 psi at 1100 rpm

Cylinder head

Warpage limit	0.005 inch
Maximum allowable cut	0.010 inch

3.1L engine

Valves and related components
Valve margin width	1/32 inch minimum
Valve seat angle	46-degrees
Valve stem-to-guide clearance	0.0010 to 0.0027 inch
Valve spring free length (intake and exhaust)	1.91 inches
Valve spring pressure	
Closed	82 lbs @ 1.58 inches
Open	191 lbs @ 1.291 inches
Installed height	1.5748 inches

Crankshaft and connecting rods
Connecting rod journal	
Diameter	1.9994 to 1.9983 inches
Bearing oil clearance	0.0011 to 0.0034 inch
Connecting rod side clearance (endplay)	0.014 to .0267 inch
Main bearing journal	
Diameter	2.6473 to 2.6483 inches
Bearing oil clearance	0.0012 to 0.0027 inch
Taper/out of round limit	0.0002 inch
Crankshaft endplay (at thrust bearing)	0.0024 to 0.0083 inch

Engine block
Cylinder bore	
Diameter	3.503 to 3.506 inches
Out-of-round limit	0.0005 inch
Taper limit (thrust side)	0.0005 inch
Block deck warpage limit	If more than 0.010 inch must be removed, replace the block

Pistons and rings
Piston-to-bore clearance	0.0009 to 0.0022 inch
Piston ring end gap	
Top compression ring	0.010 to 0.020 inch
Second compression ring	0.020 to 0.028 inch
Oil control ring	0.010 to 0.030 inch
Piston ring side clearance	
Compression rings	0.0020 to 0.0035 inch
Oil control ring	0.008 inch

Camshaft
Bearing journal diameter	1.8678 to 1.8815 inches
Bearing oil clearance	0.001 to 0.004 inch
Lobe lift	
Intake	0.2306 inch
Exhaust	0.2619 inch

Torque specifications*
	Ft-lbs
Main bearing cap bolts	72
Connecting rod cap nuts	39
Camshaft sprocket bolts	18
Driveplate-to-crankshaft bolts	50

*Note: Refer to Part A for additional torque specifications.

3800 engine

General
RPO Code	L27
VIN code	L
Displacement	231 cubic inches
Cylinder compression pressure	100 psi minimum
Maximum variation between cylinders	30-percent
Firing order	1-6-5-4-3-2
Oil pressure	15 psi at 1100 rpm

Cylinder head
Warpage limit	0.003 inch per 6 inch span/0.006 inch overall
Maximum allowable cut	0.010 inch

Chapter 2 Part C General engine overhaul procedures

Valves and related components

Margin width	1/32-inch
Seat angle	45-degrees
Stem-to-guide clearance	
Intake	0.0015 to 0.0035 inch
Exhaust	0.0015 to 0.0032 inch
Spring free length	1.981 inches
Spring pressure	
Closed	80 lbs @ 1.75 inches
Open	210 lbs @ 1.315 inches
Spring installed height	1.690 to 1.720 inches

Crankshaft and connecting rods

Connecting rod journal	
Diameter	2.2487 to 2.2499 inches
Bearing oil clearance	0.0008 to 0.0022 inch
Maximum taper	0.0003 inch
Connecting rod side clearance (endplay)	0.003 to 0.015 inch
Main bearing journal	
Diameter	2.4988 to 2.4998 inches
Bearing oil clearance	0.0008 to 0.0022 inch
Maximum taper/out-of-round	0.003 inch
Crankshaft endplay (at thrust bearing)	0.003 to 0.011 inch

Engine block

Cylinder bore	
Diameter	3.800 inches
Out-of-round limit	0.0004 inch
Taper limit	0.0005 inch

Pistons and rings

Piston-to-bore clearance limits (44 mm from top of piston)	0.0004 to 0.0022 inch
Piston ring end gap	
Compression rings	0.010 to 0.025 inch
Oil control ring	0.015 to 0.055 inch
Piston ring side clearance	
Compression rings	0.0013 to 0.0031 inch
Oil control ring	0.011 to 0.0081 inch

Camshaft

Bearing journal diameter	1.785 to 1.786 inches
Bearing oil clearance	0.0005 to 0.0035 inch
Lobe lift	
Intake	0.250 inch
Exhaust	0.255 inch

Torque specifications*

Ft-lbs (unless otherwise indicated)

Main bearing caps	
Step one	26
Step two	Rotate an additional 50-degrees
Connecting rod bolts	
Step one	20
Step two	Rotate an additional 50-degrees
Balance shaft gear bolt	
Step one	14
Step two	Rotate an additional 35-degrees
Balance shaft retainer bolts	22
Camshaft sprocket bolt	
Step one	74
Step two	Rotate an additional 105-degrees
Driveplate-to-crankshaft bolts	
Step one	11
Step two	Rotate an additional 50-degrees

*Note: Refer to Part B for additional torque specifications.

Chapter 2 Part C General engine overhaul procedures

2.4a The oil pressure sending unit on the 3.1L engine is located on the front of the block between the oil filter and the starter

2.4b The oil pressure sending unit on the 3800 engine is located on top of the oil filter housing (arrow) - remove the sending unit . . .

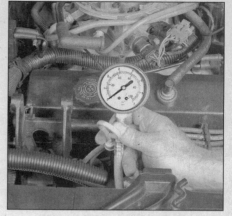

2.4c . . . and install a gauge to check the engine oil pressure

1 General information

Included in this portion of Chapter 2 are the general overhaul procedures for the cylinder heads and internal engine components. The information ranges from advice concerning preparation for an overhaul and the purchase of replacement parts to detailed, step-by-step procedures covering removal and installation of internal engine components and the inspection of parts.

The following Sections have been written based on the assumption the engine has been removed from the vehicle. For information concerning in-vehicle engine repair, as well as removal and installation of the external components necessary for the overhaul, see Parts A and B of this Chapter and Section 7 of this Part.

The Specifications included in this Part are only those necessary for the inspection and overhaul procedures which follow. Refer to Parts A and B for additional Specifications.

2 Engine overhaul - general information

Refer to illustrations 2.4a, 2.4b and 2.4c

It's not always easy to determine when, or if, an engine should be completely overhauled, as a number of factors must be considered.

High mileage isn't necessarily an indication an overhaul is needed, while low mileage doesn't preclude the need for an overhaul. Frequency of servicing is probably the most important consideration. An engine that's had regular and frequent oil and filter changes, as well as other required maintenance, will most likely give many thousands of miles of reliable service. Conversely, a neglected engine may require an overhaul very early in its life.

Excessive oil consumption is an indication that piston rings, valve seals and/or valve guides are in need of attention. Make sure oil leaks aren't responsible before deciding the rings and/or guides are bad. Perform a cylinder compression check to determine the extent of the work required (see Section 3).

Remove the oil pressure sending unit and check the oil pressure with a gauge installed in its place **(see illustrations)**. Compare the results to this Chapter's Specifications. As a general rule, engines should have ten psi oil pressure for every 1,000 rpm. If the pressure is extremely low, the bearings and/or oil pump are probably worn out.

Loss of power, rough running, knocking or metallic engine noises, excessive valve train noise and high fuel consumption rates may also point to the need for an overhaul, especially if they're all present at the same time. If a complete tune-up doesn't remedy the situation, major mechanical work is the only solution.

An engine overhaul involves restoring the internal parts to the specifications of a new engine. During an overhaul, the piston rings are replaced and the cylinder walls are reconditioned (rebored and/or honed). If a rebore is done by an automotive machine shop, new oversize pistons will also be installed. The main bearings, connecting rod bearings and camshaft bearings are generally replaced with new ones and, if necessary, the crankshaft may be reground to restore the journals. Generally, the valves are serviced as well, since they're usually in less-than-perfect condition at this point. While the engine is being overhauled, other components, such as the starter and alternator, can be rebuilt as well. The end result should be a like-new engine that will give many trouble free miles. **Note:** *Critical cooling system components such as the hoses, drivebelts, thermostat and water pump MUST be replaced with new parts when an engine is overhauled. The radiator should be checked carefully to ensure it isn't clogged or leaking (see Chapter 3, Section 5). Also, we don't recommend overhauling the oil pump on the 3.1L engine - always install a new one when an engine is rebuilt.*

Before beginning the engine overhaul, read through the entire procedure to familiarize yourself with the scope and requirements of the job. Overhauling an engine isn't particularly difficult, if you follow all of the instructions carefully, have the necessary tools and equipment and pay close attention to all specifications; however, it can be time consuming. Plan on the vehicle being tied up for a minimum of two weeks, especially if parts must be taken to an automotive machine shop for repair or reconditioning. Check on availability of parts and make sure any necessary special tools and equipment are obtained in advance. Most work can be done with typical hand tools, although a number of precision measuring tools are required for inspecting parts to determine if they must be replaced. Often an automotive machine shop will handle the inspection of parts and offer advice concerning reconditioning and replacement. **Note:** *Always wait until the engine has been completely disassembled and all components, especially the engine block, have been inspected before deciding what service and repair operations must be performed by an automotive machine shop. Since the block's condition will be the major factor to consider when determining whether to overhaul the original engine or buy a rebuilt one, never purchase parts or have machine work done on other components until the block has been thoroughly inspected. As a general rule, time is the primary cost of an overhaul, so it doesn't pay to install worn or substandard parts.*

As a final note, to ensure maximum life and minimum trouble from a rebuilt engine, everything must be assembled with care in a spotlessly clean environment.

3 Cylinder compression check

Refer to illustration 3.6

1 A compression check will tell you what mechanical condition the upper end (pistons, rings, valves, head gaskets) of the engine is in. Specifically, it can tell you if the compres-

Chapter 2 Part C General engine overhaul procedures

3.6 Use a compression gauge with a threaded fitting for the spark plug hole; don't use the type with a conical rubber tip - it relies on hand pressure to create a seal between he gauge and the combustion chamber, and usually gives an inaccurate reading

sion is down due to leakage caused by worn piston rings, defective valves and seats or a blown head gasket. **Note:** *The engine must be at normal operating temperature and the battery must be fully charged for this check.*

2 Begin by cleaning the area around the spark plugs before you remove them. Compressed air should be used, if available, otherwise a small brush or even a bicycle tire pump will work. The idea is to prevent dirt from getting into the cylinders as the compression check is being done.

3 Remove all of the spark plugs from the engine (see Chapter 1, Section 32).

4 Block the throttle wide open.

5 Disable the fuel and ignition systems by removing the ECM fuse (see Chapter 12, Section 3).

6 Install the compression gauge in the number one spark plug hole **(see illustration)**.

7 Crank the engine over at least seven compression strokes and watch the gauge. The compression should build up quickly in a healthy engine. Low compression on the first stroke, followed by gradually increasing pressure on successive strokes, indicates worn piston rings. A low compression reading on the first stroke, which doesn't build up during successive strokes, indicates leaking valves or a blown head gasket (a cracked head could also be the cause). Deposits on the undersides of the valve heads can also cause low compression. Record the highest gauge reading obtained.

8 Repeat the procedure for the remaining cylinders and compare the results to this Chapter's Specifications.

9 If the readings are below normal, add some engine oil (about three squirts from a plunger-type oil can) to each cylinder, through the spark plug hole, and repeat the test.

10 If the compression increases after the oil is added, the piston rings are definitely worn. If the compression doesn't increase significantly, the leakage is occurring at the valves or head gasket. Leakage past the valves may be caused by burned valve seats and/or faces or warped, cracked or bent valves.

11 If two adjacent cylinders have equally low compression, there's a strong possibility the head gasket between them is blown. The appearance of coolant in the combustion chambers or the crankcase would verify this condition.

12 If one cylinder is about 20-percent lower than the others, and the engine has a slightly rough idle, a worn exhaust lobe on the camshaft could be the cause.

13 If the compression is unusually high, the combustion chambers are probably coated with carbon deposits. If that's the case, the cylinder heads should be removed and decarbonized.

14 If compression is way down or varies greatly between cylinders, it would be a good idea to have a leak-down test performed by an automotive repair shop. This test will pinpoint exactly where the leakage is occurring and how severe it is.

15 Install the ECM fuse and drive the vehicle to restore the block learn memory.

4 Engine removal - methods and precautions

If you've decided the engine must be removed for overhaul or major repair work, several preliminary steps should be taken. Locating a suitable place to work is extremely important. Adequate work space, along with storage space for the vehicle, will be needed. And, as mentioned in the next Section, a vehicle lift will be required to perform the job safely.

Cleaning the engine compartment and engine before beginning the removal procedure will help keep tools clean and organized. A vehicle lift will also be necessary. Safety is of primary importance, considering the potential hazards involved in removing the engine from this vehicle.

If the engine is being removed by a novice, a helper should be available. Advice and aid from someone more experienced would also be helpful. There are many instances when one person cannot simultaneously perform all of the operations required when lifting the engine out of the vehicle.

Plan the operation ahead of time. Arrange for or obtain all of the tools and equipment you'll need prior to beginning the job. Some of the equipment necessary to perform engine removal and installation safely and with relative ease are (in addition to a vehicle lift) complete sets of wrenches and sockets as described in the front of this manual, wooden blocks and plenty of rags and cleaning solvent for mopping up spilled oil, coolant and gasoline.

Plan for the vehicle to be out of use for quite a while. A machine shop will be required to perform some of the work which the do-it-yourselfer can't accomplish without special equipment. These shops often have a busy schedule, so it would be a good idea to consult them before removing the engine in order to accurately estimate the amount of time required to rebuild or repair components that may need work.

Always be extremely careful when removing and installing the engine. Serious injury can result from careless actions. Plan ahead, take your time and a job of this nature, although major, can be accomplished successfully.

5 Engine - removal and installation

Warning 1: *Read through the following steps carefully and familiarize yourself with the procedure before beginning work. Removing and installing the engine is tricky. You must also remove the transaxle, the steering gear and the lower control arms, because they're all fastened to a subframe which must be removed as a complete assembly. The only safe way to perform the following procedure is to block up the engine/transaxle/ steering gear/subframe assembly, unbolt it from the vehicle and raise the vehicle high enough to clear the entire assembly. Therefore, we DON'T RECOMMEND attempting this procedure without a vehicle lift. Using any other means to raise the vehicle high enough to clear the engine/transaxle/subframe assembly could result in serious damage to the vehicle and/or engine/transaxle assembly - and serious injury to you - if the vehicle were to fall.*
Warning 2: *Gasoline is extremely flammable, so take extra precautions when disconnecting any part of the fuel system. Don't smoke or allow open flames or bare light bulbs in or near the work area and don't work in a garage where a natural gas appliance (such as a clothes dryer or water heater) is installed. If you spill gasoline on your skin, rinse it off immediately. Have a fire extinguisher rated for gasoline fires handy and know how to use it!*
Warning 3: *The air conditioning system is under high pressure - have a dealer service department or service station discharge the system before disconnecting any of the hoses or fittings.*
Note: *On models equipped with the Delco Loc II audio system, be sure the lockout feature is turned off before performing any procedure which requires disconnecting the battery.*

Chapter 2 Part C General engine overhaul procedures

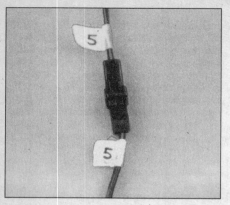

5.4 Label each wire before unplugging the connector

5.6a These two bolts (arrows) attach the power steering fluid cooling lines to the front of the subframe crossmember; they also attach the clips for a major wiring harness - forget to disconnect these clips and you'll destroy this expensive harness when the vehicle is separated from the engine/transaxle/subframe

5.6b Also make sure you disconnect these automatic transmission cooler lines (arrows)

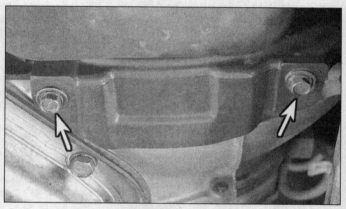

5.13a To remove the driveplate access cover, remove these two bolts on the bottom . . .

Removal

Refer to illustrations 5.4, 5.6a, 5.6b, 5.13a, 5.13b, 5.13c, 5.14, 5.18a, 5.18b, 5.18c, 5.18d, 5.19, 5.21 and 5.23

1 Relieve the fuel system pressure (see Chapter 4, Section 2), then disconnect the negative cable from the battery.

2 Cover the fenders and cowl. Special pads are available to protect the fenders, but an old bedspread or blanket will also work. It's also a good idea to remove the hood (see Chapter 11, Section 9) to give yourself more room to work. But, since you're going to lower the engine - rather than lift it - from the engine compartment, removing the hood isn't absolutely essential.

3 Remove the air cleaner assembly (see Chapter 4, Section 8).

4 To ensure correct reassembly, label each vacuum line, emission system hose, wiring connector, ground strap and fuel line. Pieces of masking tape with numbers or letters written on them prevent confusion at assembly time (see illustration). Or sketch the engine compartment routing of lines, hoses and wires. When you disconnect things, pay close attention to those that bridge the gap between the subframe and the frame. Forgetting any of these items will result in extensive damage to expensive harnesses.

5 Raise the vehicle on a hydraulic lift.

6 Working from the underside of the vehicle, finish disconnecting all hoses, lines and wire harnesses (see illustrations). Don't forget to detach the three power steering line clips from the vehicle's right frame member; all three are located adjacent to the crankshaft vibration damper pulley (No further disassembly of the power steering system is necessary - everything else is attached to the subframe and can be removed with it).

7 Drain the cooling system (see Chapter 1, Section 29) and label and detach all coolant hoses from the engine.

8 Disconnect the fuel lines running from the engine to the chassis (see Chapter 4, Section 4). Plug or cap all open fittings/lines.

9 Disconnect the throttle and cruise control cables (see Chapter 4, Section 9) and TV cable (see Chapter 7, Section 3).

10 Disconnect the air conditioning hoses from the compressor, hang them out of the way, detach the air conditioning compressor from its mounting bracket and remove it from the engine compartment (see Chapter 3, Section 15).

11 Drain the engine oil and remove the filter (see Chapter 1, Section 12).

12 Disconnect the control arms from the steering knuckles (see Chapter 10, Section 5).

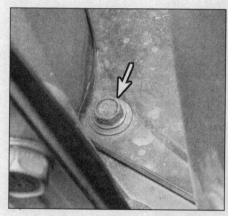

5.13b . . . one at the rear (toward the firewall) . . .

13 Remove the starter (see Chapter 5, Section 17). Remove the driveplate access cover (see illustrations).

14 Scribe or paint an alignment mark between the driveplate and the torque converter (see illustration), then rotate the crankshaft and remove the driveplate bolts.

15 Disconnect the engine strut (see Chapter 2, Part A, Section 4).

16 Disconnect the exhaust system from the engine (see Chapter 4, Section 14).

Chapter 2 Part C General engine overhaul procedures

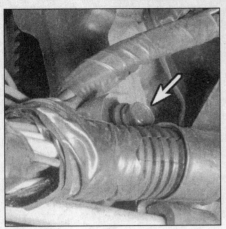

5.13c . . . and one on the front

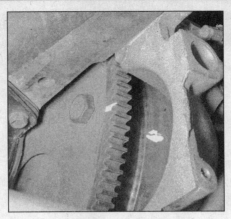

5.14 Paint or scribe alignment marks on the driveplate and the torque converter to ensure that the two components are still in balance when they're reassembled

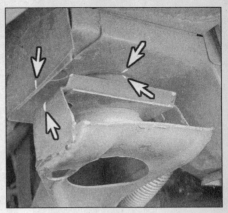

5.18a Paint or scribe alignment marks (arrows) between the corners of the subframe and the vehicle frame to ensure proper realignment of the subframe during reinstallation (right front corner shown)

5.18b Alignment marks at the right rear corner of the subframe

5.18c Front subframe-to-vehicle bolt

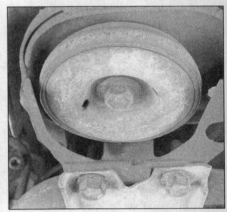

5.18d Rear subframe-to-vehicle bolt

5.19 Before lifting the vehicle too far, make one last check to verify that everything is disconnected

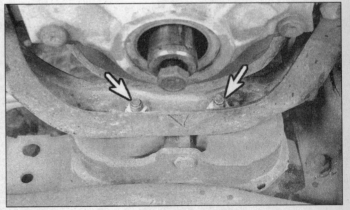

5.21 To disconnect the engine from the subframe, remove these two nuts (arrows) from the engine mount located below the timing cover

17   Support the engine/transaxle/steering gear/subframe assembly with four-by-four wood blocks arranged along both sides of the subframe.

18   Paint or scribe alignment marks at all four corners of the subframe (see illustrations), then remove the four - two front, two rear - subframe-to-frame bolts (see illustrations). Inspect all four bushings; if they're cracked or dried out, replace them.

19   Raise the vehicle just enough to make a last-minute inspection to ensure that everything if disconnected (see illustration). Raise the vehicle high enough to clear the engine/transaxle/steering gear assembly.

20   Support the engine with an engine hoist, then remove the driveplate-to-torque converter fasteners.

21   Remove the bolts and nuts from the engine mount below the timing cover (see illustration).

5.23 Lift the engine off the transaxle/subframe assembly with an engine hoist

5.27 To align the subframe assembly with the vehicle frame, insert a prybar or large screwdriver through the holes in the subframe and the frame, then lever the subframe until the marks you made at all four corners of the subframe are lined up

22 Make one last check to verify that nothing is still connecting the engine to the subframe.
23 Raise the engine slightly to disengage it from the timing cover mount. Carefully separate the engine from the transaxle. Be sure the torque converter stays in place (clamp a pair of vise-grips to the housing to keep the converter from sliding out). Slowly raise the engine off the subframe (see illustration). Check carefully to make sure nothing is hanging up as the hoist is raised.
24 Remove the driveplate (see Section 27) and mount the engine on an engine stand.

Installation

Refer to illustration 5.27
25 While the engine is out, check the engine mount and the transaxle mounts (see Chapter 7, Section 8). If they're worn or damaged, replace them.
26 Carefully lower the engine onto the subframe assembly - make sure the timing cover mount bolt holes line up. **Caution:** *DO NOT use the transaxle-to-engine bolts to force the transaxle and engine together. Take great care when mating the torque converter to the driveplate, following the procedure outlined in Chapter 7, Section 9. Make sure the alignment marks you made on the driveplate and the torque converter during removal are lined up.*
27 Lower the vehicle onto the engine/transaxle/steering gear/subframe assembly, install the subframe bolts, but don't fully tighten them yet. Using a prybar or large screwdriver, align the subframe assembly with the vehicle frame (see illustration), then tighten the subframe bolts to the torque listed in this Chapter's Specifications.
28 Install the driveplate-to-torque converter bolts and tighten them to the torque listed in this Chapter's Specifications.
29 Reinstall the remaining components in the reverse order of removal. Double-check to make sure everything is hooked up right.
30 Add coolant, oil, power steering and transmission fluid as needed.
31 Run the engine and check for leaks and proper operation of all accessories, then install the hood and test drive the vehicle.

6 Engine rebuilding alternatives

The home mechanic is faced with a number of options when performing an engine overhaul. The decision to replace the engine block, piston/connecting rod assemblies and crankshaft depends on a number of factors, with the number one consideration being the condition of the block. Other considerations are cost, access to machine shop facilities, parts availability, time required to complete the project and the extent of prior mechanical experience.

Some of the rebuilding alternatives include:

Individual parts - If the inspection procedures reveal the engine block and most engine components are in reusable condition, purchasing individual parts may be the most economical alternative. The block, crankshaft and piston/connecting rod assemblies should all be inspected carefully. Even if the block shows little wear, the cylinder bores should be surface honed.

Short block - A short block consists of an engine block with a crankshaft and piston/connecting rod assemblies already installed. All new bearings are incorporated and all clearances will be correct. The existing camshaft, valve train components, cylinder heads and external parts can be bolted to the short block with little or no machine shop work necessary.

Long block - A long block consists of a short block plus an oil pump, oil pan, cylinder heads, rocker arm covers, camshaft and valve train components, timing sprockets and chain and timing chain cover. All components are installed with new bearings, seals and gaskets incorporated throughout. The installation of manifolds and external parts is all that's necessary. Give careful thought to which alternative is best for you and discuss the situation with local automotive machine shops, auto parts dealers and experienced rebuilders before ordering or purchasing replacement parts.

7 Engine overhaul - disassembly sequence

Refer to illustrations 7.3a, 7.3b and 7.3c
1 It's much easier to disassemble and work on the engine if it's mounted on a portable engine stand. A stand can often be rented quite cheaply from an equipment rental yard. Before it's mounted on a stand, the flywheel/driveplate should be removed from the engine.
2 If a stand isn't available, it's possible to disassemble the engine with it blocked up on the floor. Be extra careful not to tip or drop the engine when working without a stand.
3 If you're going to obtain a rebuilt engine, all external components (see illustrations) must come off first, to be transferred to the replacement engine, just as they will if you're doing a complete engine overhaul yourself. These include:

Alternator and brackets
Emissions control components
Ignition coil/module assembly, spark plug wires and spark plugs
Thermostat and housing cover
Water pump
EFI components
Intake/exhaust manifolds
Oil filter
Engine mounts
Driveplate
Engine rear plate (if equipped)

Note: *When removing the external compo-*

Chapter 2 Part C General engine overhaul procedures   2C-9

7.3a Engine (3.1L) - drivebelt end

7.3b Engine (3.1L) - radiator side

7.3c Engine (3.1L) - firewall side

8.2 A small plastic bag, with an appropriate label, can be used to store the valve train components so they can be kept together and reinstalled in their original locations

nents from the engine, pay close attention to details that may be helpful or important during installation. Note the installed position of gaskets, seals, spacers, pins, brackets, washers, bolts and other small items.

4 If you're obtaining a short block, which consists of the engine block, crankshaft, pistons and connecting rods all assembled, then the cylinder heads, oil pan and oil pump will have to be removed as well. See *Engine rebuilding alternatives* for additional information regarding the different possibilities to be considered.

5 If you're planning a complete overhaul, the engine must be disassembled and the internal components removed in the following general order:

Valve covers
Intake and exhaust manifolds
Rocker arms and pushrods
Valve lifters
Cylinder heads
Timing chain cover and oil pump
Timing chain and sprockets
Camshaft
Balance shaft (3800 engine only)

Oil pan
Piston/connecting rod assemblies
Crankshaft and main bearings

6 Before beginning the disassembly and overhaul procedures, make sure the following items are available. Also, refer to *Engine overhaul - reassembly sequence* for a list of tools and materials needed for engine reassembly.

Common hand tools
Small cardboard boxes or plastic bags for storing parts
Gasket scraper
Ridge reamer
Vibration damper puller
Micrometers
Telescoping gauges
Dial indicator set
Valve spring compressor
Cylinder surfacing hone
Piston ring groove cleaning tool
Electric drill motor
Tap and die set
Wire brushes
Oil gallery brushes
Cleaning solvent

8 Cylinder head - disassembly

Refer to illustrations 8.2, 8.3 and 8.4
Note: *New and rebuilt cylinder heads are commonly available for most engines at dealerships and auto parts stores. Due to the fact that some specialized tools are necessary for the disassembly and inspection procedures, and replacement parts aren't always readily available, it may be more practical and economical for the home mechanic to purchase replacement heads rather than taking the time to disassemble, inspect and recondition the originals.*

1 Cylinder head disassembly involves removal of the intake and exhaust valves and related components. Remove the rocker arm bolts, pivots and rocker arms from the cylinder heads. Label the parts or store them separately so they can be reinstalled in their original locations.

2 Before the valves are removed, arrange to label and store them, along with their related components, so they can be kept separate and reinstalled in their original loca-

8.3 Use a valve spring compressor to compress the spring, then remove the keepers from the valve stem

tions **(see illustration)**.

3   Compress the springs on the first valve with a spring compressor and remove the keepers **(see illustration)**. Carefully release the valve spring compressor and remove the retainer, the spring and the spring seat (if used).

4   Pull the valve out of the head, then remove the oil seal from the guide. If the valve binds in the guide (won't pull through), push it back into the head and deburr the area around the keeper groove with a fine file or whetstone **(see illustration)**.

5   Repeat the procedure for the remaining valves. Remember to keep all the parts for each valve together so they can be reinstalled in the same locations.

6   Once the valves and related components have been removed and stored in an organized manner, the head should be thoroughly cleaned and inspected. If a complete engine overhaul is being done, finish the engine disassembly procedures before beginning the cylinder head cleaning and inspection process.

9 Cylinder head - cleaning and inspection

1   Thorough cleaning of the cylinder heads and related valve train components, followed by a detailed inspection, will enable you to decide how much valve service work must be done during the engine overhaul. **Note:** *If the engine was severely overheated, the cylinder head is probably warped (see Step 12).*

Cleaning

2   Scrape all traces of old gasket material and sealant off the head gasket, intake manifold and exhaust manifold mating surfaces. Be very careful not to gouge the cylinder head. Special gasket removal solvents that soften gaskets and make removal much easier are available at auto parts stores.

3   Remove all built up scale from the coolant passages.

4   Run a stiff wire brush through the various holes to remove deposits that may have formed in them.

5   Run an appropriate size tap into each of the threaded holes to remove corrosion and thread sealant that may be present. If compressed air is available, use it to clear the holes of debris produced by this operation. **Warning:** *Wear eye protection when using compressed air!*

6   Clean the rocker arm pivot stud or bolt threads with a wire brush.

7   Clean the cylinder head with solvent and dry it thoroughly. Compressed air will speed the drying process and ensure that all holes and recessed areas are clean. **Note:** *Decarbonizing chemicals are available and may prove very useful when cleaning cylinder heads and valve train components. They're very caustic and should be used with caution. Be sure to follow the instructions on the container.*

8   Clean the rocker arms, pivots, bolts and pushrods with solvent and dry them thoroughly (don't mix them up during the cleaning process). Compressed air will speed the drying process and can be used to clean out the oil passages.

9   Clean all the valve springs, spring seats, rotators, keepers and retainers with solvent and dry them thoroughly. Do the components from one valve at a time to avoid mixing up the parts.

10   Scrape off any heavy deposits that may have formed on the valves, then use a motorized wire brush to remove deposits from the valve heads and stems. Again, make sure the valves don't get mixed up.

Inspection

Note: *Be sure to perform all of the following inspection procedures before concluding machine shop work is required. Make a list of the items that need attention.*

Cylinder head

Refer to illustrations 9.12 and 9.14

11   Inspect the head very carefully for cracks, evidence of coolant leakage and other damage. If cracks are found, check with an automotive machine shop concerning repair. If repair isn't possible, a new cylinder head must be obtained.

12   Using a straightedge and feeler gauge, check the head gasket mating surface for warpage **(see illustration)**. If the warpage exceeds the limit in this Chapter's Specifications, it can be resurfaced at an automotive machine shop. **Note:** *If the heads are resurfaced, the intake manifold flanges will also require machining.*

13   Examine the valve seats in each of the combustion chambers. If they're pitted, cracked or burned, the head will require valve service that's beyond the scope of the home mechanic.

14   Check the valve stem-to-guide clearance by measuring the lateral movement of the valve stem with a dial indicator attached securely to the head **(see illustration)**. The valve must be in the guide and approximately 1/16-inch off the seat. The total valve stem

8.4 If the valve won't pull through the guide, deburr the edge of the stem end and the area around the top of the keeper groove with a file or whetstone

9.12 Check the cylinder head gasket surface for warpage by trying to slip a feeler gauge under the straightedge (see this Chapter's Specifications for the maximum warpage allowed and use a feeler gauge of that thickness)

Chapter 2 Part C General engine overhaul procedures

9.14 A dial indicator can be used to determine the valve stem-to-guide clearance (move the valve stem as indicated by the arrows)

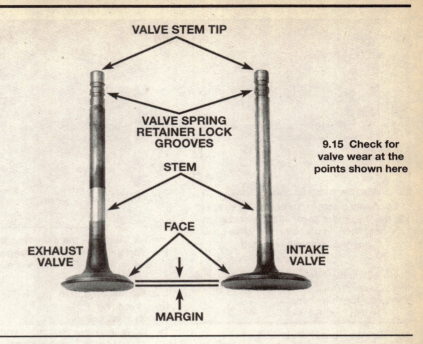

9.15 Check for valve wear at the points shown here

movement indicated by the gauge needle must be divided by two to obtain the actual clearance. After this is done, if there's still some doubt regarding the condition of the valve guides, they should be checked by an automotive machine shop (the cost should be minimal).

Valves

Refer to illustrations 9.15 and 9.16

15 Carefully inspect each valve face for uneven wear, deformation, cracks, pits and burned areas. Check the valve stem for scuffing and galling and the neck for cracks. Rotate the valve and check for any obvious indication that it's bent. Look for pits and excessive wear on the end of the stem. The presence of any of these conditions **(see illustration)** indicates the need for valve service by an automotive machine shop.

16 Measure the margin width on each valve **(see illustration)**. Any valve with a margin narrower than specified in this Chapter will have to be replaced with a new one.

Valve components

Refer to illustrations 9.17 and 9.18

17 Check each valve spring for wear (on the ends) and pits. Measure the free length and compare it to this Chapter's Specifications **(see illustration)**. Any springs that are shorter than specified have sagged and shouldn't be reused. The tension of all springs should be checked with a special fixture before deciding they're suitable for use in a rebuilt engine (take the springs to an automotive machine shop for this check).

18 Stand each spring on a flat surface and check it for squareness **(see illustration)**. If any of the springs are distorted or sagged, replace all of them with new parts.

19 Check the spring retainers and keepers for obvious wear and cracks. Any questionable parts should be replaced with new ones, as extensive damage will occur if they fail during engine operation.

Rocker arm components

20 Check the rocker arm faces (the areas that contact the pushrod ends and valve stems) for pits, wear, galling, score marks and rough spots. Check the rocker arm pivot contact areas and pivots as well. Look for cracks in each rocker arm and bolt.

21 Inspect the pushrod ends for scuffing and excessive wear. Roll each pushrod on a flat surface, like a piece of plate glass, to determine if it's bent.

22 Check the rocker arm bolt holes in the cylinder heads for damaged threads.

23 Any damaged or excessively worn parts must be replaced with new ones.

All components

24 If the inspection process indicates the valve components are in generally poor condition and worn beyond the limits specified, which is usually the case in an engine that's

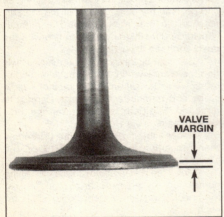

9.16 The margin width on each valve must be as specified (if no margin exists, the valve cannot be reused)

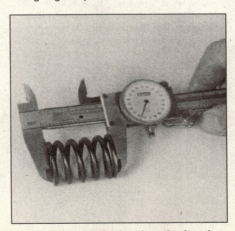

9.17 Measure the free length of each valve spring with a dial or vernier caliper

9.18 Check each valve spring for squareness

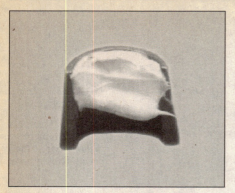

11.6 Apply a small dab of grease to each keeper as shown here before installation - it'll hold them in place on the valve stem as the spring is released

12.3 To check the camshaft lobe lift, position the dial indicator plunger directly above, and in line with, the pushrod

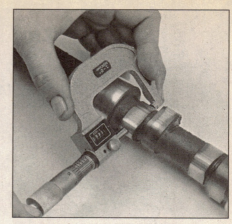

12.9 Measuring camshaft lobe lift

being overhauled, reassemble the valves in the cylinder head (see Section 10 for valve servicing recommendations).

10 Valves - servicing

1   Because of the complex nature of the job and the special tools and equipment needed, servicing of the valves, the valve seats and the valve guides, commonly known as a valve job, should be done by a professional.
2   The home mechanic can remove and disassemble the head, do the initial cleaning and inspection, then reassemble and deliver it to a dealer service department or an automotive machine shop for the actual service work. Doing the inspection will enable you to see what condition the head and valvetrain components are in and will ensure that you know what work and new parts are required when dealing with an automotive machine shop.
3   The dealer service department, or automotive machine shop, will remove the valves and springs, recondition or replace the valves and valve seats, recondition the valve guides, check and replace the valve springs, rotators, spring retainers and keepers (as necessary), replace the valve seals with new ones, reassemble the valve components and make sure the installed spring height is correct. The cylinder head gasket surface will also be resurfaced if it's warped.
4   After the valve job has been performed by a professional, the head will be in like new condition. When the head is returned, be sure to clean it again before installation on the engine to remove any metal particles and abrasive grit that may still be present from the valve service or head resurfacing operations. Use compressed air, if available, to blow out all the oil holes and passages.

11 Cylinder head - reassembly

Refer to illustration 11.6

1   Regardless of whether or not the head was sent to an automotive repair shop for valve servicing, make sure it's clean before beginning reassembly.
2   If the head was sent out for valve servicing, the valves and related components will already be in place. Begin the reassembly procedure with Step 8.
3   Install the spring seats before the valve seals.
4   Install new seals on each of the valve guides. **Note:** *3.0 and 3.8 liter engines have seals only on the intake valves.* Using a hammer and a deep socket or seal installation tool, gently tap each seal into place until it's completely seated on the guide. Don't twist or cock the seals during installation or they won't seal properly on the valve stems.
5   Beginning at one end of the head, lubricate and install the first valve. Apply molybase grease or clean engine oil to the valve stem.
6   Position the valve springs (and shims, if used) over the valves. Compress the springs with a valve spring compressor and carefully install the keepers in the groove, then slowly release the compressor and make sure the keepers seat properly. Apply a small dab of grease to each keeper to hold it in place if necessary **(see illustration)**.
7   Repeat the procedure for the remaining valves. Be sure to return the components to their original locations - don't mix them up!
8   Check the installed valve spring height with a ruler graduated in 1/32-inch increments or a dial caliper. If the head was sent out for service work, the installed height should be correct (but don't automatically assume it is). The measurement is taken from the top of each spring seat, rotator or top shim to the bottom of the retainer. If the height is greater than specified in this Chapter, shims can be added under the springs to correct it. **Caution:** *Do not, under any circumstances, shim the springs to the point where the installed height is less than specified.*
9   Apply moly-base grease to the rocker arm faces and the pivots, then install the rocker arms and pivots on the cylinder heads. Tighten the bolts finger-tight.

12 Camshaft, balance shaft and bearings - removal and inspection

Note: *Since there isn't enough room to remove the camshaft with the engine in the vehicle, the engine must be out of the vehicle and mounted on a stand for this procedure.*

Camshaft lobe lift check

With cylinder head installed

Refer to illustration 12.3

1   In order to determine the extent of cam lobe wear, the lobe lift should be checked prior to camshaft removal. Remove the valve covers (3.1L engine: see Section 5 in Part A; 3800 engine: see Section 5 in Part B).
2   Position the number one piston at TDC on the compression stroke (3.1L engine: see Section 3 in Part A; 3800 engine: see Section 3 in Part B).
3   Beginning with the number one cylinder valves, mount a dial indicator on the engine and position the plunger against the top surface of the first rocker arm. The plunger should be directly above and in line with the push-rod **(see illustration)**.
4   Zero the dial indicator, then very slowly turn the crankshaft in the normal direction of rotation (clockwise) until the indicator needle stops and begins to move in the opposite direction. The point at which it stops indicates maximum cam lobe lift.
5   Record this figure for future reference, then reposition the piston at TDC on the compression stroke.
6   Move the dial indicator to the remaining number one cylinder rocker arm and repeat the check. Be sure to record the results for each valve.
7   Repeat the check for the remaining valves. Since each piston must be at TDC on the compression stroke for this procedure, work from cylinder-to-cylinder following the

Chapter 2 Part C General engine overhaul procedures

12.12 Thread long bolts into the sprocket bolt holes to use as a handle when removing and installing the camshaft

12.13 Support the camshaft near the block to avoid damaging the bearings

firing order sequence.
8 After the check is complete, compare the results to this Chapter's Specifications. If camshaft lobe lift is less than specified, cam lobe wear has occurred and a new camshaft should be installed.

With cylinder head removed

Refer to illustration 12.9

9 If the cylinder heads have already been removed, an alternate method of lobe measurement can be used. Remove the camshaft as described later in this Section. Using a micrometer, measure the lobe at its highest point. Then measure the base circle perpendicular (90-degrees) to the lobe **(see illustration)**. Do this for each lobe and record the results.
10 Subtract the base circle measurement from the lobe height. The difference is the lobe lift. See Step 8 above.

Removal

Refer to illustrations 12.12 and 12.13

11 Remove the timing chain and sprockets, lifters and pushrods (3.1L engine: See Part A, Sections 5, 10, 13 and 14; 3800 engine: See Part B, Sections 5, 8, 13 and 16).
12 Remove the camshaft thrust plate-to-block bolts. Thread long bolts into the camshaft sprocket bolt holes to use as a handle when removing the camshaft from the block **(see illustration)**. **Note:** *The balance shaft (3800 engine only) requires special tools for removal and installation. If the bearings are bad, have the shaft assembly replaced by an automotive machine shop (see step 19).*
13 Carefully pull the camshaft out. Support the cam near the block so the lobes don't nick or gouge the bearings as it's withdrawn **(see illustration)**.

Inspection

Refer to illustration 12.15

14 After the camshaft has been removed from the engine, cleaned with solvent and dried, inspect the bearing journals for uneven wear, pitting and evidence of seizure. If the

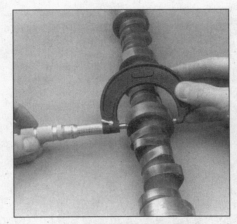

12.15 Check the camshaft bearing journal diameters to pinpoint excessive wear and out-of-round conditions

journals are damaged, the bearing inserts in the block are probably damaged as well. Both the camshaft and bearings will have to be replaced.
15 Measure the bearing journals with a micrometer **(see illustration)** to determine if they're excessively worn or out-of-round.
16 Check the camshaft lobes for heat discoloration, score marks, chipped areas, pitting and uneven wear. If the lobes are in good condition and if the lobe lift measurements are as specified, the camshaft can be reused.
17 Check the bearings in the block for wear and damage. Look for galling, pitting and discolored areas.
18 The inside diameter of each bearing can be determined with a small hole gauge and outside micrometer or an inside micrometer. Subtract the camshaft bearing journal diameter(s) from the corresponding bearing inside diameter(s) to obtain the bearing oil clearance. If it's excessive, new bearings will be required regardless of the condition of the originals.
19 Balance shaft and camshaft bearing replacement requires special tools and

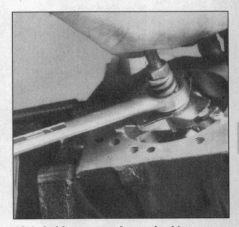

13.1 A ridge reamer is required to remove the ridge from the top of each cylinder - do this before removing the pistons!

expertise that place it outside the scope of the home mechanic. Take the block to an automotive machine shop to ensure the job is done correctly.
20 Balance shaft gear installation and timing is covered in Section 24.

13 Pistons and connecting rods - removal

Refer to illustrations 13.1, 13.3, 13.4 and 13.6
Note: *Prior to removing the piston/connecting rod assemblies, remove the cylinder heads, the oil pan and the oil pump by referring to the appropriate Sections in Part A of Chapter 2.*

1 Use your fingernail to feel if a ridge has formed at the upper limit of ring travel (about 1/4-inch down from the top of each cylinder). If carbon deposits or cylinder wear have produced ridges, they must be completely removed with a special tool **(see illustration)**. Follow the manufacturer's instructions provided with the tool. Failure to remove the ridges before attempting to remove the pis-

2C-14 Chapter 2 Part C General engine overhaul procedures

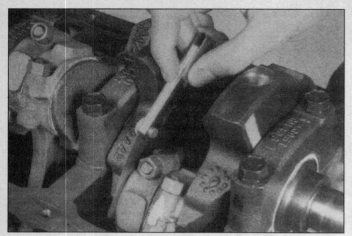

13.3 Check the connecting rod side clearance with a feeler gauge as shown

13.4 Mark the rod bearing caps in order from the front of the engine to the rear (one mark for the front cap, two for the second one and so on)

ton/connecting rod assemblies may result in piston breakage.

2 After the cylinder ridges have been removed, turn the engine upside-down so the crankshaft is facing up.

3 Before the connecting rods are removed, check the endplay with feeler gauges. Slide them between the first connecting rod and the crankshaft throw until the play is removed **(see illustration)**. The endplay is equal to the thickness of the feeler gauge(s). If the endplay exceeds the service limit, new connecting rods will be required. If new rods (or a new crankshaft) are installed, the endplay may fall under the minimum specified in this Chapter (if it does, the rods will have to be machined to restore it - consult an automotive machine shop for advice if necessary). Repeat the procedure for the remaining connecting rods.

4 Check the connecting rods and caps for identification marks. If they aren't plainly marked, use a small center-punch **(see illustration)** to make the appropriate number of indentations on each rod and cap (1, 2, 3, etc., depending on the cylinder they're associated with).

5 Loosen each of the connecting rod cap nuts 1/2-turn at a time until they can be removed by hand. Remove the number one connecting rod cap and bearing insert. Don't drop the bearing insert out of the cap.

6 Slip a short length of plastic or rubber hose over each connecting rod cap bolt to protect the crankshaft journal and cylinder wall as the piston is removed **(see illustration)**.

7 Remove the bearing insert and push the connecting rod/piston assembly out through the top of the engine. Use a wooden or plastic hammer handle to push on the upper bearing surface in the connecting rod. If resistance is felt, double-check to make sure all of the ridge was removed from the cylinder.

8 Repeat the procedure for the remaining cylinders.

13.6 To prevent damage to the crankshaft journals and cylinder walls, slip sections of rubber or plastic hose over the rod bolts before removing the pistons

9 After removal, reassemble the connecting rod caps and bearing inserts in their respective connecting rods and install the cap nuts finger tight. Leaving the old bearing inserts in place until reassembly will help prevent the connecting rod bearing surfaces from being accidentally nicked or gouged.

10 Don't separate the pistons from the connecting rods (see Section 18 for additional information).

14 Crankshaft - removal

Refer to illustrations 14.1, 14.3, 14.4a and 14.4b

Note: *The crankshaft can be removed only after the engine has been removed from the vehicle. It's assumed the driveplate, vibration damper, timing chain, oil pan, oil pump and piston/connecting rod assemblies have already been removed.*

1 Before the crankshaft is removed, check

14.1 Checking crankshaft endplay with a dial indicator

the endplay. Mount a dial indicator with the stem in line with the crankshaft and touching one of the crank throws **(see illustration)**.

2 Push the crankshaft all the way to the rear and zero the dial indicator. Next, pry the crankshaft to the front as far as possible and check the reading on the dial indicator. The distance it moves is the endplay. If it's greater than specified in this Chapter, check the crankshaft thrust surfaces for wear. If no wear is evident, new main bearings should correct the endplay.

3 If a dial indicator isn't available, feeler gauges can be used. Gently pry or push the crankshaft all the way to the front of the engine. Slip feeler gauges between the crankshaft and the front face of the thrust main bearing to determine the clearance **(see illustration)**. **Note:** *The thrust bearing is located at the number two (3800) or number 3 (3.1L) main bearing cap.*

4 Check the main bearing caps to see if they're marked to indicate their locations.

Chapter 2 Part C General engine overhaul procedures

14.3 Checking crankshaft endplay with a feeler gauge

14.4a Use a center punch or number stamping dies to mark the main bearing caps to ensure installation in their original locations on the block (make the punch marks near one of the bolt heads)

14.4b The arrow on the main bearing cap indicates the front of the engine

They should be numbered consecutively from the front of the engine to the rear. If they aren't, mark them with number stamping dies or a center-punch **(see illustrations)**. Main bearing caps generally have a cast-in arrow, which points to the front of the engine **(see illustration)**. Loosen the main bearing cap bolts 1/4-turn at a time each, until they can be removed by hand. Note if any stud bolts are used and make sure they're returned to their original locations when the crankshaft is reinstalled.

5 Gently tap the caps with a soft-face hammer, then separate them from the engine block. If necessary, use the bolts as levers to remove the caps. Try not to drop the bearing inserts if they come out with the caps.

6 Carefully lift the crankshaft out of the engine. It may be a good idea to have an assistant available, since the crankshaft is quite heavy. With the bearing inserts in place in the engine block and main bearing caps, return the caps to their respective locations on the engine block and tighten the bolts finger tight.

7 Remove the rear main oil seal from the end of the crankshaft.

15 Engine block - cleaning

Refer to illustrations 15.4a, 15.4b, 15.8 and 15.10

1 Remove the main bearing caps and separate the bearing inserts from the caps and the engine block. Tag the bearings, indicating which cylinder they were removed from and whether they were in the cap or the block, then set them aside.

2 Using a gasket scraper, remove all traces of gasket material from the engine block. Be very careful not to nick or gouge the gasket sealing surfaces.

3 Remove all of the covers and threaded oil gallery plugs from the block. The plugs are usually very tight - they may have to be drilled out and the holes retapped. Use new plugs when the engine is reassembled.

4 Remove the core plugs from the engine block. To do this, knock one side of the plugs into the block with a hammer and punch, then grasp them with large pliers and pull them out **(see illustrations)**.

5 If the engine is extremely dirty, it should be taken to an automotive machine shop to be steam cleaned or hot tanked.

6 After the block is returned, clean all oil holes and oil galleries one more time. Brushes specifically designed for this purpose are available at most auto parts stores. Flush the passages with warm water until the water runs clear, dry the block thoroughly and wipe all machined surfaces with a light, rust preventive oil. If you have access to compressed air, use it to speed the drying process and blow out all the oil holes and galleries. **Warning:** *Wear eye protection when using compressed air!*

7 If the block isn't extremely dirty or sludged up, you can do an adequate cleaning job with hot soapy water and a stiff brush. Take plenty of time and do a thorough job. Regardless of the cleaning method used, be sure to clean all oil holes and galleries very thoroughly, dry the block completely and coat all machined surfaces with light oil.

8 The threaded holes in the block must be clean to ensure accurate torque readings during reassembly. Run the proper size tap into each of the holes to remove rust, corro-

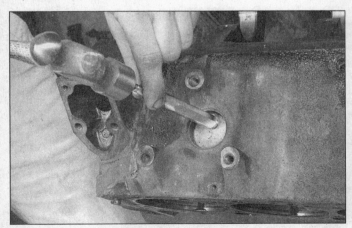

15.4a A hammer and large punch can be used to knock the core plugs sideways in their bores

15.4b Pull the core plugs from the block with pliers

Chapter 2 Part C General engine overhaul procedures

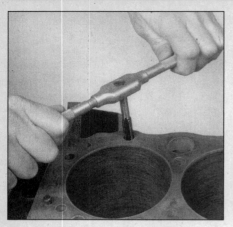

15.8 All bolt holes in the block - particularly the main bearing cap and head bolt holes - should be cleaned and restored with a tap (be sure to remove debris from the holes after this is done)

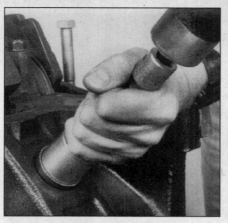

15.10 A large socket on an extension can be used to drive the new core plugs into the bores

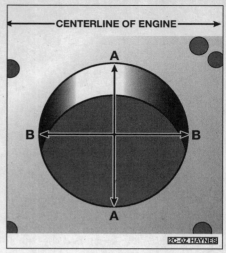

16.4a Use "A" and "B" axes for cylinder measurements as described in text

16.4b The ability to "feel" when the telescoping gauge is at the correct point will be developed over time, so work slowly and repeat the check until you're satisfied the bore measurement is accurate

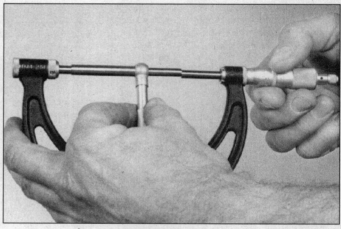

16.4c The gauge is then measured with a micrometer to determine the bore size

sion, thread sealant or sludge and restore damaged threads **(see illustration)**. If possible, use compressed air to clear the holes of debris produced by this operation. Now is a good time to clean the threads on the head bolts and the main bearing cap bolts as well.

9 Reinstall the main bearing caps and tighten the bolts finger tight.

10 After coating the sealing surfaces of the new core plugs with Permatex no. 2 sealant, install them in the engine block **(see illustration)**. Make sure they're driven in straight and seated properly or leakage could result. Special tools are available for this purpose, but a large socket, with an outside diameter that will just slip into the core plug, a 1/2-inch drive extension and a hammer will work just as well.

11 Apply non-hardening sealant (such as Permatex no. 2 or Teflon pipe sealant) to the new oil gallery plugs and thread them into the holes in the block. Make sure they're tightened securely.

12 If the engine isn't going to be reassembled right away, cover it with a large plastic trash bag to keep it clean.

16 Engine block - inspection

Refer to illustrations 16.4a, 16.4b and 16.4c

Note: *The manufacturer recommends checking the block deck and transaxle mounting bolt hole bosses for warpage and the main bearing bore concentricity and alignment. Since special measuring tools are needed, the checks should be done by an automotive machine shop.*

1 Before the block is inspected, it should be cleaned as described in Section 15.

2 Visually check the block for cracks, rust and corrosion. Look for stripped threads in the threaded holes. It's also a good idea to have the block checked for hidden cracks by an automotive machine shop that has the special equipment to do this type of work. If defects are found, have the block repaired, if possible, or replaced.

3 Check the cylinder bores for scuffing and scoring.

4 Measure the diameter of each cylinder at the top (just under the ridge area), center and bottom of the cylinder bore, parallel to the crankshaft axis **(see illustrations)**. **Note:** *These measurements should not be made with the bare block mounted on an engine stand - the cylinders will be distorted and the measurements will be inaccurate.*

5 Next, measure each cylinder's diameter at the same three locations across the crankshaft axis. Compare the results to this Chapter's Specifications.

6 If the required precision measuring tools aren't available, the piston-to-cylinder clearances can be obtained, though not quite as accurately, using feeler gauge stock. Feeler gauge stock comes in 12-inch lengths and various thicknesses and is generally available at auto parts stores.

17.3a A "bottle brush" hone will produce better results if you've never honed cylinders before

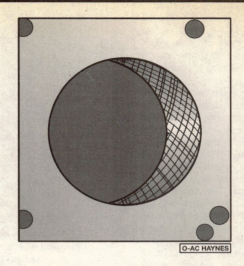

17.3b The cylinder hone should leave a smooth, crosshatch pattern with the lines intersecting at approximately a 60-degree angle

7 To check the clearance, select a feeler gauge and slip it into the cylinder along with the matching piston. The piston must be positioned exactly as it normally would be. The feeler gauge must be between the piston and cylinder on one of the thrust faces (90-degrees to the piston pin bore).

8 The piston should slip through the cylinder (with the feeler gauge in place) with moderate pressure.

9 If it falls through or slides through easily, the clearance is excessive and a new piston will be required. If the piston binds at the lower end of the cylinder and is loose toward the top, the cylinder is tapered. If tight spots are encountered as the piston/feeler gauge is rotated in the cylinder, the cylinder is out-of-round.

10 Repeat the procedure for the remaining pistons and cylinders.

11 If the cylinder walls are badly scuffed or scored, or if they're out-of-round or tapered beyond the limits given in this Chapter's Specifications, have the engine block rebored and honed at an automotive machine shop. If a rebore is done, oversize pistons and rings will be required.

12 If the cylinders are in reasonably good condition and not worn to the outside of the limits, and if the piston-to-cylinder clearances can be maintained properly, they don't have to be rebored. Honing is all that's necessary (see Section 17).

17 Cylinder honing

Refer to illustrations 17.3a and 17.3b

1 Prior to engine reassembly, the cylinder bores must be honed so the new piston rings will seat correctly and provide the best possible combustion chamber seal. **Note:** *If you don't have the tools or don't want to tackle the honing operation, most automotive machine shops will do it for a reasonable fee.*

2 Before honing the cylinders, install the main bearing caps and tighten the bolts to the torque listed in this Chapter's Specifications.

3 Two types of cylinder hones are commonly available - the flex hone or "bottle brush" type and the more traditional surfacing hone with spring-loaded stones. Both will do the job, but for the less experienced mechanic the "bottle brush" hone will probably be easier to use. You'll also need some honing oil (kerosene will work if honing oil isn't available), rags and an electric drill motor. Proceed as follows:

a) *Mount the hone in the drill motor, compress the stones and slip it into the first cylinder* **(see illustration)**. *Be sure to wear safety goggles or a face shield!*

b) *Lubricate the cylinder with plenty of honing oil, turn on the drill and move the hone up-and-down in the cylinder at a pace that will produce a fine crosshatch pattern on the cylinder walls. Ideally, the crosshatch lines should intersect at approximately a 45-degree angle* **(see illustration)**. *Be sure to use plenty of lubricant and don't take off any more material than is absolutely necessary to produce the desired finish.* **Note:** *Piston ring manufacturers may specify a different crosshatch angle - read and follow any instructions included with the new rings.*

c) *Don't withdraw the hone from the cylinder while it's running. Instead, shut off the drill and continue moving the hone up-and-down in the cylinder until it comes to a complete stop, then compress the stones and withdraw the hone. If you're using a "bottle brush" type hone, stop the drill motor, then turn the chuck in the normal direction of rotation while withdrawing the hone from the cylinder.*

d) *Wipe the oil out of the cylinder and repeat the procedure for the remaining cylinders.*

4 After the honing job is complete, chamfer the top edges of the cylinder bores with a small file so the rings won't catch when the pistons are installed. Be very careful not to nick the cylinder walls with the end of the file.

5 The entire engine block must be washed again very thoroughly with warm, soapy water to remove all traces of the abrasive grit produced during the honing operation. **Note:** *The bores can be considered clean when a lint-free white cloth - dampened with clean engine oil - used to wipe them out doesn't pick up any more honing residue, which will show up as gray areas on the cloth. Be sure to run a brush through all oil holes and galleries and flush them with running water.*

6 After rinsing, dry the block and apply a coat of light rust preventive oil to all machined surfaces. Wrap the block in a plastic trash bag to keep it clean and set it aside until reassembly.

18 Pistons and connecting rods - inspection

Refer to illustrations 18.4a, 18.4b, 18.10 and 18.11

1 Before the inspection process can be carried out, the piston/connecting rod assemblies must be cleaned and the original piston rings removed from the pistons. **Note:** *Always use new piston rings when the engine is reassembled.*

2 Using a piston ring installation tool, carefully remove the rings from the pistons. Be careful not to nick or gouge the pistons in the process.

3 Scrape all traces of carbon from the top of the piston. A hand held wire brush or a piece of fine emery cloth can be used once the majority of the deposits have been scraped away. Do not, under any circumstances, use a wire brush mounted in a drill motor to remove deposits from the pistons. The piston material is soft and may be eroded away by the wire brush.

4 Use a piston ring groove cleaning tool to remove carbon deposits from the ring

grooves. If a tool isn't available, a piece broken off the old ring will do the job. Be very careful to remove only the carbon deposits - don't remove any metal and do not nick or scratch the sides of the ring grooves **(see illustrations)**.

5 Once the deposits have been removed, clean the piston/rod assemblies with solvent and dry them with compressed air (if available). **Warning:** *Wear eye protection. Make sure the oil return holes in the back sides of the ring grooves are clear.*

6 If the pistons and cylinder walls aren't damaged or worn excessively, and if the engine block isn't rebored, new pistons won't be necessary. Normal piston wear appears as even vertical wear on the piston thrust surfaces and slight looseness of the top ring in its groove. New piston rings, however, should always be used when an engine is rebuilt.

7 Carefully inspect each piston for cracks around the skirt, at the pin bosses and at the ring lands.

8 Look for scoring and scuffing on the thrust faces of the skirt, holes in the piston crown and burned areas at the edge of the crown. If the skirt is scored or scuffed, the engine may have been suffering from overheating and/or abnormal combustion, which caused excessively high operating temperatures. The cooling and lubrication systems should be checked thoroughly. A hole in the piston crown is an indication that abnormal combustion (preignition) was occurring. Burned areas at the edge of the piston crown are usually evidence of spark knock (detonation). If any of the above problems exist, the causes must be corrected or the damage will occur again. The causes may include intake air leaks, incorrect fuel/air mixture, low octane fuel, ignition timing and EGR system malfunctions.

9 Corrosion of the piston, in the form of small pits, indicates coolant is leaking into the combustion chamber and/or the crankcase. Again, the cause must be corrected or the problem may persist in the

18.4a The piston ring grooves can be cleaned with a special tool, as shown here . . .

18.4b . . . or a section of a broken ring

rebuilt engine.

10 Measure the piston ring side clearance by laying a new piston ring in each ring groove and slipping a feeler gauge in beside it **(see illustration)**. Check the clearance at three or four locations around each groove. Be sure to use the correct ring for each groove - they are different. If the side clearance is greater than specified in this Chapter, new pistons will have to be used.

11 Check the piston-to-bore clearance by measuring the bore (see Section 16) and the piston diameter. Make sure the pistons and bores are correctly matched. Measure the piston across the skirt, at a 90-degree angle to the piston pin **(see illustration)**. The measurement must be taken at a specific point, depending on the engine type, to be accurate.

a) *The piston diameter on 3800 engines (with balance shaft) is measured directly in line with the piston pin centerline.*

b) *3.1 liter engine pistons are measured 5/8-inch (15 mm) from the bottom of the skirt.*

12 Subtract the piston diameter from the bore diameter to obtain the clearance. If it's greater than specified, the block will have to be rebored and new pistons and rings installed.

13 Check the piston-to-rod clearance by twisting the piston and rod in opposite directions. Any noticeable play indicates excessive wear, which must be corrected. The piston/connecting rod assemblies should be taken to an automotive machine shop to have the pistons and rods resized and new pins installed.

14 If the pistons must be removed from the connecting rods for any reason, they should be taken to an automotive machine shop. While they are there have the connecting rods checked for bend and twist, since automotive machine shops have special equipment for this purpose. **Note:** *Unless new pistons and/or connecting rods must be installed, do not disassemble the pistons and connecting rods.*

15 Check the connecting rods for cracks and other damage. Temporarily remove the rod caps, lift out the old bearing inserts, wipe the rod and cap bearing surfaces clean and

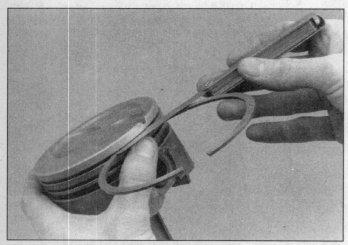

18.10 Check the ring side clearance with a feeler gauge at several points around the groove

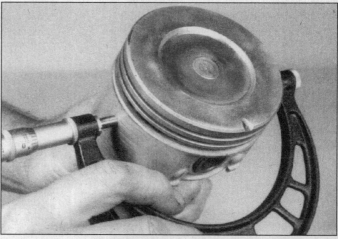

18.11 Measure the piston diameter at a 90-degree angle to the piston pin and in line with it

Chapter 2 Part C General engine overhaul procedures

19.1 The oil holes should be chamfered so sharp edges don't gouge or scratch he new bearings

19.2 Use a wire or stiff plastic bristle brush to clean the oil passages in the crankshaft

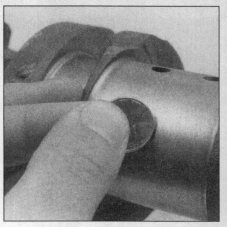

19.4 Rubbing a penny lengthwise on each journal will reveal its condition - if copper rubs off and is embedded in the crankshaft, the journal should be reground

inspect them for nicks, gouges and scratches. After checking the rods, replace the old bearings, slip the caps into place and tighten the nuts finger tight. **Note:** *If the engine is being rebuilt because of a connecting rod knock, be sure to install new rods.*

19 Crankshaft - inspection

Refer to illustrations 19.1, 19.2, 19.4, 19.6 and 19.8

1 Remove all burrs from the crankshaft oil holes with a stone, file or scraper **(see illustration)**.
2 Clean the crankshaft with solvent and dry it with compressed air (if available). **Warning:** *Wear eye protection when using compressed air.* Be sure to clean the oil holes with a stiff brush **(see illustration)** and flush them with solvent.
3 Check the main and connecting rod bearing journals for uneven wear, scoring, pits and cracks.

4 Rub a penny across each journal several times **(see illustration)**. If a journal picks up copper from the penny, it's too rough and must be reground.
5 Check the rest of the crankshaft for cracks and other damage. It should be Magnafluxed to reveal hidden cracks - an automotive machine shop will handle the procedure.
6 Using a micrometer, measure the diameter of the main and connecting rod journals and compare the results to this Chapter's Specifications **(see illustration)**. By measuring the diameter at a number of points around each journal's circumference, you'll be able to determine whether or not the journal is out-of-round. Take the measurement at each end of the journal, near the crank throws, to determine if the journal is tapered.
7 If the crankshaft journals are damaged, tapered, out-of-round or worn beyond the limits given in the Specifications, have the crankshaft reground by an automotive machine shop. Be sure to use the correct size

bearing inserts if the crankshaft is reconditioned.
8 Check the oil seal journals at each end of the crankshaft for wear and damage. If the seal has worn a groove in the journal, or if it's nicked or scratched **(see illustration)**, the new seal may leak when the engine is reassembled. In some cases, an automotive machine shop may be able to repair the journal by pressing on a thin sleeve. If repair isn't feasible, a new or different crankshaft should be installed.
9 Examine the main and rod bearing inserts (see Section 20).

20 Main and connecting rod bearings - inspection

Refer to illustration 20.1

1 Even though the main and connecting rod bearings should be replaced with new ones during the engine overhaul, the old bearings should be retained for close exami-

19.6 Measure the diameter of each crankshaft journal at several points to detect taper and out-of-round conditions

19.8 If the seals have worn grooves in the crankcase journals, or if the seal contact surfaces are nicked or scratched, the new seals will leak

nation, as they may reveal valuable information about the condition of the engine (see illustration).

2 Bearing failure occurs because of lack of lubrication, the presence of dirt or other foreign particles, overloading the engine and corrosion. Regardless of the cause of bearing failure, it must be corrected before the engine is reassembled to prevent it from happening again.

3 When examining the bearings, remove them from the engine block, the main bearing caps, the connecting rods and the rod caps and lay them out on a clean surface in the same general position as their location in the engine. This will enable you to match any bearing problems with the corresponding crankshaft journal.

4 Dirt and other foreign particles get into the engine in a variety of ways. It may be left in the engine during assembly, or it may pass through filters or the PCV system. It may get into the oil, and from there into the bearings. Metal chips from machining operations and normal engine wear are often present. Abrasives are sometimes left in engine components after reconditioning, especially when parts aren't thoroughly cleaned using the proper cleaning methods. Whatever the source, these foreign objects often end up embedded in the soft bearing material and are easily recognized. Large particles won't embed in the bearing and will score or gouge the bearing and journal. The best prevention for this cause of bearing failure is to clean all parts thoroughly and keep everything spotlessly clean during engine assembly. Frequent and regular engine oil and filter changes are also recommended.

5 Lack of lubrication (or lubrication breakdown) has a number of interrelated causes. Excessive heat (which thins the oil), overloading (which squeezes the oil from the bearing face) and oil leakage or throw off (from excessive bearing clearances, worn oil pump or high engine speeds) all contribute to lubrication breakdown. Blocked oil passages, which usually are the result of misaligned oil holes in a bearing shell, will also oil starve a bearing and destroy it. When lack of lubrication is the cause of bearing failure, the bearing material is wiped or extruded from the steel backing of the bearing. Temperatures may increase to the point where the steel backing turns blue from overheating.

6 Driving habits can have a definite effect on bearing life. Full throttle, low speed operation (lugging the engine) puts very high loads on bearings, which tends to squeeze out the oil film. These loads cause the bearings to flex, which produces fine cracks in the bearing face (fatigue failure). Eventually the bearing material will loosen in pieces and tear away from the steel backing. Short trip driving leads to corrosion of bearings because insufficient engine heat is produced to drive off the condensed water and corrosive gases. These products collect in the engine oil, forming acid and sludge. As the oil is carried to the engine bearings, the acid attacks

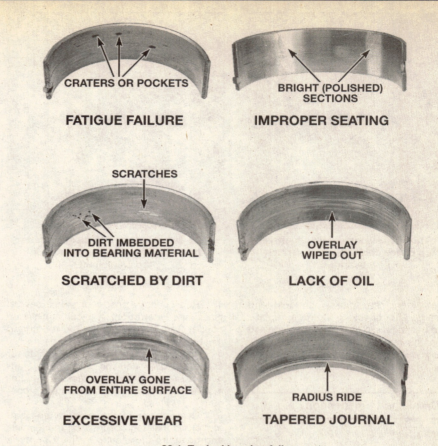

20.1 Typical bearing failures

and corrodes the bearing material.

7 Incorrect bearing installation during engine assembly will lead to bearing failure as well. Tight fitting bearings leave insufficient oil clearance and will result in oil starvation. Dirt or foreign particles trapped behind a bearing insert result in high spots on the bearing which lead to failure.

21 Engine overhaul - reassembly sequence

1 Before beginning engine reassembly, make sure you have all the necessary new parts, gaskets and seals as well as the following items on hand:

Common hand tools
Torque wrench (1/2-inch drive)
Piston ring installation tool
Piston ring compressor
Vibration damper installation tool
Short lengths of rubber or plastic hose to fit over connecting rod bolts
Plastigage
Feeler gauges
Fine-tooth file
New engine oil
Engine assembly lube or moly-base grease
Gasket sealant
Thread locking compound

2 In order to save time and avoid problems, engine reassembly must be done in the following general order:

Crankshaft and main bearings
Piston/connecting rod assemblies
Balance shaft (3800 engine only)
Camshaft
Rear main oil seal
Timing chain and sprockets
Timing chain cover and oil pump
Oil pan
Cylinder heads
Valve lifters
Valve and pushrods
Intake and exhaust manifolds
Valve covers
Driveplate

22 Piston rings - installation

Refer to illustrations 22.3, 22.4, 22.5, 22.9a, 22.9b and 22.12

1 Before installing the new piston rings, the ring end gaps must be checked. It's assumed the piston ring side clearance has been checked and verified correct (see Section 18).

Chapter 2 Part C General engine overhaul procedures

22.3 When checking piston ring end gap, the ring must be square in the cylinder bore (this is done by pushing the ring down with the top of a piston as shown)

22.4 With the ring square in the cylinder, measure the end gap with a feeler gauge

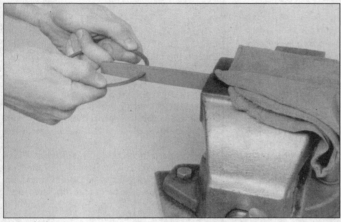

22.5 If the end gap is too small, clamp a file in a vise and file the ring ends (from the outside in only) to enlarge the gap slightly

22.9a Installing the spacer/expander in the oil control ring groove

2 Lay out the piston/connecting rod assemblies and the new ring sets so the ring sets will be matched with the same piston and cylinder during the end gap measurement and engine assembly.

3 Insert the top (number one) ring into the first cylinder and square it up with the cylinder walls by pushing it in with the top of the piston (see illustration). The ring should be near the bottom of the cylinder, at the lower limit of ring travel.

4 To measure the end gap, slip feeler gauges between the ends of the ring until a gauge equal to the gap width is found (see illustration). The feeler gauge should slide between the ring ends with a slight amount of drag. Compare the measurement to this Chapter's Specifications. If the gap is larger or smaller than specified, double-check to make sure you have the correct rings before proceeding.

5 If the gap is too small, it must be enlarged or the ring ends may come in contact with each other during engine operation, which can cause serious engine damage. The end gap can be increased by filing the ring ends very carefully with a fine file. Mount the file in a vise equipped with soft jaws, slip the ring over the file with the ends contacting the file teeth and slowly move the ring to remove material from the ends. When performing this operation, file only from the outside in (see illustration).

6 Excess end gap isn't critical unless it's greater than 0.040-inch. Again, double-check to make sure you have the correct rings for the engine.

7 Repeat the procedure for each ring that will be installed in the first cylinder and for each ring in the remaining cylinders. Remember to keep rings, pistons and cylinders matched up.

8 Once the ring end gaps have been checked/corrected, the rings can be installed on the pistons.

9 The oil control ring (lowest one on the piston) is usually installed first. It's composed of three separate components. Slip the spacer/expander into the groove (see illustration). If an anti-rotation tang is used, make sure it's inserted into the drilled hole in the ring groove. Next, install the lower side rail.

22.9b DO NOT use a piston ring installation tool when installing the oil ring side rails

Don't use a piston ring installation tool on the oil ring side rails, as they may be damaged. Instead, place one end of the side rail into the groove between the spacer/expander and the ring land, hold it firmly in place and slide a fin-

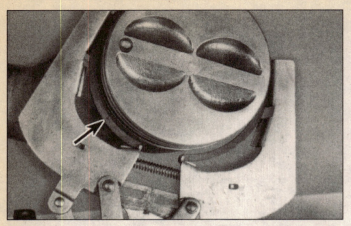

22.12 Installing the compression rings with a ring expander - the mark (arrow) must face up

23.11 Lay the Plastigage strips (arrow) on the main bearing journals, parallel to the crankshaft centerline

ger around the piston while pushing the rail into the groove **(see illustration)**. Next, install the upper side rail in the same manner.

10   After the three oil ring components have been installed, check to make sure both the upper and lower side rails can be turned smoothly in the ring groove.

11   The number two (middle) ring is installed next. It's usually stamped with a mark, which must face up, toward the top of the piston. **Note:** *Always follow the instructions printed on the ring package or box - different manufacturers may require different approaches. Don't mix up the top and middle rings, as they have different cross-sections.*

12   Use a piston ring installation tool and make sure the identification mark is facing the top of the piston, then slip the ring into the middle groove on the piston **(see illustration)**. Don't expand the ring any more than necessary to slide it over the piston.

13   Install the number one (top) ring in the same manner. Make sure the mark is facing up. Be careful not to confuse the number one and number two rings.

14   Repeat the procedure for the remaining pistons and rings.

23   Crankshaft - installation and main bearing oil clearance check

1   Crankshaft installation is the first step in engine reassembly. It's assumed at this point that the engine block and crankshaft have been cleaned, inspected and repaired or reconditioned.

2   Position the engine with the bottom facing up.

3   Remove the main bearing cap bolts and lift out the caps. Lay them out in the proper order to ensure correct installation.

4   If they're still in place, remove the original bearing inserts from the block and the main bearing caps. Wipe the bearing surfaces of the block and caps with a clean, lint-free cloth. They must be kept spotlessly clean.

Main bearing oil clearance check

Refer to illustrations 23.11 and 23.15

Note: *Don't touch the faces of the new bearing inserts with your fingers. Oil and acids from your skin can etch the bearings.*

5   Clean the back sides of the new main bearing inserts and lay one in each main bearing saddle in the block. If one of the bearing inserts from each set has a large groove in it, make sure the grooved insert is installed in the block. Lay the other bearing from each set in the corresponding main bearing cap. Make sure the tab on the bearing insert fits into the recess in the block or cap. **Caution:** *The oil holes in the block must line up with the oil holes in the bearing inserts. Do not hammer the bearing into place and don't nick or gouge the bearing faces. No lubrication should be used at this time.*

6   The flanged thrust bearing must be installed in the number three cap and saddle of a 3.1L engine or the number two cap and saddle of a 3800 engine (counting from the front of the engine).

7   Clean the faces of the bearings in the block and the crankshaft main bearing journals with a clean, lint-free cloth.

8   Check or clean the oil holes in the crankshaft, as any dirt here can go only one way - straight through the new bearings.

9   Once you're certain the crankshaft is clean, carefully lay it in position in the main bearings.

10   Before the crankshaft can be permanently installed, the main bearing oil clearance must be checked.

11   Cut several pieces of the appropriate size Plastigage (they should be slightly shorter than the width of the main bearings) and place one piece on each crankshaft main bearing journal, parallel with the journal axis **(see illustration)**.

12   Clean the faces of the bearings in the caps and install the caps in their original locations (don't mix them up) with the arrows pointing toward the front of the engine. Don't disturb the Plastigage.

13   Starting with the center main and working out toward the ends, tighten the main bearing cap bolts, in three steps, to the torque figure listed in this Chapter's Specifications. Don't rotate the crankshaft at any time during this operation.

14   Remove the bolts and carefully lift off the main bearing caps. Keep them in order. Don't disturb the Plastigage or rotate the crankshaft. If any of the main bearing caps are difficult to remove, tap them gently from side-to-side with a soft-face hammer to loosen them.

15   Compare the width of the crushed Plastigage on each journal to the scale printed on the Plastigage envelope to obtain the main bearing oil clearance **(see illustration)**. Check the Specifications to make sure it's correct.

16   If the clearance is not as specified, the bearing inserts may be the wrong size (which means different ones will be required). Before deciding different inserts are needed, make sure no dirt or oil was between the bearing inserts and the caps or block when the clearance was measured. If the Plastigage was wider at one end than the other, the journal may be tapered (see Section 19).

17   Carefully scrape all traces of the Plastigage material off the main bearing journals and/or the bearing faces. Use your fingernail or the edge of a credit card - don't nick or scratch the bearing faces.

Final crankshaft installation

18   Carefully lift the crankshaft out of the engine.

19   Clean the bearing faces in the block, then apply a thin, uniform layer of moly-base grease or engine assembly lube to each of the bearing surfaces. Be sure to coat the thrust faces as well as the journal face of the thrust bearing. Install the rear main oil seal (see Section 25).

20   Make sure the crankshaft journals are clean, then lay the crankshaft back in place in the block.

21   Clean the faces of the bearings in the caps, then apply lubricant to them.

Chapter 2 Part C General engine overhaul procedures

23.15 Measuring the width of the crushed Plastigage to determine the rod bearing oil clearance (be sure to use the correct scale - standard and metric ones are included)

22 Install the caps in their original locations with the arrows pointing toward the front of the engine.
23 Install the bolts.
24 Tighten all except the thrust bearing cap bolts to the torque listed in this Chapter's Specifications (work from the center out and approach the final torque in three steps).
25 Tighten the thrust bearing cap bolts to 10-to-12 ft-lbs.
26 Tap the ends of the crankshaft forward and backward with a lead or brass hammer to line up the main bearing and crankshaft thrust surfaces.
27 Retighten all main bearing cap bolts to the torque in this Chapter's Specifications, starting with the center main and working out toward the ends.
28 Rotate the crankshaft a number of times by hand to check for any obvious binding.
29 The final step is to check the crankshaft endplay with feeler gauges or a dial indicator as described in Section 14. The endplay should be correct if the crankshaft thrust faces aren't worn or damaged and new bearings have been installed.

24.1 Be sure to prelube the camshaft bearing journals and lobes before installation

24 Camshaft and balance shaft - installation

Camshaft

Refer to illustration 24.1

1 Lubricate the camshaft bearing journals and cam lobes with moly-base grease or special camshaft lubricant **(see illustration)**.
2 Slide the camshaft into the engine. Support the cam near the block and be careful not to scrape or nick the bearings. Install the thrust plate and tighten the bolts to the torque listed in the Part A Specifications.
3 Complete the installation of the timing chain and sprockets (3.1L engine: see Part A, Section 14; 3800 engine: see Part B, Section 16).

Balance shaft

Refer to illustration 24.6

4 If the balance shaft timing was disturbed during engine repair, turn the camshaft so that with the camshaft sprocket temporarily installed, the timing mark is straight down.
5 With the camshaft sprocket and the camshaft gear removed, turn the balance shaft so the timing mark on the gear points straight down.
6 Align the marks on the balance shaft gear and the camshaft gear **(see illustration)** by turning the balance shaft.
7 Install the timing chain and sprockets (3.1L engine: see Part A, Section 14; 3800 engine: see Part B, Section 16).

25 Rear main oil seal - replacement

Refer to illustration 25.3

1 Remove the engine/transaxle/subframe assembly and detach the engine from the transaxle, if not already done (see Section 5).
2 Remove the driveplate (see Section 27).
3 Pry out the old seal with a screwdriver or seal remover **(see illustration)**. Make sure you don't nick, gouge or scratch the crankshaft surface with the tool.
4 Clean the bore in the block and the seal contact surface on the crankshaft. Inspect the seal bore and the crankshaft surface for nicks, gouges or scratches. Even small imperfections on the crankshaft surface can damage the seal lip and cause oil leaks. If slight damage is evident, clean it up with crocus cloth. If the damage is too deep to clean up with crocus cloth, the crank will have to be remachined or replaced.
5 A special tool is recommended for installing the new seal. Lubricate the lip of the new seal with grease or clean engine oil, then slide the seal onto the mandril until the dust lip bottoms squarely against the collar of the tool. **Note:** *If the special tool isn't available, carefully work the seal over the crankshaft and tap it evenly into place with a hammer and punch.*
6 Align the dowel pin on the tool with the dowel pin hole in the crankshaft and attach the tool to the crankshaft by hand-tightening the bolts.
7 Turn the tool handle until the collar bottoms against the case, seating the seal.
8 The remainder of installation is the reverse of removal.

24.6 Align the marks on the gears as shown here

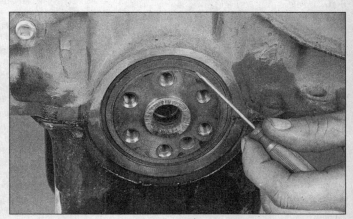

25.3 Carefully pry the old oil seal out with a screwdriver or a seal removal tool - make sure you don't scratch the seal bore or the crankshaft surface

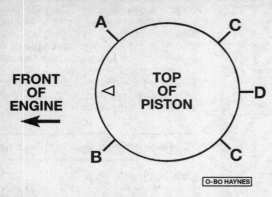

26.5 Ring end gap positions - align the oil ring spacer gap at D, the oil ring side rails at C (one inch either side of the pin centerline), and the compression rings at A and B, one inch either side of the pin centerline

26.9 The notch or arrow on each piston must face the front (timing chain) end of the engine as the pistons are installed

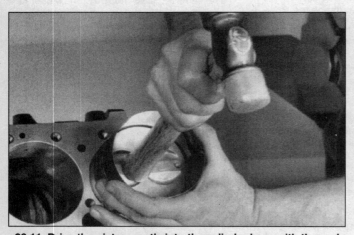

26.11 Drive the piston gently into the cylinder bore with the end of a wooden or plastic hammer handle

26.13 Lay the Plastigage strips on each rod bearing journal, parallel to the crankshaft centerline

26 Pistons and connecting rods - installation and rod bearing oil clearance check

1 Before installing the piston/connecting rod assemblies, the cylinder walls must be perfectly clean, the top edge of each cylinder must be chamfered, and the crankshaft must be in place.

2 Remove the cap from the end of the number one connecting rod (check the marks made during removal). Remove the original bearing inserts and wipe the bearing surfaces of the connecting rod and cap with a clean, lint-free cloth. They must be kept spotlessly clean.

Piston installation and rod bearing oil clearance check

Refer to illustrations 26.5, 26.9, 26.11, 26.13 and 26.17

Note: *Don't touch the faces of the new bearing inserts with your fingers. Oil and acids from your skin can etch the bearings.*

3 Clean the back side of the new upper bearing insert, then lay it in place in the connecting rod. Make sure the tab on the bearing fits into the recess in the rod. Don't hammer the bearing insert into place and be very careful not to nick or gouge the bearing face. Don't lubricate the bearing at this time.

4 Clean the back side of the other bearing insert and install it in the rod cap. Again, make sure the tab on the bearing fits into the recess in the cap, and don't apply any lubricant. It's critically important that the mating surfaces of the bearing and connecting rod are perfectly clean and oil free when they're assembled.

5 Stagger the piston ring gaps around the piston **(see illustration)**.

6 Slip a section of plastic or rubber hose over each connecting rod cap bolt.

7 Lubricate the piston and rings with clean engine oil and attach a piston ring compressor to the piston. Leave the skirt protruding about 1/4-inch to guide the piston into the cylinder. The rings must be compressed until they're flush with the piston.

8 Rotate the crankshaft until the number one connecting rod journal is at BDC (bottom dead center) and apply a coat of engine oil to the cylinder walls.

9 With the mark or notch on top of the piston **(see illustration)** facing the front of the engine, gently insert the piston/connecting rod assembly into the number one cylinder bore and rest the bottom edge of the ring compressor on the engine block.

10 Tap the top edge of the ring compressor to make sure it's contacting the block around its entire circumference.

11 Gently tap on the top of the piston with the end of a wooden or plastic hammer handle **(see illustration)** while guiding the end of the connecting rod into place on the crankshaft journal. The piston rings may try to pop out of the ring compressor just before entering the cylinder bore, so keep some pressure down on the ring compressor. Work slowly, and if any resistance is felt as the piston enters the cylinder, stop immediately. Find out what's hanging up and fix it before proceeding. Do not, for any reason, force the piston into the cylinder - you might break a ring and/or the piston.

12 Once the piston/connecting rod assembly is installed, the connecting rod bearing oil clearance must be checked before the rod cap is permanently bolted in place.

13 Cut a piece of the appropriate size Plastigage slightly shorter than the width of the

Chapter 2 Part C General engine overhaul procedures

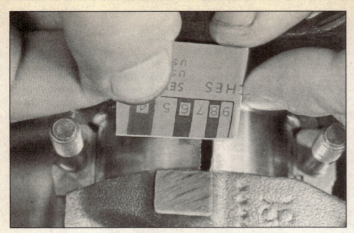

26.17 Measuring the width of the crushed Plastigage to determine the rod bearing oil clearance (be sure to use the correct scale - standard and metric ones are included)

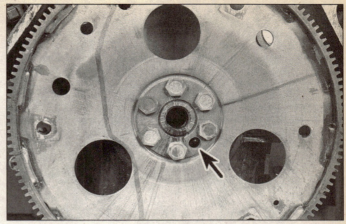

27.2 Most driveplates have locating dowels (arrow) - if the driveplate doesn't have a dowel, scribe or paint some alignment marks to ensure correct installation

connecting rod bearing and lay it in place on the number one connecting rod journal, parallel with the journal axis **(see illustration)**.

14   Clean the connecting rod cap bearing face, remove the protective hoses from the connecting rod bolts and install the rod cap. Make sure the mating mark on the cap is on the same side as the mark on the connecting rod.

15   Install the nuts and tighten them to the torque listed in this Chapter's Specifications. Work up to it in three steps. **Note:** *Use a thin-wall socket to avoid erroneous torque readings that can result if the socket is wedged between the rod cap and nut. If the socket tends to wedge itself between the nut and the cap, lift up on it slightly until it no longer contacts the cap. Do not rotate the crankshaft at any time during this operation.*

16   Remove the nuts and detach the rod cap, being very careful not to disturb the Plastigage.

17   Compare the width of the crushed Plastigage to the scale printed on the Plastigage envelope to obtain the oil clearance **(see illustration)**. Compare it to this Chapter's Specifications to make sure the clearance is correct.

18   If the clearance is not as specified, the bearing inserts may be the wrong size (which means different ones will be required). Before deciding different inserts are needed, make sure no dirt or oil was between the bearing inserts and the connecting rod or cap when the clearance was measured. Also, recheck the journal diameter. If the Plastigage was wider at one end than the other, the journal may be tapered (see Section 19).

Final connecting rod installation

19   Carefully scrape all traces of the Plastigage material off the rod journal and/or bearing face. Be very careful not to scratch the bearing - use your fingernail or the edge of a credit card.

20   Make sure the bearing faces are perfectly clean, then apply a uniform layer of clean moly-base grease or engine assembly lube to both of them. You'll have to push the piston into the cylinder to expose the face of the bearing insert in the connecting rod - be sure to slip the protective hoses over the rod bolts first.

21   Slide the connecting rod back into place on the journal, remove the protective hoses from the rod cap bolts, install the rod cap and tighten the nuts to the torque specified in this Chapter. Again, work up to the torque in three steps.

22   Repeat the entire procedure for the remaining pistons/connecting rods.

23   The important points to remember are:
a) *Keep the back sides of the bearing inserts and the insides of the connecting rods and caps perfectly clean when assembling them.*
b) *Make sure you have the correct piston/rod assembly for each cylinder.*
c) *The arrow or mark on the piston must face the front (timing chain end) of the engine.*
d) *Lubricate the cylinder walls with clean oil.*
e) *Lubricate the bearing faces when installing the rod caps after the oil clearance has been checked.*

24   After all the piston/connecting rod assemblies have been properly installed, rotate the crankshaft a number of times by hand to check for any obvious binding.

25   As a final step, the connecting rod endplay must be checked (see Section 13).

26   Compare the measured endplay to this Chapter's Specifications to make sure it's correct. If it was correct before disassembly and the original crankshaft and rods were reinstalled, it should still be right. If new rods or a new crankshaft were installed, the endplay may be inadequate. If so, the rods will have to be removed and taken to an automotive machine shop for resizing.

27 Driveplate - removal and installation

Refer to illustrations 27.2 and 27.3

1   Remove the engine/transaxle/subframe assembly, if not already done (see Section 5). Separate the engine from the transaxle.

2   If there's no dowel pin **(see illustration)** or if the bolt holes aren't staggered, scribe or paint an alignment mark on the driveplate and crankshaft to ensure correct alignment during reinstallation.

3   Remove the bolts that secure the driveplate to the crankshaft **(see illustration)**. Prevent the crankshaft from turning by inserting a screwdriver through a hole in the driveplate.

4   Remove the driveplate from the crankshaft.

5   Clean the driveplate to remove grease and oil. Inspect the surface for cracks and also check for cracked and broken ring gear teeth.

6   Clean and inspect the mating surfaces of the driveplate and the crankshaft. If the

27.3 A large screwdriver wedged in one of the holes in the driveplate can be used to keep the driveplate from turning as the mounting bolts are removed and installed

crankshaft rear seal is leaking, replace it before reinstalling the driveplate (see Section 25).

7 Position the driveplate against the crankshaft. Be sure to align the marks made during removal (if there was no alignment dowel or staggered bolt holes). Before installing the bolts, apply thread locking compound to the threads.

8 Wedge a screwdriver through a hole in the driveplate to keep it from turning as you tighten the bolts to the torque listed in this Chapter's Specifications.

9 The remainder of installation is the reverse of the removal procedure.

28 Initial start-up and break-in after overhaul

Warning: *Have a fire extinguisher handy when starting the engine for the first time.*

1 Once the engine has been installed in the vehicle, double-check the oil and coolant levels.

2 With the spark plugs out of the engine and the ECM fuse removed, crank the engine until oil pressure registers on the gauge or the light goes out.

3 Install the spark plugs, hook up the plug wires and install the ECM fuse.

4 Start the engine. It may take a few moments for the fuel system to build up pressure, but the engine should start without a great deal of effort. **Note:** *If the engine keeps backfiring, recheck the valve timing and spark plug wires.*

5 After the engine starts, it should be allowed to warm up to normal operating temperature. While the engine is warming up, make a thorough check for fuel, oil and coolant leaks.

6 Shut the engine off and recheck the engine oil and coolant levels.

7 Drive the vehicle to an area with minimum traffic, accelerate at full throttle from 30 to 50 mph, then allow the vehicle to slow to 30 mph with the throttle closed. Repeat the procedure 10 or 12 times. This will load the piston rings and cause them to seat properly against the cylinder walls. Check again for oil and coolant leaks.

8 Drive the vehicle gently for the first 500 miles (no sustained high speeds) and keep a constant check on the oil level. It's not unusual for an engine to use oil during the break-in period.

9 At approximately 500 to 600 miles, change the oil and filter.

10 For the next few hundred miles, drive the vehicle normally. Don't pamper it or abuse it.

11 After 2000 miles, change the oil and filter again and consider the engine broken in.

Chapter 3
Cooling, heating and air conditioning systems

Contents

	Section		Section
Air conditioning accumulator - removal and installation	14	Engine cooling fan - check and replacement	4
Air conditioning compressor - removal and installation	15	Evaporator core(s) - removal and installation	17
Air conditioning condenser - removal and installation	16	General information	1
Air conditioning expansion (orifice) tube - removal and installation	18	Heater and air conditioner blower motors - removal and installation	10
Air conditioning system - check and maintenance	13	Heater and air conditioner control assembly - removal and installation	12
Antifreeze - general information	2	Heater cores - removal and installation	11
Coolant level check	See Chapter 1	Radiator - removal and installation	5
Coolant reservoir - removal and installation	6	Thermostat - check and replacement	3
Coolant temperature sending unit - check and replacement	9	Water pump - check	7
Cooling system check	See Chapter 1	Water pump - removal and installation	8
Cooling system servicing (draining, flushing and refilling)	See Chapter 1		

Specifications

General

Coolant capacity	See Chapter 1
Drivebelt tension	See Chapter 1
Radiator cap pressure rating	See Chapter 1
Thermostat rating	195 to 200-degrees F
Refrigerant capacity	
Without rear air conditioning unit	3.125 lbs
With rear air conditioning unit	4.125 lbs

Torque specifications

Ft-lbs (unless otherwise indicated)

Thermostat cover bolts	
3.1L engine	18
3800 engine	20
Water pump mounting bolts/nuts	
3.1L engine	89 in-lbs
3800 engine	
Long bolts	29
Short bolts	97 in-lbs
Pulley-to-pump bolts	115 in-lbs

1 General information

Note: *On models equipped with the Delco Loc II audio system, be sure the lockout feature is turned off before performing any procedure which requires disconnecting the battery.*

Engine cooling system

All vehicles covered by this manual employ a pressurized engine cooling system with thermostatically controlled coolant circulation. An impeller type water pump mounted on the engine block pumps coolant through the engine and radiator. The coolant flows around each cylinder and back to the radiator. Cast-in coolant passages direct coolant around the intake and exhaust ports, near the spark plug areas and the exhaust valve guides.

A wax pellet type thermostat is located in a housing connected to the upper radiator hose. During warm up the closed thermostat prevents coolant from circulating through the radiator. As the engine nears normal operating temperature, the thermostat opens and allows hot coolant to travel through the radiator, where it's cooled before returning to the engine.

The cooling system is sealed by a pressure-type radiator cap, which raises the boiling point of the coolant and increases the cooling efficiency of the radiator. If the system pressure exceeds the cap pressure relief value, the excess pressure in the system forces the spring-loaded valve inside the cap off its seat and allows the coolant to escape through a hose into a coolant reservoir. When the system cools, the excess coolant is automatically drawn from the reservoir back into the radiator.

The coolant reservoir serves as both the point at which fresh coolant is added to the cooling system to maintain the proper level and as a holding tank for expelled coolant.

This type of cooling system is known as a closed design because coolant that escapes past the pressure cap is saved and reused.

Heating system

The heating system consists of a blower fan and heater core located in the heater box, the hoses connecting the heater core to the engine cooling system and the heater/air conditioning control head on the dashboard. Hot engine coolant is circulated through the heater core. When the heater mode is activated, a trap door opens to expose the heater box to the passenger compartment. A fan switch on the control head activates the blower motor, which forces air through the core, heating the air. Some models are equipped with an auxiliary heating unit which is designed to heat the rear of the passenger compartment. It is located behind the trim panel on the left hand side trim panel, behind the driver's seat.

Air conditioning system

The air conditioning system consists of a condenser mounted in front of the radiator, an evaporator mounted adjacent to the heater core, a compressor mounted on the engine, a filter-drier (accumulator), which contains a high-pressure relief valve, and the hoses and lines connecting all of the above components. Some models are equipped with a rear air conditioning system with its own control panel and blower fan.

A blower fan forces the warmer air of the passenger compartment through the evaporator core (sort of a radiator-in-reverse), transferring the heat from the air to the refrigerant. The liquid refrigerant boils off into low pressure vapor, taking the heat with it when it leaves the evaporator.

2 Antifreeze - general information

Warning: *Don't allow antifreeze to come in contact with your skin or painted surfaces of the vehicle. Rinse off spills immediately with plenty of water. Antifreeze is highly toxic if ingested. Never leave antifreeze lying around in an open container or in puddles on the floor; children and pets are attracted by its sweet smell and may drink it. Check with local authorities about disposing of used antifreeze. Many communities have collection centers which will see that antifreeze is disposed of safely. Antifreeze is also combustible, so don't store or use it near open flames.*

The cooling system should be filled with a water/ethylene glycol based antifreeze solution, which will prevent freezing down to at least -20-degrees F, or lower if local climate requires it. It also provides protection against corrosion and increases the coolant boiling point.

The cooling system should be drained, flushed and refilled at the specified intervals (see Chapter 1, Sections 2 and 29). Old or contaminated antifreeze solutions are likely to cause damage and encourage the formation of rust and scale in the system. Use distilled water with the antifreeze.

3.8 To remove the thermostat, squeeze the tabs of the hose clamp (arrows) then slide it back on the hose, detach the hose from the thermostat cover and unscrew the bolts (arrows)

Before adding antifreeze, check all hose connections, because antifreeze tends to leak through very minute openings. Engines don't normally consume coolant, so if the level goes down, find the cause and correct it.

The exact mixture of antifreeze-to-water which you should use depends on the relative weather conditions. The mixture should contain at least 50-percent antifreeze, but should never contain more than 70-percent antifreeze. Consult the mixture ratio chart on the antifreeze container before adding coolant. Hydrometers are available at most auto parts stores to test the coolant. Use antifreeze which meets the vehicle manufacturer's specifications.

3 Thermostat - check and replacement

Refer to illustrations 3.8, 3.11 and 3.13.
Warning: *DO NOT remove the radiator cap, drain the coolant or replace the thermostat until the engine has cooled completely.*

Check

1 Before assuming the thermostat is to blame for a cooling system problem, check the coolant level (see Chapter 1, Section 4), drivebelt tension (see Chapter 1, Section 21) and temperature gauge (or light) operation.
2 If the engine seems to be taking a long time to warm up (based on heater output or temperature gauge operation), the thermostat is probably stuck open. Replace the thermostat with a new one.
3 If the engine runs hot, use your hand to check the temperature of the upper radiator hose. If the hose isn't hot, but the engine is, the thermostat is probably stuck closed, preventing the coolant inside the engine from escaping to the radiator. Replace the thermostat. **Caution:** *Don't drive the vehicle without*

3.11 Note how the thermostat is installed before removing it

a thermostat. The computer may stay in open loop and emissions and fuel economy will suffer.
4 If the upper radiator hose is hot, it means the coolant is flowing and the thermostat is open. Consult the *Troubleshooting* section at the front of this manual for cooling system diagnosis.

Replacement

5 Disconnect the negative battery cable from the battery and drain the cooling system (see Chapter 1, Section 29). If the coolant is relatively new or in good condition, save it and reuse it.
6 On models with the 3800 engine, remove the air cleaner assembly for access to the thermostat housing (see Chapter 4, Section 8).
7 Follow the upper radiator hose to the engine to locate the thermostat cover.
8 Loosen the hose clamp, then detach the hose from the fitting **(see illustration)**. If the hose sticks, grasp it near the end with a pair of adjustable pliers and twist it to break the seal, then pull it off. If the hose is old or deteriorated, cut it off and install a new one.
9 If the outer surface of the large fitting that mates with the hose is deteriorated (corroded, pitted, etc.) it may be damaged further by hose removal. If it is, the thermostat cover will have to be replaced.
10 Remove the bolts/nuts and detach the thermostat cover. If the cover is stuck, tap it with a soft-face hammer to jar it loose. Be prepared for some coolant to spill as the gasket seal is broken.
11 Note how it's installed (which end is facing up), then remove the thermostat **(see illustration)**.
12 On 3800 engine models remove all traces of old gasket material and sealant from the housing and cover with a gasket scraper. Clean the gasket mating surfaces with lacquer thinner or acetone.
13 Replace the seal on the thermostat **(see illustration)**.
14 Install the new thermostat in the hous-

Chapter 3 Cooling, heating and air conditioning systems

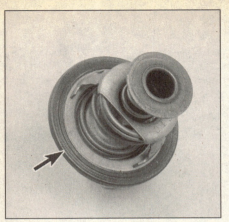

3.13 If you're working on a 3.1L engine, be sure to replace the thermostat seal (arrow)

4.1 The fan motor may be tested by running fused jumper wires directly from the battery to the motor terminals

ing. Make sure the correct end faces up - the spring end is normally directed into the engine.
15 On 3800 models replace the gasket on the thermostat, with the nubs facing up.
16 Install the cover and bolts/nuts. Tighten them to the torque figure listed in this Chapter's Specifications.
17 The remaining steps are the reverse of removal.
18 Refill the cooling system (see Chapter 1, Section 29). **Note:** *Be sure to bleed the system of air.*
19 Start the engine and allow it to reach normal operating temperature, then check for leaks and proper thermostat operation (as described in Steps 2 through 4).

4 Engine cooling fan - check and replacement

Warning: *The cooling fan is electric and can come on at any time, whether or not the engine is running. Keep hands, tools and clothing away from the fan.*
Note: *After performing any electrical checks on the fan motor circuit, make sure the Service Engine Soon light is not on. If it is, clear any stored trouble codes by following the procedure outlined in Chapter 6, Section 2.*

Fan motor check

Refer to illustration 4.1

1 To test the fan motor, unplug the electrical connector at the motor and use fused jumper wires to connect the fan directly to the battery **(see illustration)**. If the fan still doesn't work, replace the motor - see *Fan motor replacement* later in this Section.
2 If the motor works, the fault lies in the coolant temperature switch, the relay, the ECM or PCM or the wiring which connects the components. Carefully check all wiring and connections.
3 If a fan blade is bent or damaged it should be replaced with a new one. Do not attempt to repair and reuse a damaged fan blade as this will adversely effect the assembly balance.

Cooling fan control circuit diagnosis

4 This procedure should be conducted when the engine is cold. Without starting the engine turn the ignition "on" and the air conditioning "off" and then follow the diagnosis applicable to your vehicle.

3.1 Liter Engine

5 The cooling fan should be "off"; if it is, proceed to Step 7. If the cooling fan continues to operate, disconnect the cooling fan relay (see Chapter 12, Section 6 for location). The fan should stop; if it does, go to Step 6. If the fan still operates, the black/red wire between the relay and the fan motor is shorted to voltage - repair and check for proper operation.
6 With a 12-volt test lamp, probe the dark green/white wire between the ECM and the relay. If the lamp remains "off", the relay is faulty. If it is "on" check to see if the dark green/white wire is shorted to ground. If it is not, the ECM is probably faulty.
7 Ground the diagnostic terminal (see Chapter 6, illustrations 2.6a and 2.6b for location) which should activate the cooling fan. If the fan operates proceed to Step 13; if it does not, disconnect the fan relay (see Chapter 12, Section 6 for location). Probe the red wire and the pink/white wire connector terminals on the fan relay with a test lamp connected to ground. If the lamp lights on both terminals go to Step 8; if it fails to light on one or both, repair the open or short to ground in the circuit(s) that failed to light.
8 Ground the diagnostic terminal and probe the dark green/white circuit between the fan relay and the ECM with a test light connected to 12-volts.
9 If the light goes "on" proceed to Step 10; if it fails to light, check to see if the circuit is open or shorted to voltage. If the circuit checks OK there is a faulty connection at the ECM or the ECM itself is defective.
10 Using a fused jumper wire, bridge the red wire and the dark green/white wire connector terminals on the fan relay. If the fan runs, check the relay connections; if they check OK, there is a fault with the relay. If the fan fails to operate, connect a test lamp across the fan motor terminals. If the light goes "on", the motor should be replaced (see *Fan motor replacement* in this section). If it fails to come on, probe each terminal with a test light connected to ground.
11 If the light remains "Off" on both terminals repair the open in the circuit between the relay and the cooling fan motor. If the light comes "On" on one terminal there is an open in ground on the black wire.
12 If no obvious problems are found, further diagnosis should be done by a dealer service department or other repair shop.

3800 Engine

13 The fan(s) should be "off"; if they are, proceed to Step 24. If one or both of the fans are "on", proceed to Step 14.
14 If both fans are "on", check the coolant temperature sensor and the air conditioning head pressure switch. If these test OK, replace the PCM. If only the pusher fan is "on", proceed to Step 15. If only the puller fan is "on", proceed to Step 21.
15 Disconnect the pusher fan relay (see Chapter 12, Section 6 for location). If the fan is now "off", proceed to Step 16. If the fan is still "on", check the black/pink wire and the red wire connections to the fan relay and also check for shorts in these circuits.
16 Connect a test lamp between the dark blue/white wire terminal on the pusher fan relay and the positive terminal of the battery. If the light comes "on" proceed to Step 17, if it fails to come on replace the pusher fan relay.
17 Disconnect the air conditioning head pressure switch and jump across the switches connections with a test lamp. If the light comes "on" proceed to Step 18. If it does not, replace the air conditioning head pressure switch.
18 Disconnect the black multi-pin connector from the PCM. With a test light connected to 12-volts, check the dark green/white connection at the air conditioning head pressure switch. If it fails to light there is an open circuit in either the dark green/white or black/white circuit to the pressure switch.
19 If the test light is "on", disconnect the green multi-pin connector from the PCM and attach the test lamp to the dark blue/white terminal connection on the pusher fan relay. If the light is "off", the PCM is probably faulty. A short to ground in the dark blue/white circuit to the relay is suspected if the test lamp is "on".
20 If no problems are found with the pusher fan circuit, further diagnosis should be done by a dealer service department or a repair shop.
21 Disconnect the puller fan relay; if the fan continues to run, check the black/red wire

4.43a Disconnect the electrical connector by inserting a small screwdriver under the connector (arrow) and lifting it free - the connector is located below the fan motor on models with the 3.1L engine . . .

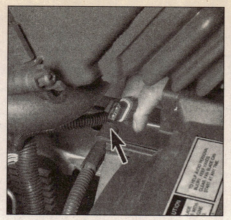

4.43b . . . and to the left of the fan on models with the 3800 engine (arrow)

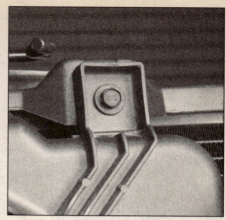

4.43c Remove the bolts located in the upper corners . . .

and red wire terminal connections to the pusher fan relay as there is a short in the circuit.

22 If the fan stops once the relay has been disconnected test the dark green/white fan relay connection with a 12-volt test lamp connected to the battery positive terminal. If it fails to light, a faulty relay is suspected. If it lights, proceed to Step 23.

23 Leave the test lamp connected to the dark green/white terminal and disconnect the green PCM connector. If the light remains "on" there is a short to ground in the dark green/white wire circuit, be sure to check its terminal connections. If the test lamp is "off", a faulty PCM is suspected and it should be replaced. If no problems are found with the puller fan circuit further diagnosis should be done by a dealer service department or a repair shop.

24 With the ignition on and the engine off, turn the air conditioning on (the engine must be cool for this check). The puller fan should activate. If it does, go to Step 31; if it does not, proceed to Step 34.

Puller fan

25 Disconnect the puller fan relay (see Chapter 12, Section 6 for location). Connect a grounded test light to the red wire puller fan relay terminal and then to the pink/white wire relay terminal. If the light turns "on" for both terminals proceed to Step 26. If it did not light for one or both terminals check the connections and repair any open circuit(s) in the circuit(s) that did not light.

26 On the puller fan relay, jump the red wire terminal to the black/red terminal. The puller fan should run, if it does proceed to Step 29. If it fails to operate connect a grounded test lamp to the black/red terminal on the puller fan motor, with the red wire and black/red wire terminals on the relay still jumped.

27 If the lamp does not light, repair the open in the black/red circuit between the puller fan and the relay, be sure to carefully check the terminals. If the lamp does light, connect the test light between the puller fan motor connections.

28 If the light is "off", repair the open ground circuit to the fan motor. If it is "on" there is either a poor connection at the fan motor or the motor is faulty. (To check the motor see Step 1.)

29 Ground the ALDL diagnostic terminal (see Chapter 6, Section 2) and connect a test light to the dark green/white wire terminal on the puller fan relay and a 12-volt power source. If the test light is "on" replace the relay. If it fails to light, check the dark green/white circuit for a short or open. If it is not shorted or open check the PCM and fan relay connection terminals. If these check OK, the PCM is probably faulty.

30 If the puller fan still does not operate take the vehicle to a dealer service department or other repair shop for diagnosis.

Pusher fan

31 Unplug the electrical connector from the coolant temperature sensor. Connect a jumper wire between the terminals in the electrical connector - after 6 seconds the pusher fan should operate. If it does, proceed to Step 32, if it does not proceed to Step 33.

32 Ensure that the air conditioner is turned off, then disconnect the air conditioning head pressure switch. After 6 seconds the pusher fan should run; if it doesn't, take the vehicle to a dealer service department or other repair shop for further diagnosis.

33 Disconnect the pusher fan relay and connect a grounded test light to the pink/white wire terminal on the fan relay and then to the red wire terminal. If the test light came "on" for both, go to Step 34. If it did not, repair the open in the circuit that did not light.

34 Connect a jumper wire across the black/pink wire and red wire terminals of the pusher fan relay. The fan should run; if it does, proceed to Step 38. If it does not run, leave the black/pink wire and red wire terminals jumped and connect a grounded test wire to the black/pink terminal at the fan motor.

35 If the light does not glow, check the black/pink terminal connections at the fan relay and the motor and also check to see if the circuit is open. If the test light is "on", remove it and connect it between the black wire and black/pink wire terminals on the fan motor.

36 If the light does not come "on", repair the open ground in the circuit from the motor. If it does come "on", there is a poor connection at the fan motor or the fan motor itself is faulty. To check the fan motor, see Step 1.

37 If the pusher fan still does not operate, further diagnosis should be done by a dealer service department or other repair shop.

38 If the fan motor ran in Step 34, ground the ALDL diagnostic terminal (see Chapter 6, Section 2) and connect a test light to the dark blue/white wire terminal on the pusher fan relay and a 12-volt power source.

39 If the light goes "on" the relay is faulty. If the light does not glow, check to see if the dark blue/white circuit is open or shorted. Also check the dark blue/white wire puller fan relay terminal and the PCM terminal. If they are sound, the PCM is probably faulty.

40 If the pusher fan still does not operate, further diagnosis should be done by a dealer service department or a repair shop.

Fan assembly replacement

Pusher fan assembly

Refer to illustrations 4.43a through 4.43d

41 Disconnect the negative battery cable from the battery.

42 Remove the air cleaner assembly (see Chapter 4, Section 8).

43 To remove the pusher fan assembly follow the photo sequence, taking care to read the accompanying captions **(see illustrations)**. Proceed to Step 47 for the motor replacement procedure, if necessary. Installation is the reverse of removal.

Puller fan assembly

Refer to illustrations 4.45a through 4.45d

44 Disconnect the negative battery cable from the battery.

Chapter 3 Cooling, heating and air conditioning systems

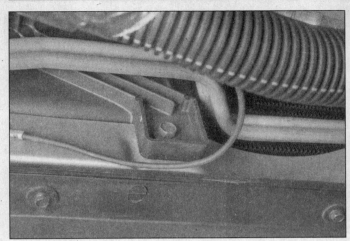

4.43d ...and lower corners of the fan assembly, then carefully pull the assembly free

4.45a The puller fan electrical connector is located below the fan motor (as seen from below) - disconnect it by prying the connector tab (arrow) away from the connector

4.45b Remove the lower...

4.45c ...and upper fan assembly retention bolts - to improve access to the middle retention bolt, remove the fan cover bolts and pull the cover forward and out...

1 Fan assembly retention bolt
2 Fan cover bolts

4.45d ...then remove the middle retention bolt and pull the assembly free

Fan motor replacement

Refer to illustrations 4.47a and 4.47b

47 Follow the photo sequence for the motor replacement procedure, taking care to read the captions **(see illustrations)**.
48 Installation is the reverse of removal.
49 After all these repairs/checks have been completed, confirm that the *Service engine soon* light is not on.

4.47a To detach the fan blades from the motor, remove the motor shaft nut. Note: **This nut unscrews clockwise**

45 Follow the photo sequence for the removal procedure **(see illustrations)**. Be sure to read all the captions. If it's necessary to replace the fan motor, proceed to Step 47.
46 Installation is the reverse of removal.

4.47b To remove the fan motor from the bracket, remove the mounting bolts/nuts

5 Radiator - removal and installation

Refer to illustrations 5.4a through 5.4i
Warning: *Wait until the engine is completely cool before beginning this procedure.*

Removal

1 Disconnect the negative battery cable from the battery.
2 Drain the cooling system (see Chapter 1, Section 29). If the coolant is relatively new or in good condition, save it and reuse it.
3 Remove the engine puller fan assembly

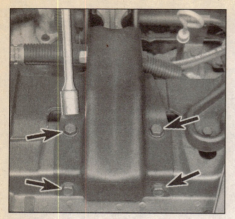

5.4a Disconnect the torque strut bracket at the radiator support by removing the bolts (arrows)

5.4b Loosen the bolt securing the rubber bushing and swing the bracket clear of the radiator

5.4c Loosen the radiator hose clamps and detach the upper and lower radiator hoses. Use a drip pan to catch spilled fluid. If the hoses are stuck, grasp each hose near the end with adjustable pliers and twist it to break the seal - be careful not to distort the radiator fitting.

5.4d Using a flare-nut wrench (if available), disconnect the cooler lines (arrow) from the upper . . .

5.4e . . . and lower corners of the radiator (cap the ends to prevent excessive fluid loss and contamination)

5.4f Disconnect the reservoir hose from the radiator neck

(see Section 4). On 3800 engine models also remove the air cleaner assembly intake duct (see Chapter 4, Section 8).

4   To remove the radiator follow the photo sequence **(see illustrations)**. Be sure to read the caption under each illustration.

5   With the radiator removed, it can be inspected for leaks and damage. If it needs repair, have a radiator shop or dealer service department perform the work, as special techniques are required.

6   Bugs and dirt can be removed from the radiator with compressed air and a soft brush. Don't bend the cooling fins as this is done.

7   Check the radiator mounts for deterioration and make sure there's nothing in them when the radiator is installed.

Installation

8   Installation is the reverse of the removal procedure.

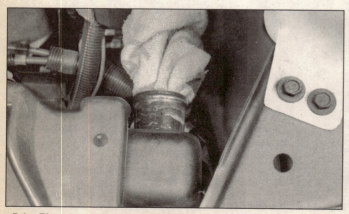

5.4g Plug or cap lines and fittings to prevent spillage of coolant or ATF as the radiator is removed

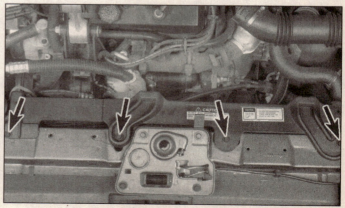

5.4h Remove the bolts from the radiator mounting panel (arrows) . . .

Chapter 3 Cooling, heating and air conditioning systems

5.4l ... then remove the mounting panel and carefully lift out the radiator. Don't spill coolant on the vehicle or scratch the paint. Take care not to dislodge and lose the lower rubber mounts

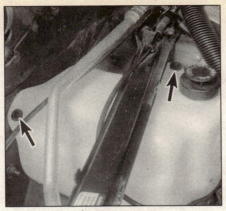

6.2 Remove the mounting screws (arrows) to detach the coolant reservoir

7.4 Check the weep hole (arrow) for leakage

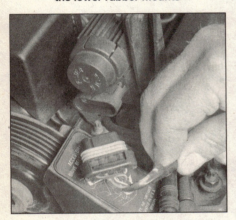

8.3a Unplug the electrical connector above the drivebelt cover for improved access

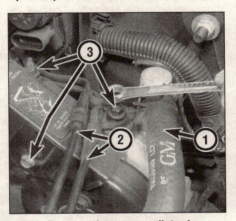

8.3b Remove the upper radiator hose, vacuum lines, the drivebelt cover bolts and lift the drivebelt cover free (3.1 engine shown, 3800 engine similar)

1 Upper radiator hose
2 Vacuum lines
3 Pulley bolts

8.3c Squeeze the hose clamp, slide it back on the hose and detach the lower radiator hose from the water pump

9 After installation, fill the cooling system with the proper mixture of antifreeze and water (see Section 2 and Chapter 1, Section 4). Be sure to bleed the system of air (see Chapter 1, Section 29).
10 Start the engine and check for leaks. Allow the engine to reach normal operating temperature, indicated by the upper radiator hose becoming hot. Re-check the coolant level and add more if required.
11 Check and add automatic transaxle fluid as needed (see Chapter 1, Section 6).

6 Coolant reservoir - removal and installation

Refer to illustration 6.2

1 Disconnect the reservoir hose from the radiator neck **(see illustration 5.4f)**.
2 Remove the mounting screws **(see illustration)** and lift the coolant reservoir from the vehicle, being careful not to spill any coolant on the vehicle's paint. Installation is the reverse of removal.

7 Water pump - check

Refer to illustration 7.4

1 A failure in the water pump can cause serious engine damage due to overheating.
2 There are three ways to check the operation of the water pump while it's installed on the engine. If the pump is defective, it should be replaced with a new or rebuilt unit.
3 With the engine running at normal operating temperature, squeeze the upper radiator hose. If the water pump is working properly, a pressure surge should be felt as the hose is released. **Warning:** *Keep your hands away from the fan blades!*
4 Water pumps are equipped with weep or vent holes. If a failure occurs in the pump seal, coolant will leak from the hole **(see illustration)**. In most cases you'll need a flashlight and mirror to find the hole on the underside of the water pump to check for leaks.
5 If the water pump shaft bearings fail there may be a howling sound coming from the drivebelt area while the engine is running. With the engine off, shaft wear can be felt if the water pump pulley is rocked up-and-down. Don't mistake drivebelt slippage, which causes a squealing sound, for water pump bearing failure.

8 Water pump - removal and installation

Refer to illustrations 8.3a through 8.3g and 8.9

Warning: *Wait until the engine is completely cool before beginning this procedure.*

Removal

1 Disconnect the negative battery cable from the battery.
2 Drain the cooling system (see Chapter 1, Section 29). If the coolant is relatively new or in good condition, save it and reuse it. Remove the drivebelt (see Chapter 1, Section 21).
3 For the removal procedure follow the photo sequence **(see illustrations)** taking care not to miss any and to read the accompanying captions.

4 Clean the fastener threads and any threaded holes in the engine to remove corrosion and sealant.
5 Compare the new pump to the old one to make sure they're identical.
6 Remove all traces of old gasket material from the engine with a gasket scraper.
7 Clean the engine and water pump mating surfaces with lacquer thinner or acetone.

Installation

8 Apply a thin coat of RTV sealant to the engine side of the new gasket.
9 Apply a thin layer of RTV sealant to the gasket mating surface of the new pump, then carefully mate the gasket and the pump. Slip a couple of bolts through the pump mounting holes to hold the gasket in place **(see illustration)**.
10 Carefully attach the pump and gasket to the engine and start the bolts/nuts finger tight.
11 Tighten the bolts in 1/4-turn increments to the torque figure listed in this Chapter's Specifications. Don't overtighten them or the pump may be distorted.
12 Reinstall all parts removed for access to the pump.
13 Refill the cooling system (see Chapter 1, Section 29). Run the engine and check for leaks.

9 Coolant temperature sending unit - check and replacement

Refer to illustration 9.1
Warning: *Wait until the engine is completely cool before beginning this procedure.*

1 The coolant temperature indicator system is composed of a light or temperature gauge mounted in the instrument panel and a coolant temperature sending unit mounted on the engine. On both the 3800 and 3.1L engines the coolant temperature sending unit is mounted at the left end (driver's side) of the rear (firewall side) cylinder head **(see illustration)**. Some vehicles have more than one sending unit, but the one used for the indicator system has only one wire.
2 If the light or gauge indicates the engine is overheating, check the coolant level in the system and then make sure the wiring between the light or gauge and the sending unit is secure and all fuses are intact.
3 When the ignition switch is turned on and the starter motor is turning, the indicator light (if equipped) should be on (overheated engine indication). Once the engine has started the light should go out.
4 If the light isn't on, the bulb may be burned out, the ignition switch may be faulty or the circuit may be open. Test the circuit by grounding the wire to the sending unit while the ignition is on (engine not running for safety). If the gauge deflects full scale or the

8.3d Prevent the pulley from turning by wedging a screwdriver between the hub and one of the pulley bolts, then loosen the pulley bolts one at a time. Once all the bolts are loose, remove the bolts and the pulley (3.1L engine shown)

8.3e Remove the mounting bolts located around the perimeter of the pump (3.1L engine shown)

8.3f Water pump mounting bolts (3800 engine)

8.3g Remove the water pump - if it is stuck, carefully pry it out with a screwdriver or prybar placed between the housing and a casting projection on the pump. DON'T pry between the sealing surfaces (leaks will develop)

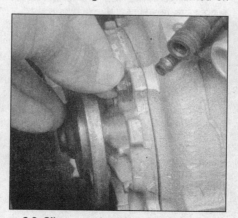

8.9 Slip a couple of bolts through the pump mounting holes to hold the new gasket in place as the pump is installed

9.1 The coolant temperature sensor (arrow) is located at the left end of the rear cylinder head

Chapter 3 Cooling, heating and air conditioning systems

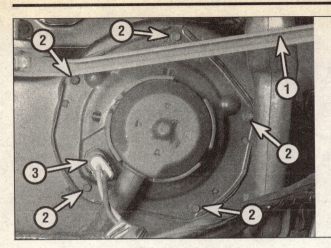

10.3 Detach the electrical connector and remove the mounting screws from the rear evaporator and blower assembly

1. Windshield wiper arm linkage
2. Blower motor mounting screws
3. Electrical connector

10.5 Unscrew the nut (arrow) from the motor shaft and pull the fan free (Note: *The retention nut unscrews clockwise*)

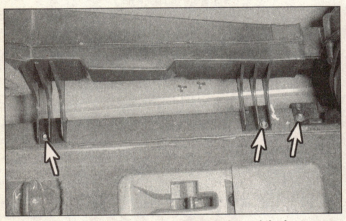

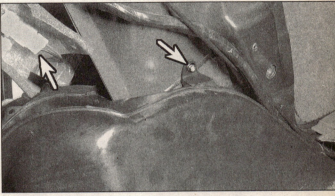

10.9a Remove the lower duct retention bolts and the lower rear evaporator and blower assembly bolt (arrows)

10.9b Remove the foam from around the electrical connector and unplug it (left arrow). Remove the upper rear evaporator and blower motor bolt (right arrow)

light comes on, replace the sending unit.
5 As soon as the engine starts, the light should go out and remain out unless the engine overheats. Failure of the light to go out may be due to a grounded wire between the light and the sending unit, a defective sending unit or a faulty ignition switch. Check the coolant to make sure it's the proper type (see Section 2). Plain water may have too low a boiling point to activate the sending unit.
6 If the sending unit must be replaced, simply unscrew it from the engine and install the replacement. Use Teflon tape or a light coat of sealant on the threads. Make sure the engine is cool before removing the defective sending unit. There will be some coolant loss as the unit is removed, so be prepared to catch it. Check the level after the replacement sending unit has been installed (see Chapter 1, Section 4).

10 Heater and air conditioner blower motor - removal and installation

Front blower motor

Refer to illustrations 10.3 and 10.5

1 Disconnect the cable from the negative battery terminal.
2 Remove the air cleaner assembly (see Chapter 4, Section 8).
3 The front blower motor is located at the rear of the engine bay. Remove the left side windshield wiper arm linkage (see Chapter 12, Section 18). Disconnect the electrical connector from the blower motor **(see illustration)**.
4 Remove the blower motor mounting bolts and separate the motor/fan assembly from the housing **(see illustration 10.3)**.
5 Remove the retaining nut and slide the blower fan off the motor shaft **(see illustration)**.
6 Installation is the reverse of the removal procedure.

Rear blower motor (without rear air conditioning)

Refer to illustrations 10.9a through 10.9e

7 Disconnect the cable from the negative battery terminal and remove the left side seat from the center row (see Chapter 11, Section 23).
8 To gain access to the rear blower motor remove the side trim panel (see Chapter 11, Section 12).
9 To remove the blower assembly follow the photo sequence **(see illustrations)** taking

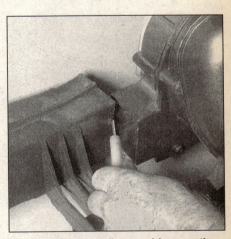

10.9c With a small screwdriver pry the rear evaporator and blower assembly away from the lower air duct and lift the assembly clear

care to read the accompanying captions. To separate the fan from the motor, unscrew the nut from the motor shaft **(see illustration 10.5)**.
10 Installation is the reverse of the removal procedure.

3-10 Chapter 3 Cooling, heating and air conditioning systems

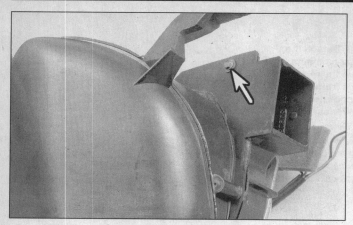

10.9d To remove the blower fan remove the front cover screw (arrow) . .

10.9e . . . and the rear (arrow) cover retention screws and separate the assembly

Rear blower motor (with rear air conditioning)

Refer to illustrations 10.14a through 10.14i

11 Disconnect the cable from the negative battery terminal.
12 Remove the middle row left hand seat (see Chapter 11, Section 23) and the middle trim panel (see Chapter 11, Section 12).
13 To remove the rear blower motor it is necessary to remove the whole assembly from the door. **Warning:** Although it is not necessary to disconnect any heater or air conditioning lines, extreme caution should be used around all such fittings and connections. Under no circumstances should they be loosened or disconnected, unless the system has been discharged by a dealer service department or a service station.
14 To remove the blower motor follow the photo procedure **(see illustrations)**, taking care to read the accompanying captions. If it's necessary to separate the fan from the motor, unscrew the nut from the motor shaft **(see illustration 10.5).**
15 Installation is the reverse of the removal procedure.

10.14a Remove the upper air duct retention bolts (arrows)

10.14b Remove the screws located on the right (arrow). . .

10.14c . . . and left (arrow) of the blower assembly

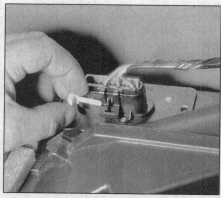

10.14d Unplug all electrical connectors - the one shown here has a lock pin which must be pulled free before disconnecting it

10.14e Disconnect the actuator vacuum line

Chapter 3 Cooling, heating and air conditioning systems

11 Heater core(s) - removal and installation

Front heater core

Refer to illustration 11.8

1 Disconnect the cable from the negative terminal of the battery.
2 Drain the cooling system (see Chapter 1, Section 29).
3 Working in the passenger compartment, remove the passenger side sound insulation panel (see Chapter 11, Section 25).
4 Remove the clamps from the heater inlet and outlet hoses and detach them from the heater core.
5 Remove the glove box (see Chapter 11, Section 26).
6 Unbolt the lower trim panel left brace (see Chapter 11, Section 22).
7 Disconnect the vacuum hose from the electric solenoid, remove the solenoid retention screws and lift it free of the cover.
8 Remove the heater core cover **(see illustration)**.
9 Remove the heater core.
10 Installation is the reverse of removal. Be sure to refill the cooling system (see Chapter 1, Section 29).

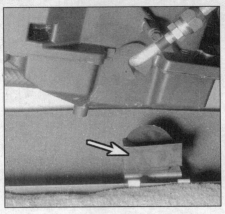

10.14f Pull the assembly upward so it disengages from the rubber drain plug (arrow)

Rear heater core

Refer to illustrations 11.15a through 11.15e

11 Disconnect the battery lead to the negative terminal. Drain the cooling system (see Chapter 1, Section 29). If the coolant is relatively new or in good condition, save it and reuse it.
12 Remove the middle row left side seat (see Chapter 11, Section 23).
13 Remove the middle trim panel (see Chapter 11, Section 12).
14 Disconnect the rear evaporator and blower assembly from the door (see Section 10).
15 Follow the photo sequence for the removal procedure **(see illustrations)**. Be

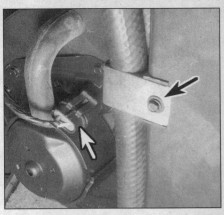

10.14g The blower motor is located at the rear of the assembly - disconnect the electrical connector and remove the hose mounting screw (arrows)

10.14h Remove the screws (arrows) that attach the blower assembly cover to the rear heating and air conditioning module

10.14i Remove the screws (arrows) and pull the blower motor free of the cover

11.8 Remove the three lower bolts and the retention clip (arrows) - three more bolts are located around the upper perimeter of the cover

11.15a Remove the lower heating core cover bolts from both sides of the assembly (arrow)

3-12 Chapter 3 Cooling, heating and air conditioning systems

11.15b Remove the two upper cover bolts (arrows) and pull the cover free

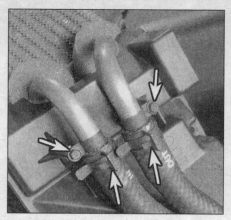
11.15c Using a pair of pliers, squeeze the hose clamps and pull the hoses free, then remove the hose connector bracket bolts (arrows)

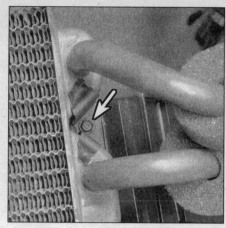
11.15d Remove the restraining clip bolt (arrow) from the bottom . . .

11.15e . . . and the top of the assembly (arrows), then pull the heater core free

12.3 Remove the retention screws (arrows) from each side of the control unit (1990 through 1992)

sure to stay in order and to read the accompanying captions.
16 Installation is the reverse of the removal procedure. Be sure to refill the cooling system (see Chapter 1, Section 29).

12 Heater and air conditioner control assembly - removal and installation

Front control assembly

Refer to illustrations 12.3 and 12.4
1 Disconnect the cable from the negative terminal of the battery.
2 Remove the trim panel from around the control assembly (see Chapter 11, Section 27).
3 Remove the screws on the left side (1990 through 1992 models) or on the front (1993 on models) of the control assembly (see illustration).
4 Disconnect the electrical connectors at the rear of the control assembly (see illustra-

12.4 Disconnect the electrical connectors at the rear of the control unit

tion). Remove the assembly from the dash.
5 Installation is the reverse of removal.

Rear control assembly

Refer to illustrations 12.6a, 12.6b and 12.6c
6 Follow the photo sequence for the

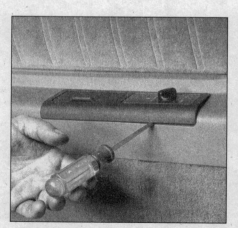
12.6a Gently pry out the control panel by pushing in the hidden tabs with a small screwdriver and prying upward

removal procedure (see illustrations) taking care to read the accompanying captions.
7 Installation is the reverse of the removal procedure.

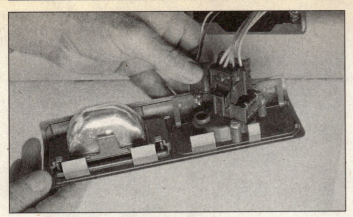
12.6b Twist the bulb connection to release it

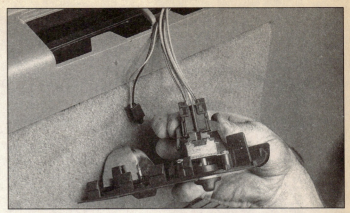
12.6c Unplug the electrical connector and pull the assembly free

13 Air conditioning system - check and maintenance

Warning: *The air conditioning system is under high pressure. DO NOT loosen any hose or line fittings or remove any components until after the system has been discharged. Air conditioning refrigerant should be properly discharged into an EPA-approved container at a dealer service department or an automotive air conditioning repair facility. Always wear eye protection when disconnecting air conditioning system fittings.*

Caution: *The air conditioning system on 1994 and later models uses the non-ozone-depleting refrigerant, referred to as R-134a. The R-134a refrigerant and its lubricating oil are not compatible with the R-12 system used on 1993 and earlier models. Under no circumstances should R-12 refrigerant or refrigerant oil be added to an R-134a system, or vice versa. If the different types of refrigerant and/or oil are mixed, it will result in damage to the air conditioning system.*

Check

1 The following maintenance checks should be performed on a regular basis to ensure the air conditioner continues to operate at peak efficiency.

a) *Check the drivebelt. If it's worn or deteriorated, replace it (see Chapter 1, Section 21).*
b) *Check the system hoses. Look for cracks, bubbles, hard spots and deterioration. Inspect the hoses and all fittings for oil bubbles and seepage. If there's any evidence of wear, damage or leaks, replace the hose(s).*
c) *Inspect the condenser fins for leaves, bugs and other debris. Use a "fin comb" or compressed air to clean the condenser.*
d) *Make sure the system has the correct refrigerant charge.*

2 It's a good idea to operate the system for about 10 minutes at least once a month, particularly during the winter. Long term non-use can cause hardening, and subsequent failure, of the seals.

3 Because of the complexity of the air conditioning system and the special equipment necessary to service it, in-depth troubleshooting and repairs are not included in this manual (refer to the Haynes *Automotive Heating & Air Conditioning Manual*). However, simple checks and component replacement procedures are provided in this Chapter.

4 The most common cause of poor cooling is simply a low system refrigerant charge. If a noticeable loss of cool air output occurs, one of the following quick checks may help you determine if the refrigerant level is low.

5 Warm the engine up to normal operating temperature.

6 Place the air conditioning temperature selector at the coldest setting and put the blower at the highest setting. Open the doors (to make sure the air conditioning system doesn't cycle off as soon as it cools the passenger compartment).

7 With the compressor engaged - the compressor clutch will make an audible click and the center of the clutch will rotate - feel the orifice tube. **Note:** *On models with 3.1L engines the orifice tube is located in the evaporator line adjacent to the right front frame rail near the radiator. On models with 3800 engines the orifice tube is located in the evaporator line next to the line mounting bracket on the right front frame rail.*

8 If a significant temperature drop is noticed, the refrigerant level is probably okay. Refer to Chapter 6, Section 4 for information on the air conditioning clutch control system. Further inspection of the system is beyond the scope of the home mechanic and should be left to a professional.

9 If the inlet has frost accumulation or feels cooler than the accumulator surface, the refrigerant charge is low. On 1993 and earlier models, have the air conditioning system charged by an automotive air conditioning specialist or dealer service department. On 1994 models, see below.

Adding refrigerant (R-134a systems only)

Note: *Because of recent Environmental Protection Agency regulations, R-12 refrigerant used in 1993 and earlier models is no longer available to the do-it-yourselfer.*

10 Buy a small can of R-134a refrigerant and a can tapper designed for use with the unique R-134a fitting. The tapper has a short section of hose that can be attached between the tapper valve and the system low side service valve. Because one can of refrigerant may not be sufficient to bring the system charge up to the proper level, it's a good idea to buy a couple of additional cans. **Warning:** *Never add more than two cans of refrigerant to the system.*

11 Connect the tapper valve to the can and the end of the hose to the system low-side fitting, following the tapper valve manufacturer's instructions. Don't open the can at this time. **Warning:** *DO NOT connect the hose to the system high side!*

12 Warm up the engine and turn on the air conditioner. Keep the hose away from the fan and other moving parts.

13 Following the can tapper manufacturer's instructions, add refrigerant to the low side of the system until both the receiver-drier surface and the evaporator inlet line feel about the same temperature. Allow stabilization time between each addition.

14 Once the receiver-drier surface and the evaporator inlet line feel about the same temperature, add the contents remaining in the can.

14 Air conditioning accumulator - removal and installation

Refer to illustrations 14.3a, 14.3b, 14.5, 14.6 and 14.7

Warning: *The air conditioning system is under high pressure. DO NOT loosen any hose or line fittings or remove any components until after the system has been discharged by a dealer service department or service station. Always wear eye protection when disconnecting air conditioning system fittings.*

1 Have the system discharged (see Warning above).

2 Disconnect the negative battery cable from the battery.

Chapter 3 Cooling, heating and air conditioning systems

14.3a The accumulator is located in the right hand front corner of the engine compartment in front of the coolant reservoir

1   Inlet line   2   Outlet line

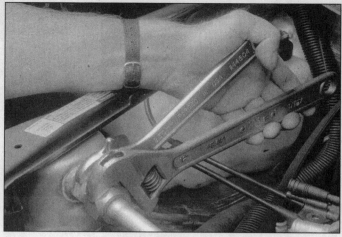

14.3b Use a back-up wrench when removing the refrigerant lines to prevent twisting the tubing

14.5 Remove the mounting bolt (arrow) and push the bracket apart to remove the accumulator

14.6 To remove the Schrader valve, unscrew the protective plastic cap (arrow)

14.7 Be sure to replace the refrigerant line O-rings

3   Disconnect the refrigerant lines from the accumulator. Use a back-up wrench to prevent twisting the tubing **(see illustrations)**.
4   Plug the open fittings to prevent the entry of dirt and moisture.
5   Loosen the mounting bracket bolt **(see illustration)** and remove the accumulator.
6   If a new accumulator is being installed, remove the Schrader valve **(see illustration)** and pour the oil out into a measuring cup, noting the amount. Add fresh refrigerant oil to the new accumulator equal to the amount removed from the old unit, plus one ounce.
7   Remove the old refrigerant line O-rings and replace with new ones. This should be done regardless of whether a new accumulator is being installed or not **(see illustration)**.
8   Installation is the reverse of removal, but be sure to lubricate the O-rings on the fittings with refrigerant oil before connecting the fittings.
9   Have the system evacuated, recharged and leak tested by the shop that discharged it.

15 Air conditioning compressor - removal and installation

Refer to illustrations 15.5, 15.6, 15.7a and 15.7b

Warning: *The air conditioning system is under high pressure. DO NOT loosen any hose or line fittings or remove any components until after the system has been discharged by a dealer service department or service station. Always wear eye protection when disconnecting air conditioning system fittings.* **Note:** *The accumulator (see Section 14) and expansion (orifice) tube (see Section 18) should be replaced whenever the compressor is replaced because of internal damage.*

Removal

1   Have the system discharged (see Warning above).
2   Disconnect the negative battery cable from the battery.
3   Set the parking brake and block the rear wheels. Raise the front of the vehicle and support it securely on jackstands.
4   Remove the drivebelt (see Chapter 1, Section 21). Remove the oil filter (see Chapter 1, Section 12) if necessary for clearance.
5   Disconnect the electrical connector from the compressor clutch **(see illustration)**.

15.5 The compressor clutch electrical connector is located on the top of the compressor

Chapter 3 Cooling, heating and air conditioning systems

3-15

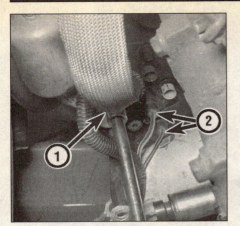

15.6 Disconnect the electrical connectors and unbolt the refrigerant lines from the rear of the compressor - 3.1L engine shown

1. Refrigerant line bolt
2. Electrical connectors

15.7a Location of the compressor rear mounting bolts (arrows)

15.7b Location of the compressor front mounting bolts (arrows)

6 Disconnect the refrigerant lines and unplug the electrical connectors from the switch(es) on the rear of the compressor. Plug the open fittings to prevent entry of dirt and moisture **(see illustration)**.

7 Unbolt the compressor from the mounting brackets and lift it out of the vehicle **(see illustrations)**.

Installation

8 If a new compressor is being installed, follow the directions with the compressor regarding the draining of excess oil prior to installation.

9 The clutch may have to be transferred from the original to the new compressor.

10 Installation is the reverse of removal. Replace all O-rings with new ones specifically made for air conditioning system use and lubricate them with refrigerant oil.

11 Have the system evacuated, recharged and leak tested by the shop that discharged it.

16 Air conditioning condenser - removal and installation

Refer to illustrations 16.4a through 16.4g
Warning: *The air conditioning system is under high pressure. DO NOT loosen any hose or line fittings or remove any components until after the system has been discharged by a dealer service department or service station. Always wear eye protection when disconnecting air conditioning system fittings.*

Removal

1 Have the air conditioning system discharged (see Warning above).

2 Disconnect the negative cable from the battery and on models so equipped, remove the auxiliary pusher fan (see Section 4).

3 Set the parking brake and block the rear wheels. Raise the front of the vehicle and support it securely on jackstands.

4 For condenser removal follow the photo sequence **(see illustrations)** taking care to read the accompanying captions.

16.4a Remove the top air deflector shield by removing the bolts (arrows) and pulling the shield free of its clips

Installation

5 If a new condenser is being installed, pour one ounce of new refrigerant oil into it prior to installation.

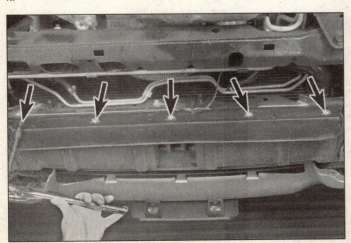

16.4b Remove the lower air deflector by removing the five bolts (arrows) . . .

16.4c . . . releasing the side flaps (arrow) . . .

Chapter 3 Cooling, heating and air conditioning systems

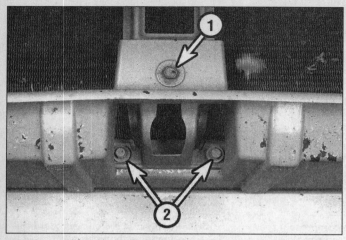

16.4d ... and removing the center retention bolt (1). Remove the center grille by removing the lower grille bolts (2) ...

16.4e ... and the upper grille bolts (arrows)

6   Reinstall the components in the reverse order of removal. Be sure the rubber pads are in place under the condenser and that the refrigerant lines have new O-rings **(see illustration 14.7)**.
7   Have the system evacuated, recharged and leak tested by the shop that discharged it.

17 Evaporator core(s) - removal and installation

Refer to illustrations 17.12a through 17.12g
Warning: *The air conditioning system is under high pressure. DO NOT loosen any hose or line fittings or remove any components until after the system has been discharged by a dealer service department or service station. Always wear eye protection when disconnecting air conditioning system fittings.*

Front evaporator

1   Have the air conditioning system discharged (see Warning above).
2   Disconnect the negative cable from the battery and drain the cooling system (see Chapter 1, Section 29).

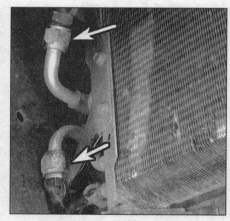

16.4f Remove the upper and lower refrigerant lines (arrows), using a back-up wrench to avoid twisting the tubing - plug the lines to keep dirt and moisture out

3   Remove the heater core cover and the core itself (see Section 11).
4   Disconnect the inlet and outlet pipes at the evaporator.
5   Remove the upper trim panel (see Chapter 11, Section 22).
6   Disconnect the temperature motor electrical connector and remove the four screws that

16.4g Remove the lower insulators (arrow), located at the lower corners of the condenser, and remove the condenser by disengaging the upper insulator and seal

attach the temperature door/slave housing.
7   Working in the engine compartment remove the four bolts that attach the evapo-

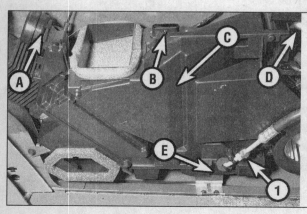

17.12a The bolts that hold the cover of the rear evaporator core together are located at the rear of the assembly - assembly viewed from the front

View A   Actuator cover bolts
View B   Upper rear bolts location
View C   Center bolt location
View D   Rear evaporator outlet
View E   Drain pan bolts
1   Evaporator core inlet

17.12b View A - Remove the bolts from around the actuator (arrows)

Chapter 3 Cooling, heating and air conditioning systems   3-17

17.12c View B - Remove the upper rear bolts . . .

17.12d View C - . . . and the bolt in the center of the assembly (arrow)

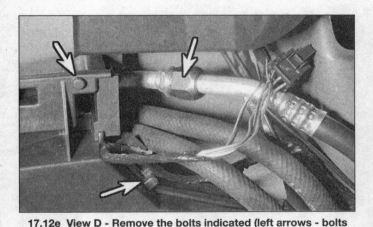

17.12e View D - Remove the bolts indicated (left arrows - bolts not visible in this photo) and the evaporator outlet (right arrow) and inlet (see illustration 17.2a) refrigerant lines. To avoid twisting the tubing use a back-up wrench

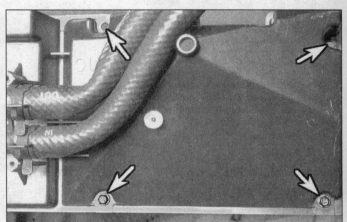

17.12f View E - Remove the four bolts securing the drain pan (arrows)

rator cover to the evaporator core. Remove the evaporator.

Rear evaporator

8 Have the air conditioning system discharged (see Warning above).
9 Disconnect the negative cable from the battery and drain the cooling system (see Chapter 1, Section 29).
10 Remove the side trim panel (see Chapter 11, Section 12)
11 Disengage the rear evaporator and blower assembly from the door (see Section 10), then remove the rear heater core (see Section 11).
12 To remove the rear evaporator follow the photo sequence (see illustrations). Be sure to stay in order and to read the accompanying captions.
13 Installation is the reverse of the removal procedure. Be sure to refill the cooling system (see Chapter 1, Section 29). Have the air conditioning system evacuated, recharged and leak tested by the shop that discharged it.

18 Air conditioning expansion (orifice) tube - removal and installation

Refer to illustration 18.5

Warning: *The air conditioning system is under high pressure. DO NOT loosen any fittings or remove any components until after the system has been properly discharged. Air conditioning refrigerant should be properly discharged into an EPA-approved container at a dealership service department or an automotive air conditioning repair facility. Always wear eye protection when disconnecting air conditioning system fittings.*

1 Have the air conditioning system discharged and the refrigerant recovered (see Warning above). Disconnect the cable from the negative terminal of the battery. **Caution:** *On models equipped with the Delco Loc II audio system, be sure the lockout feature is turned off before performing any procedure that requires disconnecting the battery.*
2 Disconnect the air conditioning con-

17.12g Lift the cover from the assembly, then carefully remove the rear evaporator core

denser tube mounting bracket from the side of the frame rail.
3 Using a pair of open-end wrenches, disconnect the air conditioning evaporator-to-

condenser line fitting and separate the lines.

4 The expansion tube is a tube with a fixed-diameter orifice and a mesh filter at each end. When you separate the pipe at the fitting you will see one end of the expansion (orifice) tube inside the pipe leading to the evaporator. Use a pair of pliers to remove the expansion tube.

5 The expansion (orifice) tube acts to meter the refrigerant, changing it from high-pressure liquid to low-pressure liquid. It is possible to reuse the expansion tube if **(see illustration):**
   a) *The screens aren't plugged with grit or foreign material*
   b) *Neither screen is torn*
   c) *The plastic housing over the screens is intact*
   d) *The brass orifice inside the plastic housing is unrestricted*

6 Installation is the reverse of removal. Be sure to insert the expansion (orifice) tube with the shorter end in first, towards the evaporator. **Caution:** *Always use a new O-ring when installing the expansion (orifice) tube.*

7 Retighten the fitting and refrigerant line, then have the system evacuated, recharged and leak-tested by the shop that discharged it.

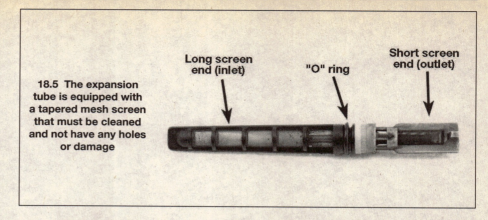

18.5 The expansion tube is equipped with a tapered mesh screen that must be cleaned and not have any holes or damage

Chapter 4
Fuel and exhaust systems

Contents

	Section		Section
Air and PCV filter replacement	See Chapter 1	Fuel pump - removal and installation	7
Air cleaner housing - removal and installation	8	Fuel tank cleaning and repair - general information	6
Electronic fuel injection system - general information	10	Fuel tank - removal and installation	5
Exhaust system check	See Chapter 1	General information	1
Exhaust system servicing - general information	14	Model 220 Throttle Body Injection (TBI) (3.1L engine) - removal, overhaul and installation	12
Fuel filter replacement	See Chapter 1		
Fuel injection system - check	11	Port Fuel Injection (PFI) (3800 engine) - component removal and installation	13
Fuel lines and fittings - repair and replacement	4		
Fuel pressure relief procedure	2	Throttle cable - removal and installation	9
Fuel pump/fuel pressure - testing	3	Throttle linkage inspection	See Chapter 1

Specifications

General
Fuel pressure
 Throttle Body Injection (TBI) .... 9 to 13 psi
 Port Fuel Injection (PFI) .... 40 to 47 psi
IAC valve pintle extension 1 1/8-inch
Minimum idle speed .... 550 to 650 rpm
Non-adjustable TPS output check .... less than 1.25 volts
Fuel injector resistance
 TBI .... 1.2 to 1.6 ohms
 PFI .... 11.4 to 12.6 ohms

Torque specifications   Ft-lbs
TBI throttle body mounting nuts 18
PFI throttle body mounting bolts
 1990 1
 1991 on 18
Plenum-to-intake manifold bolts 16

Chapter 4 Fuel and exhaust systems

1 General information

Note: *On models equipped with the Delco Loc II audio system, be sure the lockout feature is turned off before performing any procedure which requires disconnecting the battery.*

The fuel system consists of a fuel tank, an electric fuel pump, a fuel pump relay, an air cleaner assembly and either a Throttle Body Injection (TBI) system (on 3.1L engines) or a Port Fuel Injection (PFI) system (on 3800 engines). The basic difference between throttle body and port fuel injection systems is the number and location of the fuel injectors.

Throttle Body Injection (TBI) system

The throttle body system utilizes two injectors, centrally mounted in a carburetor-like housing. The injector is an electrical solenoid, with fuel delivered to the injector at a constant pressure level. To maintain the fuel pressure at a constant level, excess fuel is returned to the fuel tank.

A signal from the ECM opens the solenoid, allowing fuel to spray through the injector into the throttle body. The amount of time the injector is held open by the ECM determines the fuel/air mixture ratio.

Port Fuel Injection (PFI) system

The port system utilizes six injectors of a different type than the throttle body injection system. The fuel/air ratio is controlled in the same manner as the TBI system. Instead of dual injectors mounted in a centrally located throttle body, one injector is installed above each intake port. The throttle body on the PFI system serves only to control the amount of air passing into the system. Because each cylinder is equipped with an injector mounted immediately adjacent to the intake valve, much better control of the fuel/air mixture ratio is possible.

Fuel pump and lines

Fuel is circulated from the fuel tank to the fuel injection system, and back to the fuel tank, through a pair of metal lines running along the underside of the vehicle. An electric fuel pump is attached to the fuel sending unit inside the fuel tank. A vapor return system routes all vapors and hot fuel back to the fuel tank through a separate return line.

Exhaust system

The exhaust system includes an exhaust manifold fitted with an exhaust oxygen sensor, a catalytic converter, an exhaust pipe, and a muffler.

The catalytic converter is an emission control device added to the exhaust system to reduce pollutants. A single-bed converter is used in combination with a three-way (reduction) catalyst. Refer to Chapter 6 for more information regarding the catalytic converter.

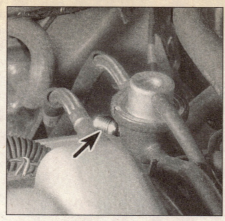

2.4 Remove the cap to the Schrader valve (arrow) and connect a fuel pressure gauge

2 Fuel pressure relief procedure

Refer to illustration 2.4

Warning 1: *Gasoline is extremely flammable, so take extra precautions when you work on any part of the fuel system. Don't smoke or allow open flames or bare light bulbs near the work area, and don't work in a garage where a natural gas-type appliance (such as a water heater or clothes dryer) with a pilot light is present. If you spill any fuel on your skin, rinse it off immediately with soap and water. When you perform any kind of work on the fuel system, wear safety glasses and have a Class B type fire extinguisher on hand.*

Warning 2: *After the fuel pressure has been relieved, wrap shop towels around any fuel connection you'll be disconnecting. They'll absorb the residual fuel that may leak out, reducing the risk of fire and preventing contact with your skin.*

1 Before servicing any fuel system component, you must relieve the fuel pressure to minimize the risk of fire or personal injury.

2 Remove the fuel filler cap - this will relieve any pressure built up in the tank.

3.1L engines (Model 220 TBI)

3 On models with Throttle Body Injection (TBI), the system contains a constant bleed feature in the fuel pressure regulator that relieves the pressure automatically. There is no special procedure that is required to relieve the fuel pressure. After turning off the ignition switch, allow about ten minutes for the fuel pressure to bleed off.

3800 engine (Port Fuel Injection [PFI])

4 On models with Port Fuel Injection, use one of the two following methods:

a) *Disconnect the fuel pump electrical connector at the fuel tank (lower the tank if necessary). Run the engine until it stops, then engage the starter again for another three seconds. With the ignition turned OFF, reconnect the fuel tank electrical connector, then disconnect the cable from the negative terminal of the battery.*

b) *Loosen the fuel filler cap. Attach fuel pressure gauge to the Schrader valve on the fuel rail* **(see illustration).** *Place the gauge bleeder hose in an approved fuel container. Open the valve on the gauge to relieve pressure, then disconnect the cable from the negative terminal of the battery.*

5 Unless this procedure is followed before servicing fuel lines or connections, fuel spray (and possible injury) may occur.

3 Fuel pump/fuel pressure - testing

Warning: *Gasoline is extremely flammable, so take extra precautions when you work on any part of the fuel system. Don't smoke or allow open flames or bare light bulbs near the work area, and don't work in a garage where a natural gas-type appliance (such as a water heater or clothes dryer) with a pilot light is present. If you spill any fuel on your skin, rinse it off immediately with soap and water. When you perform any kind of work on the fuel system, wear safety glasses and have a Class B type fire extinguisher on hand.*

Note: *In order to perform the fuel pressure test, you will need to obtain a fuel pressure gauge and adapter set for the fuel injection system being tested.*

Preliminary inspection (all models)

1 Should the fuel system fail to deliver the proper amount of fuel, or any fuel at all, to the fuel injection system, inspect it as follows.

2 Always make certain there is fuel in the tank.

3 With the engine running, inspect for leaks at the fittings at both ends of each fuel line. Repair any leaking connections. Inspect all hoses for flattening or kinks which would restrict the flow of fuel.

Pressure check

4 Relieve the fuel system pressure (see Section 2).

Models with Port Fuel Injection (PFI) (3800 engine)

5 Install a fuel pressure gauge at the Schrader valve on the fuel rail **(see illustration 2.4).**

6 Turn the ignition switch ON with the air conditioning off. The fuel pump should run for about two seconds - note the reading on the gauge. After the pump stops running the pressure should hold steady. It should be within the range listed in this Chapter's Specifications.

7 Start the engine and let it idle at normal

Chapter 4 Fuel and exhaust systems

3.14 The fuel pump test terminal is located near the battery

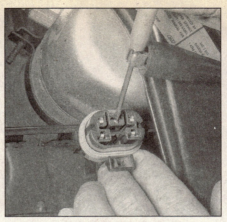

3.23 Apply battery voltage to the gray wire in the fuel pump relay connector and listen for the sound of the fuel pump

3.25 With the ignition key ON, check for battery voltage at the dark green/white terminal of the fuel pump relay connector (3.1L engine shown)

operating temperature. The pressure should be lower by 3 to 10 psi. If all the pressure readings are within the limits listed in this Chapter's Specifications, the system is operating properly.

8   If the pressure did not drop by 3 to 10 psi after starting the engine, apply 12 to 14 inches of vacuum to the pressure regulator. If the pressure drops, repair the vacuum source to the regulator. If the pressure does not drop, replace the regulator.

9   If the fuel pressure is not within specifications, check the following:
  a) *If the pressure is higher than specified, check for a faulty regulator or a pinched or clogged fuel return hose or pipe.*
  b) *If the pressure is lower than specified:*
  c) *Inspect the fuel filter - make sure it's not clogged.*
  d) *Look for a pinched or clogged fuel hose between the fuel tank and the fuel rail.*
  e) *Check the pressure regulator for a malfunction.*
  f) *Look for leaks in the fuel line.*
  g) *Look for a pinched, broken or disconnected regulator vacuum hose.*
  h) *Check for leaking injectors.*
  i) *Check the in-tank fuel pump check valve.*

10 After the testing is done, relieve the fuel pressure (see Section 2) and remove the fuel pressure gauge.

11 If there are no problems with any of the above-listed components, check the fuel pump (see below).

Models with Throttle Body Injection (TBI) (3.1L engine)

Refer to illustration 3.14

12 Relieve fuel system pressure (see Section 2).

13 Install a fuel pressure gauge between the fuel feed hose and the inlet fitting of the throttle body. The fuel hoses will be under high pressure during this check, so make your connections secure, with no possibility of leaks. **Note:** *If the fuel line fittings to adapt your fuel pressure gauge are not available,*

use a special tool available at an auto parts store or automotive tool company.

14 With the ignition OFF, use a fused jumper wire from a 12 volt source (battery voltage), jump the fuel pump test terminal (red wire) and note the pressure reading **(see illustration)**.

15 If the pressure is within the limits listed in this Chapter's Specifications, no further testing is necessary.

16 If the pressure was higher than specified, check for a restricted fuel return line. If the line is OK, then replace the pressure regulator (see Section 12).

17 If the pressure was less than specified, slowly pinch the hose between the gauge and the TBI unit and note the pressure. If the pressure goes above 9 psi, then replace the pressure regulator. If there is not any pressure, then check for a plugged fuel filter, plugged fuel pump inlet filter or a restricted fuel line.

18 After testing is completed, relieve the fuel pressure and remove the fuel pressure gauge.

19 If no problems are found with any of the above listed components, check the fuel pump (see below).

Fuel pump check (all models)

Refer to illustration 3.23

20 If you suspect a problem with the fuel pump, verify the pump actually runs. Have an assistant turn the ignition switch to ON - you should hear a brief whirring noise as the pump comes on and pressurizes the system. Have the assistant start the engine. This time you should hear a constant whirring sound from the pump (but it's more difficult to hear with the engine running).

21 If the pump does not come on (makes no sound), proceed to the next Step.

22 Check the fuel pump fuse located in the passenger compartment **(see illustration 3.1 in Chapter 12, Section 3)**. If the fuse is blown, replace the fuse and see if the pump works. If the pump still does not work, go to the next step.

23 With the ignition OFF, apply 12 volts (battery voltage) to the fuel pump relay terminal direct to the fuel pump **(see illustration)** and listen for the fuel pump running. **Note:** *Refer to illustrations 6.2a and 6.2b in Chapter 12, Section 6 for the locations of the relays.*

24 If the pump runs, then check the fuel pump relay. If the pump does not run, check for an open circuit between the relay and the fuel pump. If the circuit is OK, disconnect the electrical connector(s) at the fuel pump. Connect battery voltage to the gray wire terminal of the pump. Connect the black wire terminal to ground. If the fuel pump now runs, it's OK; check all connections and wiring again. If it does not run, it's faulty. Replace the fuel pump.

Fuel pump relay check (all models)

Refer to illustration 3.25

25 To test the fuel pump relay, disconnect the fuel pump relay **(see illustration 6.2a and 6.2b in Chapter 12)** and connect a test light to the dark green/white terminal and have an assistant turn the ignition key ON **(see illustration)**. The test light should come on for two or three seconds and cycle itself off.

26 If the test result is correct, but the fuel pump relay does not come on (if it comes on, you should hear/feel a click at the relay when an assistant turns on the ignition key when the relay is plugged in), replace the fuel pump relay with a new part. If there is no presence of voltage, check the ECM/PCM and the relating fuel pump circuits for a short or open circuit.

4 Fuel lines and fittings - repair and replacement

Refer to illustrations 4.10, 4.12 and 4.13
Warning: *Gasoline is extremely flammable, so take extra precautions when you work on*

Chapter 4 Fuel and exhaust systems

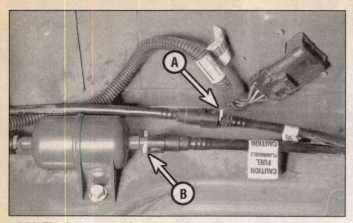

4.10 The fuel return (A) and feed lines (B) are easily released by squeezing the tabs and separating each connector (3.1L engine shown)

4.12 Use special fuel line un-locking tools, available at most auto parts stores, and push them into the connector (arrows) to separate the fuel lines (3.1L engine shown)

any part of the fuel system. Don't smoke or allow open flames or bare light bulbs near the work area, and don't work in a garage where a natural gas-type appliance (such as a water heater or clothes dryer) with a pilot light is present. If you spill any fuel on your skin, rinse it off immediately with soap and water. When you perform any kind of work on the fuel system, wear safety glasses and have a Class B type fire extinguisher on hand.

1   Always relieve the fuel pressure before servicing fuel lines or fittings (see Section 2).
2   The fuel feed, return and vapor lines extend from the fuel tank to the engine compartment. The lines are secured to the underbody with clip and screw assemblies. These lines must be occasionally inspected for leaks, kinks and dents.
3   If evidence of dirt is found in the system or fuel filter during disassembly, the line should be disconnected and blown out. Check the fuel strainer on the fuel gauge sending unit (see Section 7) for damage and deterioration.

Steel tubing

4   If replacement of a fuel line or emission line is called for, use welded steel tubing meeting the manufacturer's specifications.
5   Don't use copper or aluminum tubing to replace steel tubing. These materials cannot withstand normal vehicle vibration.
6   Because fuel lines used on fuel-injected vehicles are under high pressure, they require special consideration.
7   Most fuel lines have threaded fittings with O-rings. Any time the fittings are loosened to service or replace components:

  a) Use a backup wrench while loosening and tightening the fittings.
  b) Check all O-rings for cuts, cracks and deterioration. Replace any that appear worn or damaged.
  c) If the lines are replaced, always use original equipment parts, or parts that meet the manufacturer's standards specified in this Section.

Rubber hose

Warning: *On models equipped with port fuel injection, use only original equipment replacement hoses or their equivalent. Others may fail from the high pressures of this system.*
8   When rubber hose is used to replace a metal line, use reinforced, fuel resistant hose with the word "Fluoroelastomer" imprinted on it. Hose(s) not clearly marked like this could fail prematurely and could fail to meet Federal emission standards. Hose inside diameter must match line outside diameter.
9   Don't use rubber hose within four inches of any part of the exhaust system or within ten inches of the catalytic converter. Metal lines and rubber hoses must never be allowed to chafe against the frame. A minimum of 1/4-inch clearance must be maintained around a line or hose to prevent contact with the frame.

Removal and installation

Note: *The following procedure and accompanying illustrations are typical for vehicles covered by this manual. On quick-disconnect (non-threaded) fittings, clean off the fittings before disconnection to prevent dirt from getting in the fittings. After disconnection, clean the fittings with compressed air and apply a few drops of oil.*
10 Relieve the fuel pressure (see Section 2) and disconnect the fuel feed, return or vapor line at the fuel tank **(see illustration)**. **Note:** *Some fuel line connections may be threaded. Be sure to use a back-up wrench when separating the connections.*
11 Remove all fasteners attaching the lines to the vehicle body.
12 Detach the fitting(s) that attach the fuel hoses to the engine compartment metal lines **(see illustration)**. Twisting them back and forth will allow them to separate more easily.
13 Installation is the reverse of removal. Be sure to install new O-rings at the threaded fittings **(see illustration)**. Apply a light coat of clean engine oil to the O-rings.

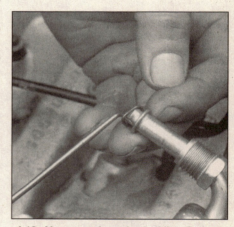

4.13 Always replace the fuel line O-rings (if equipped)

Repair

14 In repairable areas, cut a piece of fuel hose four inches longer than the portion of the line removed. If more than a six inch length of line is removed, use a combination of steel line and hose so hose lengths won't be more than ten inches. Always follow the same routing as the original line.
15 Cut the ends of the line with a tube cutter. Using the first step of a double flaring tool, form a bead on the end of both line sections. If the line is too corroded to withstand bead operation without damage, the line should be replaced.
16 Use a screw type hose clamp. Slide the clamp onto the line and push the hose on. Tighten the clamps on each side of the repair.
17 Secure the lines properly to the frame to prevent chafing.

5 Fuel tank - removal and installation

Refer to illustrations 5.6 and 5.8
Warning: *Gasoline is extremely flammable,*

Chapter 4 Fuel and exhaust systems

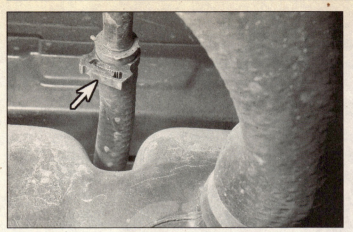

5.6 To release the clamp (arrow) from the fuel vapor line, pry up on one side of the locking tab with a small screwdriver

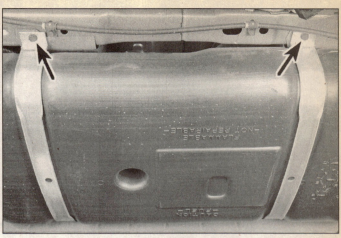

5.8 Remove the bolts (arrows) from the tank straps

so take extra precautions when you work on any part of the fuel system. Don't smoke or allow open flames or bare light bulbs near the work area, and don't work in a garage where a natural gas-type appliance (such as a water heater or clothes dryer) with a pilot light is present. If you spill any fuel on your skin, rinse it off immediately with soap and water. When you perform any kind of work on the fuel system, wear safety glasses and have a Class B type fire extinguisher on hand.

Note: *Don't begin this procedure until the fuel gauge indicates the tank is empty or nearly empty. If the tank must be removed when it's full (for example, if the fuel pump malfunctions), siphon any remaining fuel from the tank prior to removal.*

1 Unless the vehicle has been driven far enough to completely empty the tank, it's a good idea to siphon the residual fuel out before removing the tank from the vehicle. **Warning:** *DO NOT start the siphoning action by mouth! Use a siphoning kit, available at most auto parts stores.*
2 Relieve the fuel pressure (see Section 2).
3 Detach the cable from the negative terminal of the battery.
4 Raise the vehicle and place it securely on jackstands.
5 Locate the electrical connector for the electric fuel pump and fuel gauge sending unit in front of the tank, and unplug it. If the vehicle doesn't have an electrical connector, see Step 9 below.
6 Disconnect the fuel feed and return lines, the vapor return line and the filler neck and vent tubes **(see illustration)**.
7 Support the fuel tank with a piece of wood and a floor jack.
8 Disconnect both fuel tank retaining straps **(see illustration)**.
9 Lower the tank enough to disconnect the wires and ground strap from the fuel pump/fuel gauge sending unit, if you haven't already done so.
10 Remove the tank from the vehicle.
11 Installation is the reverse of removal.

6 Fuel tank cleaning and repair - general information

1 All repairs to the fuel tank or filler neck should be carried out by a professional who has experience in this critical and potentially dangerous work. Even after cleaning and flushing of the fuel system, explosive fumes can remain and ignite during repair of the tank.
2 If the fuel tank is removed from the vehicle, it should not be placed in an area where sparks or open flames could ignite the fumes coming out of the tank. Be especially careful inside garages where a natural gas-type appliance is located, because the pilot light could cause an explosion.

7 Fuel pump - removal and installation

Refer to illustrations 7.5 and 7.6

Warning: *Gasoline is extremely flammable, so take extra precautions when you work on any part of the fuel system. Don't smoke or allow open flames or bare light bulbs near the work area, and don't work in a garage where a natural gas-type appliance (such as a water heater or clothes dryer) with a pilot light is present. If you spill any fuel on your skin, rinse it off immediately with soap and water. When you perform any kind of work on the fuel system, wear safety glasses and have a Class B type fire extinguisher on hand.*

Removal

1 Relieve the fuel pressure (see Section 2).
2 Remove the cable from the negative battery terminal.
3 Remove the fuel tank (see Section 6).
4 The fuel pump/sending unit assembly is located inside the fuel tank. It's held in place by a cam lock ring mechanism consisting of an inner ring with three locking cams and an outer ring with three retaining tangs.
5 To unlock the fuel pump/sending unit assembly, turn the inner ring counterclockwise until the locking cams are free of the retaining tangs **(see illustration)**. **Note:** *If the rings are locked together too tightly to release by hand, tap them gently with a hammer and brass punch or wood dowel.* **Warning:** *Do not use a steel punch - a spark could cause an explosion!*

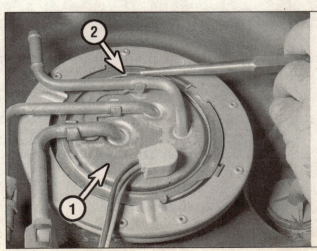

7.5 To remove the fuel gauge unit and fuel pump assembly (1), turn the inner cam lock ring (2) counterclockwise and pull the pump and O-ring gasket out of the tank – be extremely careful not to bang the float mechanism on the sides of the hole

4-6 Chapter 4 Fuel and exhaust systems

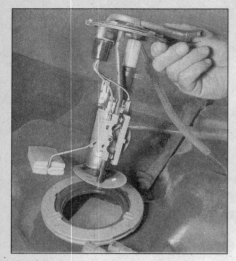

7.6 Lift the fuel pump assembly from the fuel tank

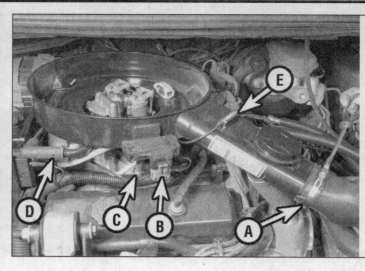

8.4 Air cleaner assembly details on the 3.1L engine

A Clamp
B MAP sensor connector
C Vacuum line
D Crankcase vent hose
E Vacuum connector

6   Pull the fuel pump/sending unit assembly out of the tank **(see illustration)**. **Caution:** *The fuel level float and sending unit are delicate. Don't bump them into the lock ring during removal or the accuracy of the sending unit may be affected.*
7   Replace the rubber O-ring around the mouth of the lock ring mechanism.
8   Inspect the filter on the lower end of the fuel pump. If it's dirty, remove it, clean it with solvent and blow it out with compressed air. If it's too dirty to be cleaned, replace it.
9   If you have to separate the fuel pump and sending unit, pull the fuel pump assembly into the rubber connector and slide the pump away from the bottom support. Care should be taken to prevent damage to the rubber insulator and fuel strainer during removal. After the pump is clear of the bottom support, pull it out of the rubber connector.

Installation

10 Position the rubber gasket around the opening in the fuel tank and guide the fuel pump/sending unit assembly into the tank.
11 Turn the inner lock ring clockwise until the locking cams are fully engaged by the retaining tangs. **Note:** *Since you've installed a new O-ring, it may be necessary to push down on the inner lock ring until the locking cams slide under the retaining tangs.*
12 Install the fuel tank (see Section 5).

8 Air cleaner housing - removal and installation

1   Detach the cable from the negative terminal of the battery.
2   Remove the battery from the vehicle if necessary (see Chapter 5).

3.1L engine

Refer to illustration 8.4
3   Remove the air cleaner element (see Chapter 1).
4   Unclamp the air intake duct from the air cleaner housing **(see illustration)**.

5   Unplug the electrical connector from the MAP sensor and remove any vacuum lines from the housing and detach any other fasteners retaining the air duct.
6   Remove the crankcase vent hose.
7   Disconnect the heated air inlet hose and lift the housing from the vehicle.
8   Installation is the reverse of removal.

3800 engine

Refer to illustrations 8.9, 8.10 and 8.11
9   Remove the bolts **(see illustration)** that retain the air intake assembly to the front of the engine compartment.
10 Remove the bolt and the clamp from the engine brace **(see illustration)** and separate the air intake duct from the engine compartment.
11 Loosen the clamp from the throttle body/air intake duct **(see illustration)**.
12 Disconnect the Air Intake Temperature (AIT) sensor **(see illustration 8.11)** and lift the complete air cleaner housing from the engine compartment.
13 Installation is the reverse of removal.

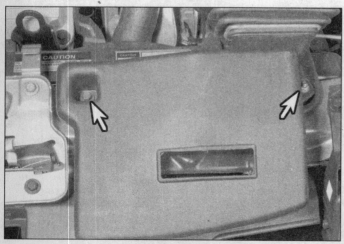

8.9 Remove the bolts (arrows) that retain the air intake duct to the engine compartment

8.10 Remove the bracket that holds the air intake brace

Chapter 4 Fuel and exhaust systems

8.11 Loosen the clamp (A), disconnect the Air Intake Temperature sensor (B) and lift off the cleaner housing as a unit

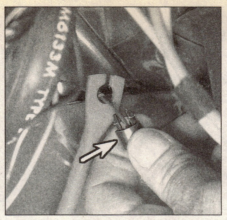

9.2 To detach the throttle cable at the pedal, pull the spring cup (arrow) toward the end of the cable and slide the cable out of the slot

9.6 Pop off the retaining clip from the throttle lever arm with a small screwdriver

9 Throttle cable - removal and installation

3.1L engine

Refer to illustrations 9.2 and 9.6

Removal

1 Detach the screws and the clip retaining the lower instrument trim panel and partially lower the trim.
2 Detach the accelerator cable from the accelerator pedal **(see illustration)**.
3 Squeeze the accelerator cable cover tangs and push the cable through the firewall into the engine compartment.
4 Remove the air cleaner assembly (see Section 8).
5 Detach the routing clip and the accelerator cable (if equipped).
6 Detach the accelerator cable-to-throttle lever retainer and separate the accelerator cable from the throttle body lever **(see illustration)**.
7 Squeeze the accelerator cable retaining tangs and push the cable through the accelerator cable bracket.

9.11 Be careful not to bend the clip when applying pressure

Installation

8 Installation is the reverse of removal. **Note:** *To prevent possible interference, flexible components (hoses, wires, etc.) must not be routed within two inches of moving parts, unless routing is controlled.*
9 Operate the accelerator pedal and check for any binding condition by completely opening and closing the throttle.
10 At the engine compartment side of the firewall, apply sealant around the accelerator cable.

3800 engine

Refer to illustration 9.11

11 Use a small screwdriver to disconnect the throttle cable circlip from the cable end **(see illustration)**.
12 Remove the spring clip from the housing bracket and slide the cable out.
13 Disconnect the cable from inside the passenger compartment and remove the cable from the vehicle.
14 Installation is the reverse of removal.

10 Electronic fuel injection system - general information

Electronic fuel injection provides optimum fuel/air mixture ratios at all stages of combustion and offers immediate throttle response characteristics. It also enables the engine to run at the leanest possible fuel/air mixture ratio, reducing exhaust gas emissions.

On models equipped with a 3.1L V6 engine, the Throttle Body Injection (TBI) unit replaces a conventional carburetor atop the intake manifold. Models equipped with a 3800 engine are fitted with a Port Fuel Injection (PFI) system. TBI systems are controlled by an Electronic Control Module (ECM) while PFI systems are controlled by a Powertrain Control Module (PCM). The ECM/PCM monitors engine performance and adjusts the air/fuel mixture accordingly (see Chapter 6 for a complete description of the ECM/PCM control system).

An electric fuel pump located in the fuel tank with the fuel gauge sending unit pumps fuel to the fuel injection system through the fuel feed line and an in-line fuel filter. **Note:** *The 3.1L engine is equipped with a twin turbine pump (medium pressure) while the 3800 engine is equipped with the roller vane pump (high pressure). Also, the 3800 engine is equipped with a pulsator* **(see illustration 7.8)** *to reduce noise generated by the pulsation of the fuel directly flowing from the pump.* A pressure regulator keeps fuel available at a constant pressure. Fuel in excess of injector needs is returned to the fuel tank by a separate line.

The basic TBI unit is made up of two major casting assemblies - a throttle body with an Idle Air Control (IAC) valve controls air flow and a Throttle Position Sensor (TPS) monitors throttle angle. The fuel body consists of a fuel meter with a built-in pressure regulator and a fuel injector to supply fuel to the engine.

The fuel injector is a solenoid operated device controlled by the ECM. The ECM turns on the solenoid, which lifts a normally closed ball valve off its seat. The fuel, which is under pressure, is injected in a conical spray pattern at the walls of the throttle body bore above the throttle valve. The fuel which is not used by the injector passes through the pressure regulator before being returned to the fuel tank.

On Port Fuel Injection (PFI) systems, the throttle body has a throttle valve to control the amount of air delivered to the engine. The Throttle Position Sensor (TPS) and Idle Air Control (IAC) valves are located on the throttle body.

The fuel rail is mounted on the top of the engine. It distributes fuel to the individual injectors.

Fuel is delivered to the input end of the rail by the fuel feed line, goes through the rail and then to the pressure regulator. The regulator keeps the pressure to the injectors at a constant level.

The remaining fuel is returned to the fuel tank.

11.8 Measure the resistance of each injector with a digital volt-ohmmeter

12.6 Carefully peal away the old fuel meter outlet passage gasket and fuel meter cover gasket with a razor blade

11 Fuel injection system - check

Warning: *Gasoline is extremely flammable, so take extra precautions when you work on any part of the fuel system. Don't smoke or allow open flames or bare light bulbs near the work area, and don't work in a garage where a natural gas-type appliance (such as a water heater or clothes dryer) with a pilot light is present. If you spill any fuel on your skin, rinse it off immediately with soap and water. When you perform any kind of work on the fuel system, wear safety glasses and have a Class B type fire extinguisher on hand.*

Note: *The following procedure is based on the assumption that the fuel pump is working and the fuel pressure is adequate (see Section 3).*

Preliminary checks

1 Check all electrical connectors that are related to the system. Loose electrical connectors and poor grounds can cause many problems that resemble more serious malfunctions.
2 Check to see that the battery is fully charged, as the control unit and sensors depend on an accurate supply voltage in order to properly meter the fuel.
3 . Check the air filter element - a dirty or partially blocked filter will severely impede performance and economy (see Chapter 1).
4 If a blown fuse is found, replace it and see if it blows again. If it does, search for a grounded wire in the harness to the fuel pump.

Port Fuel Injection only

Refer to illustration 11.8

5 Check the air intake duct from the Mass Airflow Sensor (MAF) (if equipped) to the intake manifold for leaks, which will result in an excessively lean mixture. Also check the condition of the vacuum hoses connected to the intake manifold.
6 Remove the air intake duct from the throttle body and check for dirt, carbon or other residue build-up in the throttle body, particularly around the throttle plate. If it's dirty, clean it with carburetor cleaner and a toothbrush.
7 With the engine running, place a screwdriver against each injector, one at a time, and listen through the handle for a clicking sound, indicating operation.
8 With the engine OFF and the fuel injector electrical connectors disconnected, measure the resistance of each injector **(see illustration)**. Each injector should measure about 11.4 to 12.6 ohms. If not, the injector is probably faulty.
9 The remainder of the system checks should be left to a dealer service department or other qualified repair shop, as there is a chance that the control unit may be damaged if not performed properly.

TBI systems only

10 Set the parking brake, remove the air cleaner top plate and, with the engine idling in Park, observe the operating fuel injectors. The spray pattern should be even and conical in shape. The spray should touch the throttle body bore.
  a) If the spray is weak or uneven, the injector is clogged or faulty. Gasoline additives designed to clean fuel injectors can sometimes clear a clogged injector. If not, a dealer service department or other qualified shop has more effective cleaning equipment.
  b) If an injector is not operating at all, check its electrical connector. If the connection is good and the injector is receiving voltage, but the injector still doesn't work, the injector is faulty.

12 Model 220 Throttle Body Injection (TBI) (3.1L engine) - removal, overhaul and installation

Refer to illustrations, 12.6, 12.7, 12.14, 12.16, 12.21 through 12.25, 12.37, 12.40, 12.41, 12.50, 12.52, 12.65 and 12.66

Note: *Because of its relative simplicity, the throttle body assembly does not need to be removed from the intake manifold or disassembled for component replacement. However, for the sake of clarity, the following procedures are shown with the TBI assembly removed from the vehicle.*

1 Relieve system fuel pressure (see Section 2).
2 Detach the cable from the negative terminal of the battery.
3 Remove the air cleaner housing assembly, adapter and gaskets.

Fuel meter cover/fuel pressure regulator assembly

Note: *The fuel pressure regulator is housed in the fuel meter cover. Whether you are replacing the meter cover or the regulator itself, the entire assembly must be replaced. The regulator must not be removed from the cover.*

4 Unplug the electrical connectors to the fuel injectors.
5 Remove the long and short fuel meter cover screws and remove the fuel meter cover.
6 Remove the fuel meter outlet passage gasket, cover gasket and pressure regulator seal. Carefully remove any old gasket material that is stuck with a razor blade **(see illustration)**. **Caution:** *Do not attempt to re-use either of these gaskets.*
7 Inspect the cover for dirt, foreign material and casting warpage. If it is dirty, clean it with a clean shop rag soaked in solvent. Do not immerse the fuel meter cover in cleaning solvent - it could damage the pressure regulator diaphragm and gasket. **Warning:** *Do not remove the four screws* **(see illustration)** *securing the pressure regulator to the fuel meter cover. The regulator contains a large spring under compression which, if accidentally released, could cause injury. Disassembly might also result in a fuel leak between the diaphragm and the regulator housing. The new fuel meter cover assembly will include a new pressure regulator.*
8 Install the new pressure regulator seal, fuel meter outlet passage gasket and cover gasket.

Chapter 4 Fuel and exhaust systems

12.7 Never remove the four pressure regulator screws (arrows) from the fuel meter cover

12.14 To remove either injector electrical connector, depress the two tabs on the front and rear of each connector and lift straight up

12.16 To remove an injector, slip the tip of a flat-bladed screwdriver under the lip of the lug on top of the injector and, using another screwdriver as a fulcrum, carefully pry the injector up and out

12.21 Slide the new filter onto the nozzle of the fuel injector

9 Install the fuel meter cover assembly using Loctite 262 or equivalent on the screws. **Note:** *The short screws go next to the injectors.*
10 Attach the electrical connectors to both injectors.
11 Attach the cable to the negative terminal of the battery.
12 With the engine off and the ignition on, check for leaks around the gasket and fuel line couplings.
13 Install the air cleaner, adapter and gaskets.

Fuel injector assembly

14 To unplug the electrical connectors from the fuel injectors, squeeze the plastic tabs and pull straight up **(see illustration)**.
15 Remove the fuel meter cover/pressure regulator assembly. **Note:** *Do not remove the fuel meter cover assembly gasket - leave it in place to protect the casting from damage during injector removal.*
16 Use a screwdriver and fulcrum **(see illustration)** to pry out the injector.
17 Remove the upper (larger) and lower (smaller) O-rings and filter from the injector.
18 Remove the steel backup washer from the top of the injector cavity.
19 Inspect the fuel injection filter for evi-

12.22 Lubricate the lower O-ring with transmission fluid, then place it on the shoulder in the bottom of the injector cavity

12.23 Place the steel back-up washer on the shoulder near the top of the injector cavity

dence of dirt and contamination. If present, check for the presence of dirt in the fuel lines and fuel tank.
20 Be sure to replace the fuel injector with an identical part. Injectors from other models can fit in the Model 220 TBI assembly but are calibrated for different flow rates.
21 Slide the new filter into place on the nozzle of the injector **(see illustration)**.
22 Lubricate the new lower (smaller) O-ring with automatic transmission fluid and place it on the small shoulder at the bottom of the fuel injector cavity in the fuel meter body **(see illustration)**.
23 Install the steel back-up washer in the injector cavity **(see illustration)**.

Chapter 4 Fuel and exhaust systems

12.24 Lubricate the upper O-ring with transmission fluid, then install it on top of the steel washer

12.25 Make sure that the lug is aligned with the groove in the bottom of the fuel injector cavity

12.37 The IAC valve can be removed with an adjustable wrench (shown) or a 1-1/4 inch wrench

12.40 The IAC pintle must not extend more than 1-1/8 inches - also replace the O-ring if it is damaged or brittle

- A Distance of pintle extension
- B O-ring

12.41 To reduce the IAC valve pintle extension, grasp the valve and depress the pintle using a slight side-to-side motion

24 Lubricate the new upper (larger) O-ring with automatic transmission fluid and install it on top of the steel back-up washer **(see illustration)**. **Note:** *The backup washer and the large O-ring must be installed before the injector. If they aren't, improper seating of the large O-ring could cause fuel leakage.*
25 To install the injector, align the raised lug on the injector base with the notch in the fuel meter body cavity **(see illustration)**. Push down on the injector until it is fully seated in the fuel meter body. **Note:** *The electrical terminals should be parallel with the throttle shaft.*
26 Install the fuel meter cover assembly and gasket.
27 Attach the cable to the negative terminal of the battery.
28 With the engine off and the ignition on, check for fuel leaks.
29 Attach the electrical connectors to the fuel injector(s).
30 Install the air cleaner housing assembly, adapter and gaskets.

Throttle Position Sensor (TPS)

31 Remove the two TPS attaching screws and retainers and remove the TPS from the throttle body.
32 If you intend to re-use the same TPS, do not attempt to clean it by soaking it in any liquid cleaner or solvent. The TPS is a delicate electrical component and can be damaged by solvents.
33 Install the TPS on the throttle body while lining up the TPS lever with the TPS drive lever.
34 Install the two TPS attaching screws and retainers.
35 Install the air cleaner housing assembly, adapter and gaskets.
36 Attach the cable to the negative terminal of the battery. **Note:** *See non-adjustable TPS output check at end of this Section.*

Idle Air Control (IAC) valve

37 Unplug the electrical connector from the IAC valve and remove the IAC valve **(see illustration)**.
38 Remove and discard the old IAC valve gasket. Clean any old gasket material from the surface of the throttle body assembly to insure proper sealing of the new gasket.
39 All pintles in IAC valves on Model 220 TBI units have the same dual taper. However, the pintles on some units have a 12mm diameter and the pintles on others have a 10mm diameter. A replacement IAC valve must have the appropriate pintle taper and diameter for proper seating of the valve in the throttle body.
40 Measure the distance between the tip of the pintle and the housing mounting surface with the pintle fully extended **(see illustration)**. If this dimension is greater than the dimension listed in this Chapter's Specifications, it must be reduced to prevent damage to the valve.
41 If the pintle must be adjusted, determine whether your valve is a Type I (collar around the electrical terminal) or a Type II (no collar around the electrical terminal).

 a) *To adjust the pintle of an IAC valve with a collar, grasp the valve and exert firm pressure on the pintle with the thumb. Use a slight side-to-side movement on the pintle as you press it in with your thumb* **(see illustration)**.

12.50 Remove the fuel inlet and outlet nuts from the fuel meter body

12.52 Once the fuel inlet and outlet nuts are off, pull the fuel meter body straight up to separate it from the throttle body

12.65 When disconnecting the fuel feed and return lines from the fuel inlet and outlet nuts, be sure to use a back-up wrench to prevent damage to the lines

12.66 To remove the Model 220 throttle body from the intake manifold, remove three nuts (arrows) (third nut not visible)

b) To adjust the pintle of an IAC valve without a collar, compress the retaining spring while turning the pintle clockwise. Return the spring end to its original position with the straight portion aligned in the slot under the flat surface of the valve.

42 Install the IAC valve and tighten the screws securely. Attach the electrical connector.
43 Install the air cleaner housing assembly, adapter and gaskets.
44 Attach the cable to the negative terminal of the battery.
45 Start the engine and allow it to reach operating temperature, then turn it off. No adjustment of the IAC valve is required after installation. The IAC valve is reset by the ECM when the engine is turned off.

Fuel meter body assembly

46 Unplug the electrical connectors from the fuel injectors.
47 Remove the fuel meter cover/pressure regulator assembly, fuel meter cover gasket, fuel meter outlet gasket and pressure regulator seal.

48 Remove the fuel injectors.
49 Unscrew the fuel inlet and return line threaded fittings, detach the lines and remove the O-rings.
50 Remove the fuel inlet and outlet nuts and gaskets from the fuel meter body assembly **(see illustration)**. Note the locations of the nuts to ensure proper reassembly. The inlet nut has a larger passage than the outlet nut.
51 Remove the gasket from the inner end of each fuel nut.
52 Remove the fuel meter body-to-throttle body attaching screws and remove the fuel meter body from the throttle body **(see illustration)**.
53 Install the new throttle body-to-fuel meter body gasket. Match the cut-out portions in the gasket with the openings in the throttle body.
54 Install the fuel meter body on the throttle body. Coat the fuel meter body-to-throttle body attaching screws with thread locking compound before installing them.
55 Install the fuel inlet and outlet nuts, with new gaskets, in the fuel meter body and tighten the nuts to the specified torque. Install the fuel inlet and return line threaded fittings with new O-rings. Use a backup wrench to prevent the nuts from turning.
56 Install the fuel injectors.
57 Install the fuel meter cover/pressure regulator assembly.
58 Attach the electrical connectors to the fuel injectors.
59 Attach the cable to the negative terminal of the battery.
60 With the engine off and the ignition on, check for leaks around the fuel meter body, the gasket and around the fuel line nuts and threaded fittings.
61 Install the air cleaner housing assembly, adapters and gaskets.

Throttle body assembly

62 Unplug all electrical connectors - the IAC valve, TPS and fuel injectors. Detach the grommet with the wires from the throttle body.
63 Detach the throttle linkage, return spring(s), transmission control cable (automatics) and, if equipped, cruise control.
64 Clearly label, then detach, all vacuum hoses.
65 Using a backup wrench, detach the inlet and outlet fuel line nuts **(see illustration)**. Remove the fuel line O-rings from the nuts and discard them.
66 Remove the TBI mounting nuts **(see illustration)** and lift the TBI unit from the intake manifold. Remove and discard the TBI manifold gasket.
67 Place the TBI unit on a holding fixture, available at most auto parts stores. **Note:** *If you don't have a holding fixture, and decide to place the TBI directly on a work bench surface, be extremely careful when servicing it. The throttle valve can be easily damaged.*
68 Remove the fuel meter body-to-throttle body attaching screws and separate the fuel meter body from the throttle body.
69 Remove the throttle body-to-fuel meter body gasket and discard it.
70 Remove the TPS.
71 Invert the throttle body on a flat surface for greater stability and remove the IAC valve.
72 Clean the throttle body assembly in a

13.7a Remove the bolts (arrows) from the throttle cable bracket and . . .

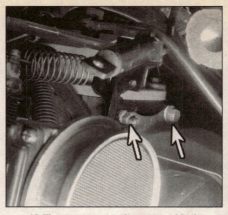

13.7b . . . remove the nut and bolt (arrows) from the bracket assembly near the transaxle

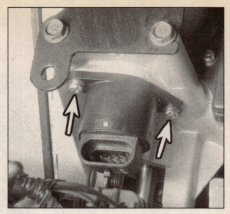

13.12 You'll need a Phillips screwdriver to remove the IAC retaining screws (arrows)

cold immersion cleaner. Clean the metal parts thoroughly and blow dry with compressed air. Be sure that all fuel and air passages are free of dirt or burrs. **Caution:** *Do not place the TPS, IAC valve, pressure regulator diaphragm, fuel injectors or other components containing rubber in the solvent or cleaning bath. If the throttle body requires cleaning, soaking time in the cleaner should be kept to a minimum. Some models have throttle shaft dust seals that could lose their effectiveness by extended soaking.*
73 Inspect the mating surfaces for damage that could affect gasket sealing. Inspect the throttle lever and valve for dirt, binds, nicks and other damage.
74 Invert the throttle body on a flat surface for stability and install the IAC valve and the TPS.
75 Install a new throttle body-to-fuel meter body gasket and place the fuel meter body assembly on the throttle body assembly. Coat the fuel meter body-to-throttle body attaching screws with thread locking compound and tighten them securely.
76 Install the TBI unit and tighten the mounting nuts to the torque listed in this Chapter's Specifications. Use a new TBI-to-manifold gasket.
77 Install new O-rings on the fuel line nuts. Install the fuel line and outlet nuts by hand to prevent stripping the threads. Using a backup wrench, tighten the nuts securely once they have been correctly threaded into the TBI unit.
78 Attach the vacuum hoses, throttle linkage, return spring(s), transmission control cable (automatics) and, if equipped, cruise control cable. Attach the grommet, with wire harness, to the throttle body.
79 Plug in all electrical connectors, making sure that the connectors are fully seated and latched.
80 Check to see if the accelerator pedal is free by depressing the pedal to the floor and releasing it with the engine off.
81 Connect the negative battery cable, and, with the engine off and the ignition on, check for leaks around the fuel line nuts.
82 Adjust the minimum idle speed and check

the TPS output (see Steps 84 through 93).
83 Install the air cleaner housing assembly, adapter and gaskets.

Minimum idle speed adjustment

Note: *This adjustment should be performed only when the throttle body has been replaced. The engine should be at normal operating temperature before making the adjustment.*
84 Remove the air cleaner housing assembly, adapter and gaskets.
85 Plug any vacuum ports as required by the VECI label.
86 With the IAC valve connected, ground the diagnostic terminal of the ALDL electrical connector (see Chapter 6). Turn on the ignition but do not start the engine. Wait at least 30 seconds to allow the IAC valve pintle to extend and seat in the throttle body. Disconnect the IAC valve electrical connector. Remove the ground from the diagnostic terminal and start the engine.
87 Remove the plug by first piercing it with an awl (see item "28 in illustration 12.5), then applying leverage.
88 Adjust the idle stop screw to obtain the specified rpm in Park.
89 Turn the ignition off and reconnect the IAC valve electrical connector.
90 Unplug any plugged vacuum line ports.
91 Install the air cleaner housing assembly, adapter and new gaskets (see Section 8).

Non-adjustable TPS output check

Note: *This check should be performed only when the throttle body or the TPS has been replaced or after the minimum idle speed has been adjusted.*
92 Connect a digital voltmeter from the TPS electrical connector center terminal "B" to outside terminal "A" (you'll have to fabricate jumpers for terminal access).
93 With the ignition on and the engine off, TPS voltage should be less than the voltage listed in this Chapter's Specifications. If it's

more than the specified voltage, check the minimum idle speed before replacing the TPS.

13 Port Fuel Injection (PFI) (3800 engine) - component removal and installation

Warning: *Before servicing an injector, fuel pump, fuel line, fuel rail or pressure regulator, relieve the pressure in the fuel system to minimize the risk of fire and injury (refer to the fuel pressure relief procedure described in Section 2). After servicing the fuel system, cycle the ignition between ON and OFF several times (wait 10 seconds between cycles) and check the system for leaks.*

Throttle Body

Refer to illustrations 13.7a and 13.7b
1 Disconnect the cable from the negative terminal of the battery.
2 Unplug the Idle Air Control (IAC) valve and the Throttle Position Sensor (TPS) electrical connectors **(see illustration 2.1b in Chapter 6)**.
3 Disconnect any vacuum hoses to the throttle body.
4 Disconnect the throttle cable (see Section 9).
5 Remove the breather hose.
6 Detach the air intake duct (see Section 8).
7 Remove the bolts that retain the throttle cable assembly **(see illustrations)** to the throttle body.
8 Remove the throttle body bolts and detach the throttle body.
9 Install the throttle body and gasket and tighten the bolts to the torque listed in this Chapter's Specifications.
10 The rest of the procedure is the reverse of removal.

Idle Air Control (IAC) valve

Refer to illustration 13.12
11 Unplug the electrical connector from the Idle Air Control (IAC) valve assembly.
12 Remove the two IAC valve attaching

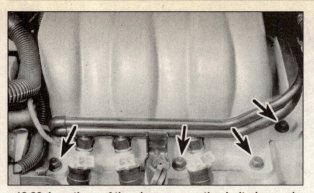

13.26 Locations of the plenum mounting bolts (arrows) for the 3800 engine

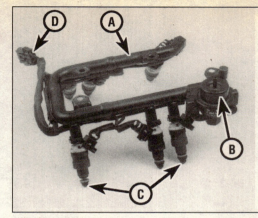

13.38 3800 engine fuel rail assembly

A Fuel rail
B Fuel pressure regulator
C Fuel injectors
D Fuel injector electrical connector

screws and withdraw the valve **(see illustration)**.
13 Remove the IAC valve assembly and replace the rubber O-ring.
14 Clean the sealing surface and the bore of the idle air/vacuum signal housing assembly to ensure a good seal. **Caution:** *The IAC valve assembly itself is an electrical component and must not be soaked in any liquid cleaner or solvent or damage may result.*
15 Before installing the IAC valve assembly, the position of the pintle must be checked. If the pintle is extended too far, damage to the assembly may occur.
16 Measure the distance from the gasket mounting surface of the IAC valve assembly to the tip of the pintle.
17 If the distance is greater than 1-1/8 inch, reduce it by applying a firm hand pressure on the pintle to retract it (a slight side-to-side motion may help).
18 Position the new O-ring seal on the IAC valve assembly. Lubricate the O-ring with engine oil.
19 Install the IAC valve in the idle air/vacuum signal housing assembly and tighten it securely.
20 Plug in the electrical connector at the IAC valve assembly. **Note:** *No adjustment is made to the IAC assembly after reinstallation. The IAC resetting is controlled by the ECM when the engine is started.*

Throttle Position Sensor (TPS)

21 Replacement of this component should be performed by a dealer service department or repair shop equipped with a "SCAN" tool to properly adjust the TPS.

Plenum

Refer to illustration 13.26
22 Remove the cable from the negative terminal of the battery.
23 Mark and remove all of the vacuum lines that may interfere, then remove the throttle cable bracket nuts.
24 Remove the throttle body.
25 Remove the bolts which secure the plastic spark plug wire shield.
26 Remove the plenum bolts **(see illustration)**.
27 Remove the plenum and gaskets. If the plenum sticks, use a block of wood and a rubber mallet to dislodge it. Do not pry between the sealing flanges, as this will damage the machined surfaces and vacuum leaks may develop.
28 Remove all traces of old gasket material from the plenum and intake manifold mating surfaces. It is a good idea to stuff rags into the intake manifold openings to prevent debris and old gasket material from falling in.
29 Install the new gaskets and set the plenum into position.
30 Install the plenum bolts and tighten them to the torque listed in this Chapter's Specifications.
31 The rest of the procedure is the reverse of removal.

Fuel rail and related components

Refer to illustration 13.38
Warning: *Before any work is performed on the fuel lines, fuel rail or injectors, the fuel system pressure must be relieved (refer to the fuel pressure relief procedure in Section 2).*
Note: *An eight-digit identification number is stamped on the side of the fuel rail assembly. Refer to this number if servicing or parts replacement is required.*
32 Detach the negative battery cable from the battery.
33 Remove the air intake plenum (see Step 22).
34 Remove the fuel lines at the fuel rail (see Section 4).
35 Remove the vacuum line at the regulator.
36 Label and unplug the injector electrical connectors.
37 Remove the fuel rail retaining bolts.
38 Carefully remove the fuel rail with the injectors **(see illustration)**. **Caution:** *Use care when handling the fuel rail assembly to avoid damaging the injectors.*

Fuel injectors

Refer to illustration 13.40
Caution: *To prevent dirt from entering the engine, the area around the injectors should be cleaned before servicing.*
39 To remove the fuel injectors, spread open the end of the injector clip slightly and remove it from the fuel rail, then extract the injector.
40 Inspect the injector O-ring seal(s). These should be replaced whenever the fuel rail is removed **(see illustration)**.
41 Install the new O-ring seal(s) on the injector(s) and lubricate them with engine oil.
42 Install the injectors on the fuel rail.
43 Secure the injectors with the retainer clips.

Fuel pressure regulator

Refer to illustration 13.45
44 Relieve the fuel pressure from the system (see Section 2).
45 Clean any dirt deposits from the fuel

13.40 If you plan to reuse the same injector, always replace the O-ring with a new one

13.45 The retaining ring (arrow) is located on the top of the fuel pressure regulator

pressure regulator ring located directly on the top **(see illustration)**.
46 Remove the vacuum line.
47 Compress the ring with snap ring pliers and lift the ring from the regulator.
48 Lift while simultaneously twisting the regulator and remove it from the housing.
49 Cover the regulator housing with a clean shop towel to prevent any dirt from entering the system.
50 Install new O-rings and lubricate them with a light coat of oil.
51 Installation is the reverse of removal.

Mass Air Flow (MAF) sensor

Refer to illustration 13.53
52 Disconnect the electrical connector.
53 Remove the MAF sensor bolts **(see illustration)** from the assembly.
54 Lift the MAF sensor out of the vehicle.
Caution: *the MAF sensor is delicate - if you plan to reinstall the existing unit, handle it carefully.*
55 Installation is the reverse of the removal procedure.

---

14 Exhaust system servicing - general information

Refer to illustrations 14.1, 14.4a and 14.4b
Warning: *Inspection and repair of exhaust system components should be done only after enough time has elapsed after driving the vehicle to allow the system components to cool completely. Also, when working under the vehicle, make sure it is securely supported on jackstands.*

1 The exhaust system consists of the exhaust manifold(s), the catalytic converter, the muffler, the tailpipe and all connecting pipes, brackets, hangers and clamps. The exhaust system is attached to the body with mounting brackets and rubber hangers **(see**

13.53 Remove the screws (arrows) form the MAF sensor

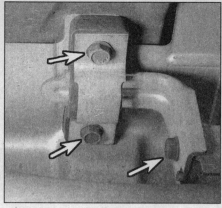

14.1 To detach this rubber block type hanger, remove the bolts (arrows)

illustration). If any of the parts are improperly installed, excessive noise and vibration will be transmitted to the body.
2 Conduct regular inspections of the exhaust system to keep it safe and quiet. Look for any damaged or bent parts, open seams, holes, loose connections, excessive corrosion or other defects which could allow exhaust fumes to enter the vehicle. Deteriorated exhaust system components should not be repaired; they should be replaced with new parts.
3 If the exhaust system components are extremely corroded or rusted together, welding equipment will probably be required to remove them. The convenient way to accomplish this is to have a muffler repair shop remove the corroded sections with a cutting torch. If, however, you want to save money by doing it yourself (and you don't have a welding outfit with a cutting torch), simply cut off the old components with a hacksaw. If you have compressed air, special pneumatic cutting chisels can also be used. If you decide to tackle the job at home, be sure to

wear safety goggles to protect your eyes from metal chips and work gloves to protect your hands.
4 Here are some simple guidelines to follow when repairing the exhaust system:
 a) *Work from the back to the front when removing exhaust system components.*
 b) *Apply penetrating oil to the exhaust system component fasteners to make them easier to remove* **(see illustration)**.
 c) *Use new gaskets, hangers and clamps when installing exhaust systems components.*
 d) *Apply anti-seize compound to the threads of all exhaust system fasteners during reassembly.*
 e) *Be sure to allow sufficient clearance between newly installed parts and all points on the underbody to avoid overheating the floor pan and possibly damaging the interior carpet and insulation. Pay particularly close attention to the catalytic converter and heat shield* **(see illustration)**.

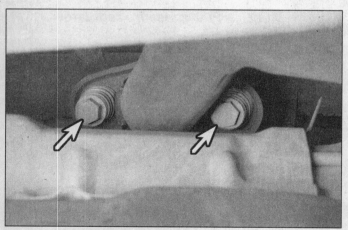

14.4a Lubricate the exhaust manifold-to-flange bolts with penetrating spray before attempting to remove them

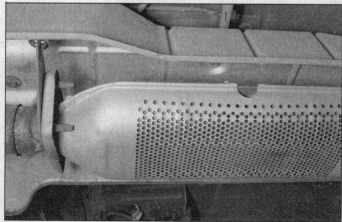

14.4b Be sure the catalytic converter is completely cooled down before repairing the assembly

Chapter 5
Engine electrical systems

Contents

	Section		Section
Alternator - removal and installation	13	Ignition coil - removal, check and installation	9
Battery cables - check and replacement	4	Ignition module and distributorless ignition sensors - check and replacement	7
Battery check and maintenance	See Chapter 1	Ignition pick-up coil (3.1L engine) - check and replacement	8
Battery - emergency jump starting	3	Ignition system - check	6
Battery - removal and installation	2	Ignition system - general information	5
Charging system - check	12	Spark plug replacement	See Chapter 1
Charging system - general information	10	Starter motor - removal and installation	16
Charging system - precautions	11	Starter motor - testing in vehicle	15
Distributor (3.1L engine) - removal and installation	6	Starter solenoid - removal and installation	17
Drivebelt check, adjustment and replacement	See Chapter 1	Starting system - general information	14
General information	1		

Specifications

Coil
3800 engine
- Primary coil resistance .... 0.3 to 0.6 ohms
- Secondary coil resistance .... 5,000 to 6,500 ohms

1 General information

Note: *On models equipped with the Delco Loc II audio system, be sure the lockout feature is turned off before performing any procedure which requires disconnecting the battery.*

Ignition systems

The 3.1L engines covered by this manual are equipped with an HEI (High Energy Ignition) and the 3800 engines are equipped with a Direct (distributorless) Ignition System (DIS).

HEI (High Energy Ignition) systems (3.1L engines)

HEI equipped vehicles use a special HEI distributor with Electronic Spark Timing (EST). HEI distributors covered by this manual are equipped with a separate coil.

All spark timing changes in the HEI/EST distributor are carried out by the Electronic Control Module (ECM), which monitors data from various engine sensors, computes the desired spark timing and signals the distributor to change the timing accordingly. No vacuum or mechanical advance is used.

Electronic Spark Control (ESC)

3.1L and 3800 engines are equipped with an Electronic Spark Control (ESC), which uses a knock sensor in connection with the ECM/PCM to control spark timing to allow the engine to have maximum spark advance without spark knock. This improves driveability and fuel economy.

Secondary (spark plug) wiring

The secondary (spark plug) wire used with the HEI system is a carbon impregnated cord conductor encased in an 8 mm diameter rubber jacket with an outer silicone jacket. This type of wire will withstand very high temperatures and provides an excellent insulator for the HEI's high voltage. **Warning:** *Because of the very high voltage generated by the HEI system, extreme care should be taken whenever an operation involving ignition components is performed. This not only includes the distributor, coil, control module and spark plug wires, but related items that are connected to the systems as well, such as the plug connections, tachometer and testing equipment.*

DIS (Direct Ignition System)

The distributorless ignition is controlled by the PCM (Powertrain Control Module). This system offers no moving parts, less maintenance, more coil cool-down between spark plug firing and the elimination of the mechanical timing adjustments.

The DIS ignition systems use a "waste spark" method of spark distribution. Each cylinder is paired with its opposing cylinder in the firing order (1-4, 2-5, 3-6) so one cylinder under compression fires simultaneously with its opposing cylinder, where the piston is on the exhaust stroke. Since the cylinder on exhaust requires very little of the available voltage to fire its plug, most of the voltage is used to fire the cylinder on compression.

The DIS system includes three coil packs, an ignition module, a crankshaft position sensor (dual Hall Effect), an engine crankshaft balancer with crankshaft sensor interrupter rings attached to the rear, a camshaft sensor, spark plug wires, and the PCM (Powertrain Control Module).

Conventional ignition coils have one end of the secondary winding connected to the engine ground. On DIS, neither end of the secondary winding is grounded; instead, each end of the coil's secondary winding is directly attached to the spark plug and its companion.

2.2 Remove the bolts (arrows) and lift the brace from the engine compartment

2.4 Remove the battery hold-down clamp bolt (arrow) from the battery carrier

The dual magnetic crankshaft sensors are mounted in an aluminum mounting bracket bolted to the front left side of the engine timing cover. The interrupter rings mounted on the crankshaft balancer will disrupt the signal voltage causing an ON OFF ON OFF etc. reaction much the same as the opening and closing of the ignition points on the older ignition systems. The DIS system is also equipped with an Electronic Spark Timing (EST) and control wires from the PCM, just like conventional distributor systems.

Charging system

The charging system consists of a belt-driven alternator with an integral voltage regulator and the battery. These components work together to supply electrical power for the ignition system, the lights and all accessories.

There are three models of alternators used. Earlier vehicles use the CS 130 model (100 amp) and later models are equipped with the CS 130 (105 amp w/ rear air conditioning) and CS 144 (140 amp, 3800 engines). All types use a conventional pulley and fan.

Starting system

The starting circuit consists of the battery, starter motor, ignition switch and other related electrical wiring. The starter motor is equipped with a solenoid mounted directly to the assembly. Parts are available for repair of the motor and solenoid as separate units.

2 Battery - removal and installation

Refer to illustrations 2.2 and 2.4

1 **Caution:** *Always disconnect the negative cable first and hook it up last or the battery may be shorted by the tool being used to loosen the cable clamps.* Disconnect both cables from the battery terminals.
2 Remove windshield washer reservoir and the upper engine compartment brace from chassis **(see Illustration)**.
3 On some models it may be necessary to remove the air cleaner assembly intake duct (see Chapter 4, Section 8) and the battery heat shield (if equipped).
4 Remove the clamp from the bottom of the battery **(see illustration)**.
5 Lift out the battery. Be careful - it's heavy.
6 While the battery is out, inspect the carrier (tray) for corrosion.
7 If you are replacing the battery, make sure that you get one that's identical, with the same dimensions, amperage rating, cold cranking rating, etc.
8 Installation is the reverse of removal.

3 Battery - emergency jump starting

Refer to the Booster battery (jump) starting *procedure at the front of this manual.*

4 Battery cables - check and replacement

1 Periodically inspect the entire length of each battery cable for damage, cracked or burned insulation and corrosion. Poor battery cable connections can cause starting problems and decreased engine performance.
2 Check the cable-to-terminal connections at the ends of the cables for cracks, loose wire strands and corrosion. The presence of white, fluffy deposits under the insulation at the cable terminal connection is a sign that the cable is corroded and should be replaced. Check the terminals for distortion, missing mounting bolts and corrosion.
3 When removing the cables, always disconnect the negative cable first and hook it up last or the battery may be shorted by the tool used to loosen the cable clamps. Even if only the positive cable is being replaced, be sure to disconnect the negative cable from the battery first (see Chapter 1 for further information regarding battery cable removal).
4 Disconnect the old cables from the battery, then trace each of them to their opposite ends and detach them from the starter solenoid and ground terminals. Note the routing of each cable to ensure correct installation.
5 If you are replacing either or both of the old cables, take them with you when buying new cables. It is vitally important that you replace the cables with identical parts. Cables have characteristics that make them easy to identify: positive cables are usually red, larger in cross-section and have a larger diameter battery post clamp; ground cables are usually black, smaller in cross-section and have a slightly smaller diameter clamp for the negative post.
6 Clean the threads of the solenoid or ground connection with a wire brush to remove rust and corrosion. Apply a light coat of battery terminal corrosion inhibitor, or petroleum jelly, to the threads to prevent future corrosion.
7 Attach the cable to the solenoid or ground connection and tighten the mounting nut/bolt securely.
8 Before connecting a new cable to the battery, make sure that it reaches the battery post without having to be stretched.
9 Connect the positive cable first, followed by the negative cable.

5 Ignition system - check

Refer to illustration 5.2
Warning: *Because of the very high voltage generated by the ignition system, extreme care should be taken whenever an operation is performed involving ignition components. This not only includes the distributor, coil(s), control module and spark plug wires, but related items that are connected to the system as well, such as the plug connections, tachometer and any test equipment.*
1 If the engine turns over but won't start, disconnect the spark plug wire from any

Chapter 5 Engine electrical systems

5.2 To use a calibrated ignition tester (available at most auto parts stores), simply disconnect a spark plug wire, attach the wire to the tester, clip the tester to a convenient ground (like a valve cover bolt) and operate the starter - if there's enough power to fire the plug, sparks will be visible between the electrode tip and the tester body

6.4 Mark the position of the rotor to the distributor before removing the distributor - also mark the relationship of the distributor to the engine block

6.5 The distributor hold-down clamp and bolt must be removed before the distributor can be removed from the engine (four-cylinder engine shown)

spark plug and attach it to a calibrated ignition tester (available at most auto parts stores).
2 Connect the clip on the tester to a bolt or metal bracket on the engine **(see illustration)**. If you're unable to obtain a calibrated ignition tester, remove the wire from one of the spark plugs and, using an insulated tool, hold the end of the wire about 1/4-inch from a good ground.
3 Crank the engine and watch the end of the tester or spark plug wire to see if bright blue, well-defined sparks occur. If you're not using a calibrated tester, have an assistant crank the engine for you. **Warning:** *Keep clear of drivebelts and other moving engine components that could injure you.*
4 If sparks occur, sufficient voltage is reaching the plug to fire it (repeat the check at the remaining plug wires to verify the wires, distributor cap and rotor [if equipped] or other coils [distributorless systems] are OK). However, the plugs themselves may be fouled, so remove them and check them as described in Chapter 1.
5 On distributor (HEI) systems, if no sparks or intermittent sparks occur, remove the distributor cap and check the cap and rotor as described in Chapter 1. If moisture is present, dry out the cap and rotor, then reinstall the cap.
6 On distributor systems, if there's still no spark, detach the coil secondary wire from the distributor cap and hook it up to the tester (reattach the plug wire to the spark plug), then repeat the spark check. Again, if you don't have a tester, hold the end of the wire about 1/4-inch from a good ground. If sparks occur now, the distributor cap, rotor or plug wire(s) may be defective.
7 On all models, if no sparks occur, check the spark plug wires and the wire connections at the coil(s) to make sure they're clean and tight. Check for voltage to the coil. Make any necessary repairs, then repeat the check again.
8 On distributor systems, if there's still no spark, the coil-to-cap wire may be bad (check the resistance with an ohmmeter - it should be 7000 ohms per foot or less). If a known good wire doesn't make any difference in the test results, the ignition module may be defective.

6 Distributor (3.1 engine) - removal and installation

Removal

Refer to illustrations 6.4 and 6.5
1 Disconnect the cable from the negative battery terminal. **Caution:** *If the vehicle is equipped with a Delco Loc II audio system, make sure you have the correct activation code before disconnecting the battery. See the information at the front of this manual for the radio re-activation procedure.*
2 Remove the coil wire or small-wire electrical connections from the distributor cap.
3 Remove the distributor cap (see Chapter 1).
4 Note the position of the rotor and the distributor-to-block alignment. Make an alignment mark on the distributor to indicate the position of the rotor **(see illustration)**. Also make a mark to indicate distributor-to-block relationship.
5 Remove the distributor hold-down clamp bolt and clamp **(see illustration)**. Remove the distributor from the engine. **Caution:** *Do not turn the crankshaft while the distributor is removed from the engine. If the crankshaft is turned, the position of the rotor will be altered and the engine will have to be re-timed.*

Installation (crankshaft not turned after distributor removal)

6 Insert the distributor into the engine in exactly the same relation to the block in which it was removed. To mesh the gears, it may be necessary to turn the rotor slightly. At this point the distributor may not seat down against the block completely. This is due to the lower end of the distributor shaft not mating properly with the oil pump shaft. If this is the case, check again to make sure the distributor is aligned with the block in the same position it was in before removal and that the rotor is correctly aligned with the distributor body. The gear on the distributor shaft is engaged with the gear on the camshaft, and this relationship cannot change as long as the distributor is not lifted from the engine. Use a socket and breaker bar on the crankshaft bolt to turn the engine over in the normal direction of rotation. The rotor will turn, but the oil pump shaft will not because the two shafts are not engaged. When the proper alignment is reached the distributor will drop down over the oil pump shaft, and the distributor body will seat properly against the block.
7 Install the hold-down clamp and tighten the bolt securely.
8 Install the distributor cap and coil wire.
9 Connect the cable to the negative terminal of the battery.

Installation (crankshaft turned after distributor removal)

10 Remove the number one spark plug.
11 Place your finger over the spark plug hole while turning the crankshaft in the normal direction of rotation with a wrench on the pulley bolt at the front of the engine.
12 When you feel compression, continue turning the crankshaft slowly until the timing mark on the crankshaft pulley is aligned with the "0" on the engine timing indicator.

Chapter 5 Engine electrical systems

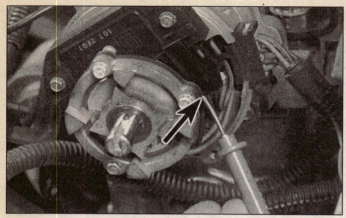

7.4 To check the module for voltage, insert the voltmeter probe into the module positive terminal (arrow)

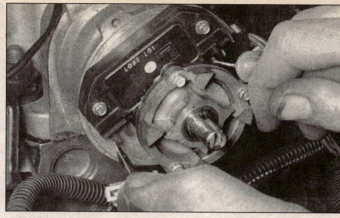

7.6 With a test light connected at terminal "P," check the voltage with the meter probe at the "C" terminal

13 Position the rotor to point at the number one distributor terminal.
14 Insert the distributor into the engine in exactly the same relation to the block in which it was removed. To mesh the gears, it may be necessary to turn the rotor slightly. If the distributor does not seat fully against the block it is because the oil pump shaft has not seated in the distributor shaft. Make sure the distributor drive gear is fully engaged with the camshaft gear, then use a socket on the crankshaft bolt to turn the engine over in the normal direction of rotation until the two shafts engage and the distributor seats against the block.
15 Install the hold-down clamp and tighten the bolt securely.
16 Install the distributor cap and coil wire.
17 Connect the cable to the negative terminal of the battery.

7 Ignition module and distributorless ignition system sensors - check and replacement

Distributor ignition system (3.1L engine)

Note: *It is not necessary to remove the distributor to check or replace the module.*

Check

Refer to illustrations 7.4, 7.6 and 7.11

1 Disconnect the tachometer (if so equipped) at the distributor.
2 Check for a spark at the coil and spark plug wires (Section 5).
3 If there is no spark, remove the distributor cap. Remove the ignition module from the distributor but leave the connector plugged in.
4 With the ignition switch turned On, check for voltage at the module positive terminal **(see illustration)**.
5 If the reading is less than ten volts, there is a fault in the wire between the module positive (+) terminal and the ignition coil positive connector or the ignition coil and primary cir-

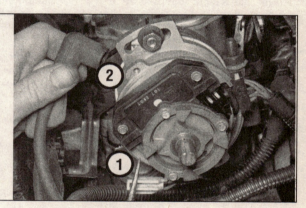

7.11 As the test light probe (1) is removed, check for a spark at the coil wire (2)

cuit-to-ignition switch.
6 If the reading is ten volts or more, check the "C" terminal on the module **(see illustration)**.
7 If the reading is less than one volt, there is an open or grounded lead in the distributor-to-coil "C" terminal connection or ignition coil or an open primary circuit in the coil itself.
8 If the reading is one to ten volts, replace the module with a new one and check for a spark (Section 5). If there is a spark the module was faulty and the system is now operating properly. If there is no spark, there is a fault in the ignition coil.
9 If the reading in Step 4 is 10 volts or more, unplug the pickup coil connector from the module. Check the "C" terminal voltage with the ignition switch On and watch the voltage reading as a test light is momentarily (five seconds or less) connected between the battery positive (+) terminal and the module "P" terminal **(see illustration 7.6)**.
10 If there is no drop in voltage, check the module ground and, if it is good, replace the module with a new one.
11 If the voltage drops, check for spark at the coil wire as the test light is removed from the module terminal. If there is no spark, the module is faulty and should be replaced with a new one. If there is a spark, the pick-up coil or connections are faulty or not grounded **(see illustration)**.

Replacement

Refer to illustration 7.16

12 Detach the cable from the negative terminal of the battery. **Caution:** *If the vehicle is equipped with a Delco Loc II audio system, make sure you have the correct activation code before disconnecting the battery. See the information at the front of this manual for the radio re-activation procedure.*
13 Remove the distributor cap and rotor (see Chapter 1).
14 Remove both module attaching screws and lift the module up and away from the distributor.
15 Disconnect both electrical leads from the module. Note that the leads cannot be interchanged.
16 Do not wipe the grease from the module or the distributor base if the same module is to be reinstalled. If a new module is to be installed, a package of silicone grease will be included with it. Wipe the distributor base and the new module clean, then apply the silicone grease on the face of the module and on the distributor base where the module seats **(see illustration)**. This grease is necessary for heat dissipation.
17 Install the module and attach both electrical leads.
18 Install the distributor rotor and cap (see Chapter 1).
19 Attach the cable to the negative terminal of the battery.

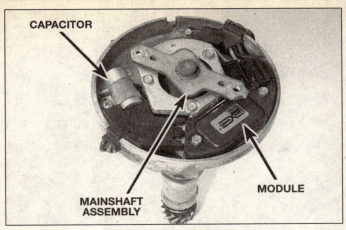

7.16 Silicone lubricant applied to the distributor base in the area under the ignition module dissipates heat and prevents module failure

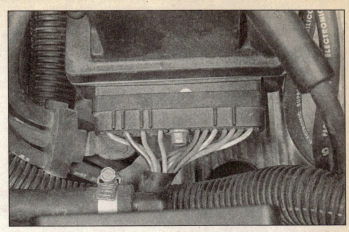

7.23 Unbolt and unplug the connector from the right end of the C3I ignition coil/module assembly

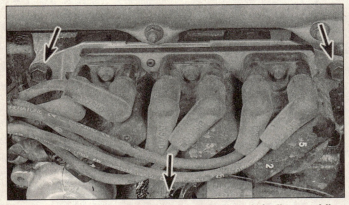

7.25 Mounting bolt locations (arrows) for module/coil assemblies (V6 engine shown)

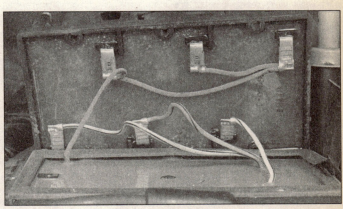

7.28 After removing the screws that hold the coil and module assemblies together, label, then detach, the wires from the module

Distributorless ignition systems (DIS)

Check

20 Because of the complex nature of these systems and their interrelationship with the engine management system, complete troubleshooting is beyond the scope of the home mechanic. Check for trouble codes (see Chapter 6), check the spark plug wires and spark plugs (see Chapter 1), check for a loose electrical connection at the coils, perform a basic ignition system check (see Section 5) and check the coils (see Section 9). If you are not able to identify the problem, take the vehicle to a dealer service department or other qualified shop for complete diagnosis.

Ignition module replacement

Refer to illustrations 7.23, 7.25, 7.28 and 7.29
21 Detach the cable from the negative terminal of the battery. **Caution:** *If the vehicle is equipped with a Delco Loc II audio system, make sure you have the correct activation code before disconnecting the battery. See the information at the front of this manual for the radio re-activation procedure.*
22 Clearly label, then disconnect, all spark plug wires from the DIS or C3I assembly.
23 Unbolt or unplug the electrical connector at the module **(see illustration)**.
24 If equipped, detach the vacuum lines and the electrical connector from the EGR valve solenoid on the left end of the coil/module assembly.
25 Unbolt and remove the DIS or C3I and support bracket assembly **(see illustration)**.
26 Using a Torx screwdriver or bit (most

7.29 To detach the C3I module from the support bracket, remove the three nuts

models), remove the coil-to-module attaching screws.
27 Separate the coil and module assemblies.
28 Open the coil and module halves as shown **(see illustrations)**. Label, then detach, the wires between the module and the coil assemblies from the spade terminals on the underside of the coils.
29 Unbolt the module **(see illustration)**

Chapter 5 Engine electrical systems

7.32 On most models, the electrical connector for the crankshaft sensor is located near the starter motor

7.36 The camshaft sensor is located near the water pump (arrow)

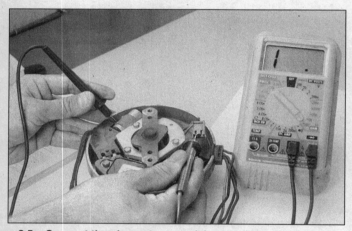

8.5a Connect the ohmmeter to a pick-up coil terminal and the distributor body. If continuity is indicated, there is a short from the pick-up coil wiring to the distributor body

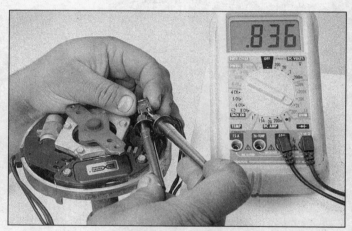

8.5b Connect the ohmmeter to the pick-up coil terminals as shown and measure the resistance of the pick-up coil. It should be between 500 and 1,500 ohms

from the support bracket.
30   Installation is the reverse of removal. Be sure to attach the wires of the new module to the coil assembly spade terminals in exactly the same order in which they were removed.

Crankshaft or combination sensor replacement

3800 engine

Refer to illustration 7.32

31   Remove the crankshaft balancer (vibration damper) (see Chapter 2, Part B).
32   Disconnect the electrical connector for the sensor **(see illustration)**.
33   Unbolt and remove the sensor.
34   Check the interrupter rings on the back of the crankshaft balancer for damage and bends. Replace the balancer as an assembly if any problems exist. A special tool which fits over the end of the crankshaft should be used to check the interrupter rings and adjust the new sensor.
35   Installation is the reverse of removal. Before tightening the mounting bolts, adjust the sensor with the special tool.

Camshaft sensor replacement (3800 engine)

Refer to illustration 7.36

36   The sensor is located just below the water pump pulley, towards the front of the vehicle **(see illustration)**. Disconnect the electrical connector, remove the bolt and detach the sensor. Installation is the reverse of removal.

8   Ignition pickup coil (3.1L engine) - check and replacement

Refer to illustrations 8.5a, 8.5b, 8.9 and 8.10

Check

1   Detach the cable from the negative terminal of the battery. **Caution:** *If the vehicle is equipped with a Delco Loc II audio system, make sure you have the correct activation code before disconnecting the battery. See the information at the front of this manual for the radio re-activation procedure.*
2   Remove the distributor cap and rotor (see Chapter 1).
3   Remove the distributor from the engine (see Section 6). **Note:** *This Step is necessary on most models, since it is difficult to gain access to the pick-up coil wires in the distributor. If access on your vehicle is relatively unrestricted, this Step may not be necessary.*
4   Detach the pickup coil wires from the module.
5   Connect one lead of an ohmmeter to the terminal of the pickup coil lead and the other to ground as shown **(see illustrations)**. Flex the leads by hand to check for intermittent opens. The ohmmeter should indicate infinite resistance at all times. If it doesn't, the pickup coil is defective and must be replaced.
6   Connect the ohmmeter leads to both terminals of the pickup coil wires. Flex the wires by hand to check for intermittent opens. The ohmmeter should read one steady value between 500 and 1500 ohms as the leads are flexed by hand. If it doesn't, the pickup coil is defective and must be replaced.

Chapter 5 Engine electrical systems

8.9 To remove the distributor shaft assembly, knock out the roll pin with a punch and hammer

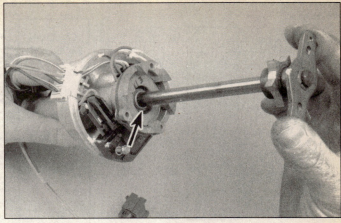

8.10 Remove the shaft from the distributor, then remove the "C" clip (arrow) to remove the pickup coil assembly

Replacement

7 Remove distributor, if not already done, then remove the spring from the distributor shaft.
8 Mark the distributor gear drive and shaft so that they can be reassembled in the same position.
9 Carefully mount the distributor in a soft-jawed vise or on a block of wood and, using a hammer and punch, remove the roll pin from the distributor shaft and gear **(see illustration)**. Pull the shaft out of the distributor body. **Caution:** *If the shaft binds when being pulled out, you may need to lightly sand the lower shaft area so it will pull through without binding.*
10 To remove the pick up coil, remove the thin "C" clip **(see illustration)**. **Note:** *On some models, you may have to unbolt a shield to gain access to the "C" clip.*
11 Lift the pickup coil assembly straight up and remove it from the distributor.
12 Reassembly is the reverse of disassembly.
13 Installation is the reverse of removal.

9 Ignition coil - removal, testing and installation

1 Disconnect the cable from the negative terminal of the battery. **Caution:** *If the vehicle is equipped with a Delco Loc II audio system, make sure you have the correct activation code before disconnecting the battery. See the information at the front of this manual for the radio re-activation procedure.*

Distributor ignition system (3.1L engine)

Removal

2 On models with a separately mounted coil, unplug the coil high tension wire and both electrical leads from the coil.
3 Remove both mounting nuts and remove the coil from the engine.

4 On models with the coil in the distributor cap, remove the coil cover screws and lift off the cover. **Note:** *If you are just checking the coil, it is not necessary to remove the coil from the cap. Refer to the checking procedure below.*
5 Push the coil electrical leads through the top of the hood with a small screwdriver.
6 Remove the coil mounting screws and lift the coil, with the leads, from the cap.

Check

Models with a separately mounted coil
Refer to illustration 9.7

7 On models with a separately mounted coil, check the coil for opens and grounds by performing the following three tests with an ohmmeter **(see illustration).**

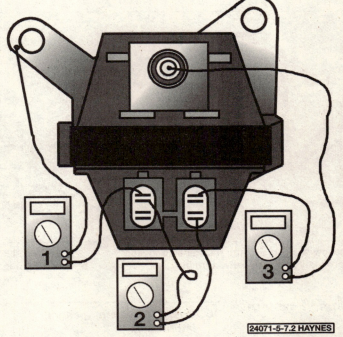

9.7 To check the ignition coil, use an ohmmeter to perform the following three checks

1 On high scale, the ohmmeter should read infinity
2 On low scale, the ohmmeter should read very low or zero
3 On high scale, the ohmmeter should read high but not infinite (if the coil fails any of these tests, replace it)

8 Using the ohmmeter's high scale, hook up the ohmmeter leads as illustrated (see test #1 in illustration 9.13). The ohmmeter should indicate a very high, or infinite, resistance value. If it doesn't, replace the coil.
9 Using the low scale, hook up the leads as illustrated (see test #2 in illustration 9.13). The ohmmeter should indicate a very low, or zero, resistance value. If it doesn't, replace the coil.
10 Using the high scale, hook up the leads as illustrated (see test #3 in illustration 9.13). The ohmmeter should not indicate an infinite resistance. If it does, replace the coil.

Installation

11 Installation of the coil is the reverse of the removal procedure.

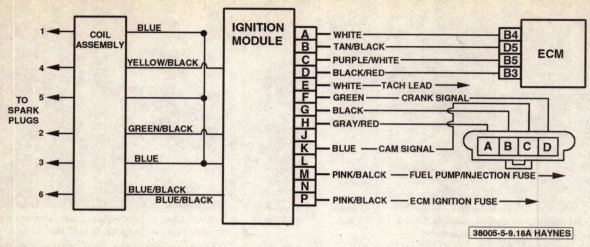

9.13a Although a distributorless ignition coil assembly looks different from a conventional coil, its condition is determined by performing the same three tests described in the previous illustration - this wiring diagram will help you identify the correct terminals

9.13b Test #2: Using the low scale, check the resistance between the positive terminal and the negative terminal

9.13c Test #3: Using the high scale, check the resistance between the negative terminal of the coil and the high tension tower terminal - it should be infinite

Distributorless ignition system (3800 engine)

Removal

12   Refer to Section 7 and remove the DIS or C3I coil/module assembly, then separate the coils from the module/bracket.

Check

Refer to illustrations 9.13a, 9.13b and 9.13c

13   The three-test checking procedure for the DIS coil is the same as the one outlined above for the HEI-type coil. However, the primary terminals are not marked; study the accompanying photos and wiring diagram before attempting to check the coil **(see illustrations)**.

Installation

14   Installation is the reverse of removal. Be careful when inserting the module/coil assembly to avoid damage to the sensor.

10   Charging system - general information

The charging system consists of a belt-driven alternator with an integral voltage regulator and the battery. These components work together to supply electrical power for the ignition system, the lights and all accessories.

There are three models of alternators used. Earlier vehicles use the CS 130 model (100 amp) and later models are equipped with the CS 130 (105 amp w/ rear air conditioning systems) and CS 144 (140 amp, 3800 engines). All types use a conventional pulley and fan.

To determine which type of alternator is installed on your vehicle, look at the fasteners employed to attach the two halves of the alternator housing. All CS 130 models use rivets instead of screws. CS 130 alternators are rebuildable once the rivets are drilled out. However, we don't recommend this practice. CS 144 alternators use conventional through-bolts to secure the two halves. The CS 144 incorporates a slightly different bearing design in the Slip Ring End casting. For all intents and purposes, CS types should be considered non-serviceable and, if defective, should be exchanged as cores for new or rebuilt units.

The purpose of the voltage regulator is to limit the alternator's voltage to a preset value. This prevents power surges, circuit overloads, etc., during peak voltage output. On all models with which this manual is concerned, the voltage regulator is contained within the alternator housing.

The charging system does not ordinarily require periodic maintenance. The drivebelts, electrical wiring and connections should, however, be inspected at the intervals suggested in Chapter 1.

13.5 After detaching the electrical connectors from the back of the alternator, remove the front mounting bolt (arrow) . . .

13.6 . . . then remove the rear mounting bolt and brace bolt (arrows)

11 Charging system - precautions

Take extreme care when making circuit connections to a vehicle equipped with an alternator and note the following. When making connections to the alternator from a battery, always match correct polarity. Before using arc welding equipment to repair any part of the vehicle, disconnect the wires from the alternator and the battery terminal. Never start the engine with a battery charger connected. Always disconnect both battery leads before using a battery charger.

The charging indicator lamp on the dash lights when the ignition switch is turned on and goes out when the engine starts. If the lamp stays on or comes on once the engine is running, a charging system problem has occurred. See Section 12 for the proper diagnosis procedure for each type of alternator.

12 Charging system - check

1   If a malfunction occurs in the charging circuit, do not immediately assume that the alternator is causing the problem. First check the following items:
   a) *The battery cables where they connect to the battery. Make sure the connections are clean and tight.*
   b) *The battery electrolyte specific gravity. If it is low, charge the battery.*
   c) *Check the external alternator wiring and connections. They must be in good condition.*
   d) *Check the drivebelt condition and tension (see Chapter 1).*
   e) *Make sure the alternator mounting bolts are tight.*
   f) *Run the engine and check the alternator for abnormal noise (may be caused by a loose drive pulley, loose mounting bolts, worn or dirty bearings, defective diode or defective stator).*
2   Using a voltmeter, check the battery voltage with the engine off. It should be approximately 12 volts.
3   Start the engine and check the battery voltage again. It should now be approximately 14 to 15 volts.

13 Alternator - removal and installation

Refer to illustrations 13.5 and 13.6

Note: *There are three different model CS alternators installed on the vehicles covered by this manual. These units are NOT serviceable. In the event the alternator becomes defective, exchange the alternator for a new or rebuilt part at your local auto parts store or dealership parts department.*

1   Detach the cable from the negative terminal of the battery.
2   Remove the serpentine drivebelt (see Chapter 1, Section 21).
3   On models with the 3.1L engine, remove the air cleaner housing (see Chapter 4, Section 8).
4   Detach the electrical connector from the back of the alternator. Also remove the nut and detach the battery positive cable from the back of the alternator.
5   Remove the alternator front mounting bolt **(see illustration)**.
6   Remove the rear mounting bolt and the alternator brace, if equipped **(see illustration)**.
7   Detach any wiring harness routing clips and remove the alternator from the vehicle.
8   Installation is the reverse of the removal procedure.

14 Starting system - general information

The function of the starting system is to crank the engine. The starting system is composed of a starting motor, solenoid and battery. The battery supplies the electrical energy to the solenoid, which then completes the circuit to the starting motor, which does the actual work of cranking the engine.

The solenoid and starting motor are mounted together at the lower front side of the engine. No periodic lubrication or maintenance is required.

The electrical circuitry of the vehicle is arranged so that the starter motor can only be operated when the clutch pedal is depressed (manual transmission) or the transmission selector lever is in Park or Neutral (automatic transmission).

Never operate the starter motor for more than 30 seconds at a time without pausing to allow it to cool for at least two minutes. Excessive cranking can cause overheating, which can seriously damage the starter.

15 Starter motor - testing in vehicle

Refer to illustration 15.6

1   If the starter motor does not turn at all when the switch is operated, make sure that the shift lever is in Neutral or Park (automatic transmission) or that the clutch pedal is depressed (manual transmission).
2   Make sure that the battery is charged and that all cables, both at the battery and starter solenoid terminals, are secure.
3   If the starter motor spins but the engine is not cranking, the overrunning clutch in the starter motor is slipping and the motor must be removed from the engine for replacement.
4   If, when the switch is actuated, the starter motor does not operate at all but the solenoid clicks, then the problem lies with either the battery, the main solenoid contacts or the starter motor itself. **Note:** *Before diagnosing starter problems, make sure that the battery is fully charged.*
5   If the solenoid plunger cannot be heard when the switch is actuated, the solenoid itself is defective or the solenoid circuit is open.
6   To check the solenoid, connect a jumper lead between the battery (+) and the

15.6 Typical starter "S" terminal location (arrow)

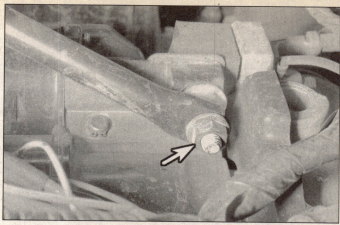

16.2 Remove the nut (arrow) and then the starter motor brace from the engine block

16.5 One of the starter mounting bolts is accessed behind and under the frame

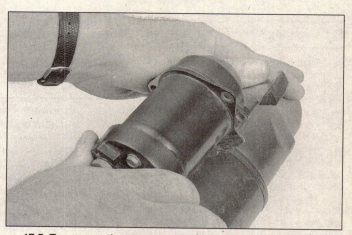

17.5 To remove the starter solenoid from the starter motor, remove the solenoid-to-starter in a counterclockwise direction and pull

"S" terminal on the solenoid **(see illustration)**. If the starter motor now operates, the solenoid is OK and the problem is in the ignition switch, neutral start switch or in the wiring.

7   If the starter motor still does not operate, remove the starter/solenoid assembly for disassembly, testing and repair.

8   If the starter motor cranks the engine at an abnormally slow speed, first make sure that the battery is charged and that all terminal connections are tight. If the engine is partially seized, or has the wrong viscosity oil in it, it will crank slowly.

9   Run the engine until normal operating temperature is reached, then disconnect the coil wire from the distributor cap and ground it on the engine.

10   Connect a voltmeter positive lead to the starter motor terminal of the solenoid and then connect the negative lead to ground.

11   Crank the engine and take the voltmeter readings as soon as a steady figure is indicated. Do not allow the starter motor to turn for more than 30 seconds at a time. A reading of 9 volts or more, with the starter motor turning at normal cranking speed, is normal. If the reading is 9 volts or more but the cranking speed is slow, the motor is faulty. If the reading is less than 9 volts and the cranking speed is slow, the solenoid contacts are probably burned.

16   Starter motor - removal and installation

Refer to illustration 16.2 and 16.5

1   Detach the cable from the negative terminal of the battery.
2   Remove the starter brace **(see illustration)**.
3   Raise the vehicle and place it securely on jackstands.
4   Detach the wires from the starter solenoid.
5   Remove the two bolts, nuts, washers and shims holding the starter to the engine **(see illustration)**.
6   Remove the starter from the engine.
7   Installation is the reverse of removal.

17   Starter solenoid - removal and installation

Refer to illustration 17.5

1   Disconnect the cable from the negative terminal of the battery.
2   Remove the starter motor (Section 16).

Removal

3   Disconnect the strap from the solenoid to the starter motor terminal.
4   Remove the two screws which secure the solenoid to the starter motor.
5   Twist the solenoid in a counterclockwise direction to disengage the flange from the starter body **(see illustration)**.

Installation

6   To install, first make sure the return spring is in position on the plunger, then insert the solenoid body into the starter housing and turn the solenoid clockwise to engage the flange.
7   Install the two solenoid screws and connect the motor strap.

Chapter 6
Emissions control systems

Contents

	Section		Section
Catalytic converter	12	Exhaust Gas Recirculation (EGR) system	
Computer Command Control (CCC) system and		(all 3.1L engines and 1993 and later 3800 engines)	7
trouble codes	2	General information	1
Electronic Control Module (ECM) PROM (3.1L engine)/Powertrain		Information sensors	4
Control Module (PCM) MEM-CAL (3800 engine)	3	Oxygen sensor	See Section 4
Electronic Spark Control (ECS) system (1990 through 1992)		Positive Crankcase Ventilation (PCV) system	9
or Knock Sensor (KS) system (1993 on)	6	Thermostatic air cleaner (THERMAC) (3.1L engine)	10
Electronic Spark Timing (EST)	5	Transmission Converter Clutch (TCC)	11
Evaporative Emission Control System (EECS)	8		

Specifications

Torque specifications
Camshaft sensor .... 35 to 53 in-lbs

1 General information

Refer to illustration 1.6

To prevent pollution of the atmosphere from incompletely burned and evaporating gases, and to maintain good driveability and fuel economy, a number of emission control devices are incorporated. They include the:

Electronic Spark Timing (EST) system
Electronic Spark Control (ESC) system
Exhaust Gas Recirculation (EGR) system
Evaporative Emission Control System (EECS)
Positive Crankcase Ventilation (PCV) system
Transmission Converter Clutch (TCC)
Catalytic converter
Thermostatic air cleaner
Air conditioning control

All of these systems are linked, directly or indirectly, to the Computer Command Control (CCC or C3) system.

The Sections in this Chapter include general descriptions, checking procedures within the scope of the home mechanic and component replacement procedures (when possible) for each of the systems listed above.

Before assuming that an emissions control system is malfunctioning, check the fuel and ignition systems carefully. The diagnosis of some emission control devices requires specialized tools, equipment and training. If checking and servicing become too difficult or if a procedure is beyond the scope of your skills, consult your dealer service department.

This doesn't mean, however, that emission control systems are particularly difficult to maintain and repair. You can quickly and easily perform many checks and do most (if not all) of the regular maintenance at home with common tune-up and hand tools. **Note:** *The most frequent cause of emissions problems is simply a loose or broken vacuum hose or wiring connection, so always check the hose and wiring connections first.*

Pay close attention to any special precautions outlined in this Chapter. It should be noted that the illustrations of the various systems may not exactly match the system installed on your vehicle because of changes made by the manufacturer during production or from year-to-year.

A Vehicle Emissions Control Information (VECI) label is located in the engine compartment **(see illustration)**. This label contains important emissions specifications and ignition timing procedures, as well as a vacuum hose schematic and emissions components

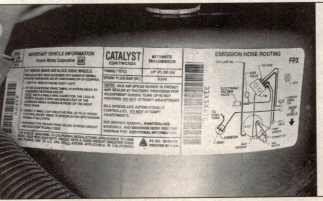

1.6 The Vehicle Emission Control Information (VECI) label is located on the strut tower in the engine compartment and contains information on idle sped adjustment, ignition timing, location of emission devices on your vehicle, vacuum line routing, etc.

Chapter 6 Emissions control systems

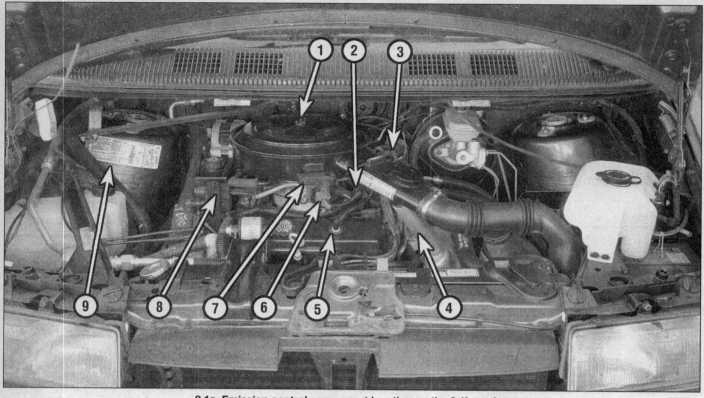

2.1a Emission control component location on the 3.1L engine

1. Air cleaner housing
2. Throttle Position Sensor (TPS)
3. Distributor
4. Warm air duct for the THERMAC system
5. PCV valve
6. Idle Air Control (IAC) valve
7. MAP sensor
8. Electronic Vacuum Regulator Valve (EVRV) EGR
9. VECI label

identification guide. When servicing the engine or emissions systems, the VECI label in your particular vehicle should always be checked for up-to-date information.

Note: *On models equipped with the Delco Loc II audio system, be sure the lockout feature is turned off before performing any procedure which requires disconnecting the battery.*

2 Computer Command Control (CCC) system and trouble codes

Note: *To access the code information on 1995 models, it is necessary to use a scan tool to access the OBD II codes.*

Refer to illustrations 2.1a, 2.1b, 2.7a and 2.7b

1 The Computer Command Control (CCC) system consists of an Electronic Control Module (ECM on 3.1L engines) or Powertrain Control Module (PCM on 3800 engines) and information sensors which monitor various functions of the engine and send data back to the ECM/PCM **(see illustrations)**.

2 There are two types of computers used. Vehicles equipped with a 3.1L engine have an ECM with a PROM and a CALPAK. Vehicles equipped with a 3800 engine have a PCM with a MEM-CAL (Memory and Calibration) unit. **Note:** *The CALPAK on the 3.1L model is not replaceable. The ECM must be replaced as a unit in the event of CALPAK failure.*

3 The ECM/PCM controls the following systems:

Fuel control
Electronic spark timing
Electronic spark control
Exhaust gas recirculation
Canister purge
Engine cooling fan
Idle Air Control (IAC)
Transmission converter clutch
Air conditioning clutch control

4 The CCC system is analogous to the central nervous system in the human body. The sensors (nerve endings) constantly relay information to the ECM/PCM (brain), which processes the data and, if necessary, sends out a command to change the operating parameters of the engine (body).

5 Here's a specific example of how one portion of this system operates: An oxygen sensor, located in the exhaust manifold, constantly monitors the oxygen content of the exhaust gas. If the percentage of oxygen in the exhaust gas is incorrect, an electrical signal is sent to the ECM/PCM. The ECM/PCM takes this information, processes it and then sends a command to the fuel injection system, telling it to change the air/fuel mixture. This happens in a fraction of a second and it goes on continuously when the engine is running. The end result is an air/fuel mixture ratio which is constantly maintained at a predetermined ratio, regardless of driving conditions.

6 One might think that a system which uses an on-board computer and electrical sensors would be difficult to diagnose. This is not necessarily the case. The CCC system has a built-in diagnostic feature which indicates a problem by flashing a SERVICE ENGINE SOON light on the instrument panel. When this light comes on during normal vehicle operation, a fault in one of the information sensor circuits or the ECM/PCM itself has been detected. More importantly, a trouble code is stored in the ECM/PCM's memory.

7 To retrieve this information from the ECM/PCM memory, you must use a short jumper wire to ground a diagnostic terminal. This terminal is part of an electrical connector known as the Assembly Line Data Link (ALDL) **(see illustrations)**. The ALDL is located underneath the dashboard, just below the instrument panel and to the left of the center console. To use the ALDL, remove the plastic cover (if equipped) by sliding it toward you. With the electrical connector exposed to view, push one end of the jumper

Chapter 6 Emissions control systems

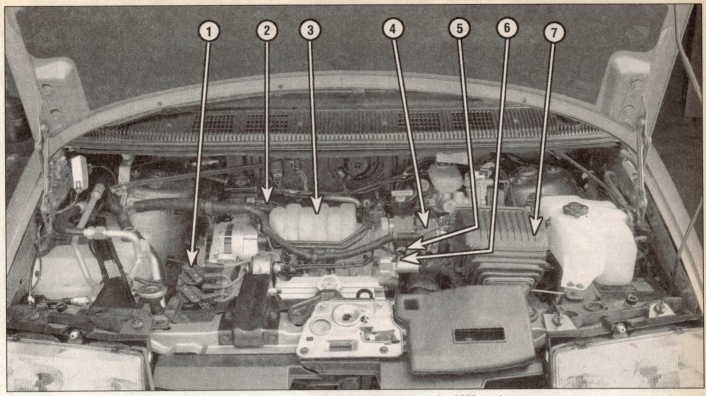

2.1b Emission control component location on the 3800 engine

1. Direct Ignition System (DIS)
2. PCV valve and housing
3. Air intake plenum
4. Mass Air Flow (MAF) sensor
5. Idle Air Control (IAC) valve
6. Throttle Position Sensor (TPS)
7. Air cleaner housing

2.7a The Assembly Line Data Link (ALDL) (arrow) is located under the dash near the fuse panel

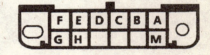

2.7b The Assembly Line Data Link (ALDL) terminal identification

- A Ground
- B Diagnostic terminal
- C AIR (if used)
- D Service Engine Soon Light
- E Serial data (special tool required – do not use)
- F TCC (if used)
- G Fuel pump (if used)
- M Serial data (if used)

wire into the diagnostic terminal and the other end into the ground terminal.

8 When the diagnostic terminal is grounded with the ignition on and the engine stopped, the system will enter the Diagnostic Mode. **Caution:** *Don't start or crank the engine with the diagnostic terminal grounded.* In this mode the ECM/PCM will display a "Code 12" by flashing the SERVICE ENGINE SOON light, indicating that the system is operating. A code 12 is simply one flash, followed by a brief pause, then two flashes in quick succession. This code will be flashed three times. If no other codes are stored, Code 12 will continue to flash until the diagnostic terminal ground is removed.

9 After flashing Code 12 three times, the ECM/PCM will display any stored trouble codes. Each code will be flashed three times, then Code 12 will be flashed again, indicating that the display of any stored trouble codes has been completed.

10 When the ECM/PCM sets a trouble code, the SERVICE ENGINE SOON light will come on and a trouble code will be stored in memory. If the problem is intermittent, the light will go out after 10 seconds, when the fault goes away. However, the trouble code will stay in the ECM/PCM memory until the battery voltage to the ECM/PCM is inter-

rupted. Removing battery voltage for 10 seconds will clear all stored trouble codes. Trouble codes should always be cleared after repairs have been completed. **Caution:** *To prevent damage to the ECM/PCM, the ignition switch must be OFF when disconnecting or connecting power to the ECM/PCM.*

11 Following is a list of the typical trouble codes which may be encountered while diagnosing the Computer Command Control System. Also included are simplified troubleshooting procedures. If the problem persists after these checks have been made, more detailed service procedures will have to be done by a dealer service department.

Chapter 6 Emissions control systems

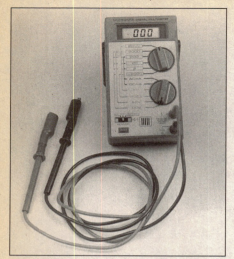

2.12 Digital multimeters can be used for testing all types of circuits; because of their high impedance, they are much more accurate than analog meters for measuring millivolts in low-voltage computer circuits

2.13 Scanners like the Actron Scantool and the AutoXray XP240 are powerful diagnostic aids - programmed with comprehensive diagnostic information, they can tell you just about anything you want to know about your engine management system

Diagnostic tool information

Refer to illustrations 2.12, 2.13 and 2.15

12 A digital multimeter is a necessary tool for checking fuel injection and emission related components **(see illustration)**. A digital volt-ohmmeter is preferred over the older style analog multimeter for several reasons. The analog multimeter cannot display the volts-ohms or amps measurement in hundredths and thousandths increments. When working with electronic circuits which are often very low voltage, this accurate reading is most important. Another good reason for the digital multimeter is the high impedance circuit. The digital multimeter is equipped with a high resistance internal circuitry (10 million ohms). Because a voltmeter is hooked up in parallel with the circuit when testing, it is vital that none of the voltage being measured should be allowed to travel the parallel path set up by the meter itself. This dilemma does not show itself when measuring larger amounts of voltage (9 to 12 volt circuits) but if you are measuring a low voltage circuit such as the oxygen sensor signal voltage, a fraction of a volt may be a significant amount when diagnosing a problem.

13 Hand-held scanners are the most powerful and versatile tools for analyzing engine management systems used on later model vehicles **(see illustration)**. Early model scanners handle codes and some diagnostics for many OBD I systems. Each brand scan tool must be examined carefully to match the year, make and model of the vehicle you are working on. Often interchangeable cartridges are available to access the particular manufacturer; Ford, GM, Chrysler, etc.). Some manufacturers will specify by continent; Asia, Europe, USA, etc. Seek the advice of your local auto parts retailer.

14 With the arrival of the Federally mandated emission control system (OBD II), a specially designed scanner must also be developed. At this time, several manufacturers plan to release OBD II scan tools for the home mechanic. Ask the parts salesman at a local auto parts store for additional information concerning dates and costs. **Note:** *Although OBD II codes cannot be accessed on 1996 models without a Scan tool, follow the simple component checks in Section 4.*

15 Another type of code reader and less expensive is available at parts stores **(see illustration)**. These tools simplify the procedure for extracting codes from the engine management computer by simply "plugging in" to the diagnostic connector on the vehicle wiring harness. **Note:** *Some diagnostic connectors are located under the dash, kick panel or glovebox while others are located in the engine compartment.*

2.15 Trouble code tools simplify the task of extracting the trouble codes

OBD I trouble codes

Trouble code	Circuit or system	Probable cause
Code 12 (1 flash, pause, 2 flashes)	No distributor reference pulses to ECM/PCM	This code will flash whenever the diagnostic terminal is grounded with the ignition turned ON and the engine not running. If additional trouble codes are stored in the ECM/PCM they will appear after this code has flashed three times. If this code appears while the engine is running, no reference pulses from the distributor are reaching the ECM/PCM.
Code 13 (1 flash, pause, 3 flashes)	Oxygen sensor circuit	Check for a sticking or misadjusted throttle position sensor. Check the wiring and connectors from the oxygen sensor. Replace the oxygen sensor.
Code 14 (1 flash, pause, 4 flashes)	Coolant sensor/high temp	If the engine is experiencing cooling system problems the problem must be rectified before continuing. Check all wiring and connectors associated with the coolant temperature sensor. Replace the coolant temperature sensor.*
Code 15 (1 flash, pause, 5 flashes)	Coolant sensor/low temp	See above, then check the wiring connections at the ECM.

Chapter 6 Emissions control systems

Trouble code	Circuit or system	Probable cause
Code 16 (3800 engine only) (1 flash, pause, 6 flashes)	System voltage/high or low	System voltage above 17-volts or below 9-volts is detected.
Code 18 (3800 engine only) (1 flash, pause, 8 flashes)	Cam/Crank error	Intermittent problem with the cam/crank signal or fuel control signal.
Code 21 (2 flashes, pause, 1 flash)	Throttle position sensor/voltage high	Check for a sticking or misadjusted TPS plunger. Check all wiring and connections between the TPS and the ECM/PCM. Adjust or replace the TPS (see Chapter 4).*
Code 22 (2 flashes, pause, 2 flashes)	Throttle position sensor/voltage low	Check the TPS adjustment (Chapter 4). Check the ECM/PCM connector. Replace the TPS (Chapter 4).*
Code 24 (2 flashes, pause, 4 flashes)	Vehicle speed sensor	A fault in this circuit should be indicated only when the vehicle is in motion. Disregard Code 24 if it is set when the drive wheels are not turning. Check the connections at the ECM. Check the TPS setting.
Code 26 (3800 engine only) (2 flashes, pause, 6 flashes)	QDM "A" failure	The PCM detects an improper voltage level on the circuit that is connected to the QDM (Quad Driver Module) "A".
Code 32 (3.1L engine only) (3 flashes, pause, 2 flashes)	EGR	Vacuum switch shorted to ground on start-up, switch not closed after the ECM has commanded the EGR for a specified period of time or the EGR solenoid circuit is open for specified period of time. Replace the EGR valve.*
Code 33 (3.1L engine only) (3 flashes, pause, 3 flashes)	MAP sensor	Check the vacuum hoses from the MAP sensor. Check the electrical connections at the ECM. Replace the MAP sensor.*
Code 34 (3.1L engine only) (3 flashes, pause, 4 flashes)	Vacuum sensor or MAP sensor	Code 34 will set when the signal voltage from the MAP sensor is too low. Instead the ECM will substitute a fixed MAP value and use the TPS to control fuel delivery. Replace the MAP sensor.*
Code 34 (3800 engine only) (3 flashes, pause, 4 flashes)	MAF sensor	Code 34 will set when the signal voltage from the MAF sensor is not detected for a period over 4 seconds.
Code 41 (3800 engine only) (4 flashes, pause, 1 flashes)	Cam sensor circuit	Code 41 will set when the signal voltage from the cam sensor is not detected for a period over 5 seconds.
Code 42 (4 flashes, pause, 2 flashes)	Electronic Spark Timing	Electronic Spark Timing bypass circuit or EST circuit is grounded or open. (EST) A malfunctioning HEI module can cause this code.
Code 43 (4 flashes, pause, 3 flashes)	Electronic spark control	The ESC retard signal has been on for too long or the system has failed a unit (ESC) functional check.
Code 44 (4 flashes, pause, 4 flashes)	O2 sensor indicates lean exhaust	Check the ECM wiring connections, particularly terminals 15 and 8. Check for vacuum leakage at the TBI base gasket, vacuum hoses or the intake manifold gasket. Replace the oxygen sensor.*
Code 45 (4 flashes, pause, 5 flashes)	O2 sensor indicates rich exhaust	Possible rich or leaking injector, high fuel pressure or faulty TPS or MAF sensor. Also, check the evaporative charcoal canister and its components for the presence of fuel. Replace the oxygen sensor.*
Code 51 (5 flashes, pause, 1 flash)	PROM (3.1L engine) or MEMCAL (3800 engine)	Make sure that the PROM or MEM-CAL is properly installed in the ECM. Replace the PROM or MEM-CAL.*
Code 52 (3.1L engine only) (5 flashes, pause, 2 flashes)	CALPAK	Check the ECM to insure proper installation. Replace the ECM.*
Code 54 (3.1L engine only) (5 flashes, pause, 4 flashes)	Fuel pump	Low fuel pump voltage. Sets when the fuel pump voltage is less than 2-volts when reference pulses are being received.
Code 55 (5 flashes, pause, 5 flashes)	ECM/PCM	Be sure that the ECM ground connections are tight. If they are, replace the ECM/PCM.*
Code 56 (3800 engine only) (5 flashes, pause, 6 flashes)	QDM "B" failure	The PCM detects an improper voltage level on the circuit that is connected to the QDM (Quad Driver Module) "B".

* Component replacement may not cure the problem in all cases. For this reason, you may want to seek professional advice before purchasing replacement parts.

Note: on 3800 engines, there are a few codes that cannot be displayed on the SERVICE ENGINE SOON light on the instrument panel. A special diagnostic tool is necessary in order to read these special codes.

Chapter 6 Emissions control systems

5-digit trouble codes (OBD II models)

Code	Code definition
P0101	Mass Air Flow (MAF) sensor error
P0102	Mass Air Flow (MAF) sensor circuit - low frequency detected
P0103	Mass Air Flow (MAF) sensor - high frequency detected
P0106	Manifold Absolute Pressure (MAP) sensor performance fault
P0107	Manifold Absolute Pressure (MAP) sensor circuit low input
P0108	Manifold Absolute Pressure (MAP) sensor circuit high input
P0112	Intake Air Temperature (IAT) sensor circuit low input
P0113	Intake Air Temperature (IAT) sensor circuit high input
P0117	Engine Coolant Temperature (ECT) sensor circuit low input
P0118	Engine Coolant Temperature (ECT) sensor circuit high input
P0121	Throttle Position Sensor (TPS) range/performance fault
P0122	Throttle Position Sensor (TPS) circuit low input
P0123	Throttle Position Sensor (TPS) circuit high input
P0125	Engine Coolant Temperature (ECT) takes too long to enter closed loop
P0131	Upstream heated O2 sensor circuit low voltage (Bank 1, Sensor 1)
P0132	Upstream heated O2 sensor circuit high voltage (Bank 1, Sensor 1)
P0133	Upstream heated O2 sensor slow response (lazy sensor) (Bank 1, Sensor 1)
P0134	Upstream heated O2 sensor insufficient activity (Bank 1, Sensor 1)
P0135	Upstream heated O2 sensor - heater circuit fault (Bank 1, Sensor 1)
P0137	Downstream heated O2 sensor circuit low voltage (Bank 1, Sensor 2)
P0138	Downstream heated O2 sensor circuit high voltage (Bank 1, Sensor 2)
P0140	Downstream heated O2 sensor insufficient activity (Bank 1, Sensor 2)
P0141	O2 sensor heater circuit fault (Bank 1, Sensor 2)
P0171	System Adaptive fuel too lean
P0172	System Adaptive fuel too rich
P0191	Injector Pressure sensor system performance
P0192	Injector Pressure sensor circuit low input
P0193	Injector Pressure sensor circuit high input
P0300	Cylinder misfire detected (random)
P0301	Cylinder number 1 misfire detected
P0302	Cylinder number 2 misfire detected
P0303	Cylinder number 3 misfire detected
P0304	Cylinder number 4 misfire detected
P0305	Cylinder number 5 misfire detected
P0306	Cylinder number 6 misfire detected
P0321	Crankshaft Position sensor circuit fault

Code	Code definition
P0325	Knock sensor circuit fault
P0326	Knock sensor circuit performance
P0327	Knock sensor noise channel - low voltage
P0335	Crankshaft Position sensor circuit
P0336	Crankshaft reference signal circuit
P0341	Camshaft Position sensor circuit
P0342	Camshaft Position sensor circuit
P0401	Exhaust Gas Recirculation (EGR) system flow insufficient
P0420	Three Way Catalyst (TWC) system - low efficiency
P0441	EVAP system - no flow during purge cycle
P0351	COP ignition coil 1 primary circuit fault
P0352	COP ignition coil 2 primary circuit fault
P0353	COP ignition coil 3 primary circuit fault
P0354	COP ignition coil 4 primary circuit fault
P0400	EGR flow fault
P0401	EGR insufficient flow detected
P0402	EGR excessive flow detected
P0420	Catalyst system efficiency below threshold (Bank 1)
P0421	Catalyst system efficiency below threshold (Bank 1)
P0430	Catalyst system efficiency below threshold (Bank 2)
P0431	Catalyst system efficiency below threshold (Bank 2)
P0441	EVAP incorrect purge flow
P0443	EVAP VMV circuit fault
P0452	EVAP fuel tank pressure sensor low input
P0453	EVAP fuel tank pressure sensor high input
P0502	VSS circuit low input
P0503	VSS circuit range performance
P0506	IAC system rpm lower than expected
P0507	IAC system rpm higher than expected
P0530	A/C refrigerant pressure sensor circuit
P0560	System voltage out of range
P0600	PCM serial data communication link fault
P0601	PCM memory problem
P0602	PCM control module programming error
P0650	Quad driver module circuit
P0703	TCC brake switch input circuit fault
P0705	Transaxle Range sensor circuit malfunction

5-digit trouble codes (OBD II models) (continued)

Code	Code definition
P0706	Transaxle Range sensor performance
P0712	Transaxle fluid temperature sensor circuit low input
P0713	Transaxle fluid temperature sensor circuit high input
P0740	TCC circuit fault
P0755	Shift solenoid circuit fault

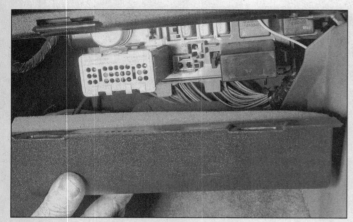

3.4 Remove the lower trim panel under the glovebox for access to the relay panel and the ECM/PCM

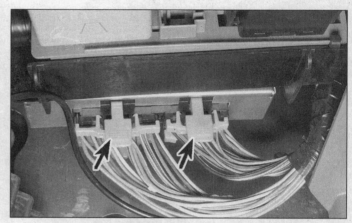

3.5 Press firmly on the tabs (arrows) to release the electrical connectors from the ECM/PCM

3 Electronic Control Module (ECM) PROM (3.1L engine)/Powertrain Control Module (PCM) MEM/CAL (3800 engine)

Refer to illustrations 3.4, 3.5, 3.6, 3.8a, 3.8b, 3.13, 3.14 and 3.17

ECM/PCM removal

1 The Electronic Control Module (ECM)/Powertrain Control Module (PCM) is located inside the passenger compartment under the dashboard behind the relay panel (right side).
2 Disconnect the negative battery cable from the battery.
3 Remove the screws from the trim panel under the right end of the dash.
4 Remove the trim panel to expose the relay panel (see illustration).
5 Unplug both electrical connectors (see illustration) from the ECM/PCM. Caution: *The ignition switch must be turned OFF when pulling out or plugging in the electrical connectors to prevent damage to the ECM/PCM.*
6 Remove the retaining nut (see illustration) from the left side of the ECM/PCM.
7 Carefully slide the ECM/PCM down far enough to clear the fuse panel. Note: *Avoid any static electricity damage to the computer by using gloves and a special anti-static pad to store the ECM/PCM once it is removed.*

3.6 Remove the nut (arrow) that holds the computer to the carrier

General description

PROM (3.1L engine)

8 To allow one model of ECM to be used for many different vehicles (see illustration), a device called a PROM (Programmable Read Only Memory) is used. To access the PROM remove the cover. The PROM (see illustration) is located inside the ECM and contains information on the vehicle's weight, engine, transmission, axle ratio, etc. One ECM part number can be used by many GM vehicles but the PROM is very specific and must be used only in the vehicle for which it was designed. For this reason, it is essential to check the latest parts book and Service Bulletin information for the correct part number when replacing a PROM. An ECM purchased at the dealer is purchased without a PROM. The PROM from the old ECM must be carefully removed and installed in the new ECM.

3.8a If you are replacing the ECM, compare the service numbers on the label on the old unit to the numbers on the new one

3.8b Location of the PROM (arrow) inside an ECM for a 3.1L engine (PCM's for the 3800 engine are similar but have a MEM-CAL instead of a PROM and CALPAK)

3.13 Using a PROM removal tool, grasp the PROM carrier at the narrow ends and gently rock the removal tool until the PROM is unplugged from the socket

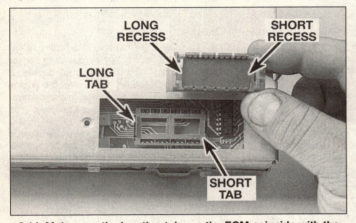

3.14 Make sure the locating tabs on the ECM coincide with the recess grooves in the PROM

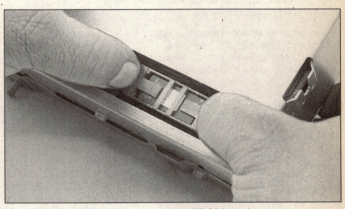

3.17 Press only on the ends of the PROM carrier – pressure on the area between could result in bent or broken pins or damage to the PROM

CALPAK (3.1L engine)

9   A device known as a CALPAK is used to allow fuel delivery if other parts of the ECM are damaged. The CALPAK is not replaceable. Install a new ECM in the event of CALPAK failure.

MEM-CAL (3800 engine)

10   The MEM-CAL contains the functions of the PROM, CALPAK and ESC module used on other GM applications. Like the PROM, it contains the calibrations needed for a specific vehicle as well as the back-up fuel control circuitry required if the rest of the PCM becomes damaged or faulty.

Component replacement

PROM (3.1L engine)

11   Turn the ECM so that the bottom cover is facing up and place it on a clean work surface.
12   Remove the PROM/CALPAK access cover.
13   Using a PROM removal tool (available at your dealer), grasp the PROM carrier at the narrow ends **(see illustration)**. Gently rock the carrier from end to end while applying firm force up. The PROM carrier and PROM should lift off the PROM socket easily. **Caution:** *The PROM carrier should only be removed with the special PROM removal tool. Removal without this tool or with any other type of tool may damage the PROM or the PROM socket.*
14   Note the reference end of the PROM carrier **(see illustration)** before setting it aside.
15   If you are replacing the ECM, remove the new ECM from its container and check the service number to make sure that it is the same as the number on the old ECM **(see illustration 3.8a)**.
16   If you are replacing the PROM, remove the new PROM from its container and check the service number to make sure that it is the same as the number of the old PROM.
17   Position the PROM/PROM carrier assembly squarely over the PROM socket with the small notched end of the carrier aligned with the small notch in the socket at the pin 1 end. Press on the PROM carrier until it seats firmly in the socket **(see illustration)**.
18   If the PROM is new, make sure that the notch in the PROM is matched to the small notch in the carrier. **Caution:** *If the PROM is installed backwards and the ignition switch is turned on, the PROM will be destroyed.*
19   Using the tool, install the new PROM carrier in the PROM socket of the ECM. The small notch of the carrier should be aligned with the small notch in the socket. Press on the PROM carrier until it is firmly seated in the socket. **Caution:** *Do not press on the PROM - press only on the carrier.*
20   Attach the access cover to the ECM and tighten the two screws.
21   Install the ECM in the support bracket, plug in the electrical connectors to the ECM and install the under-dash panel.
22   Start the engine.
23   Enter the diagnostic mode by grounding the diagnostic lead of the ALDL (see Section 2). If no trouble codes occur, the PROM is correctly installed.
24   If Trouble Code 51 occurs, or if the SERVICE ENGINE SOON light comes on and remains constantly lit, the PROM is not fully seated, is installed backwards, has bent pins or is defective.
25   If the PROM is not fully seated, pressing firmly on both ends of the carrier should correct the problem.
26   If the pins have been bent, remove the PROM, straighten the pins and reinstall the PROM. If the bent pins break or crack when

4.3a The coolant temperature sensor on the 3.1L engine is located in the intake manifold near the TBI unit

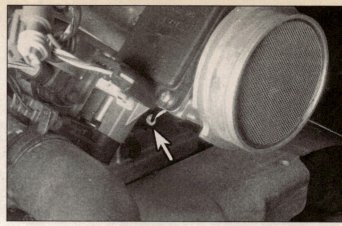
4.3b The coolant temperature sensor on the 3800 engine is located in the intake manifold directly under the throttle body

you attempt to straighten them, discard the PROM and replace it with a new one.

27 If careful inspection indicates that the PROM is fully seated, has not been installed backwards and has no bent pins, but the SERVICE ENGINE SOON light remains lit, the PROM is probably faulty and must be replaced.

MEM-CAL (3800 engine)

Note: *With respect to removal of the PCM from the vehicle, MEM-CAL replacement is similar to the procedure described for the PROM. However, the actual removal and installation steps for the MEM-CAL and the functional check to ensure that the MEM-CAL is installed properly differ in detail from the Steps for the PROM and CALPAK.*

28 Remove the MEM-CAL access cover.

29 Using two fingers, push both retaining clips back away from the MEM-CAL. At the same time, grasp the MEM-CAL at both ends and lift it up out of its socket. Do not remove the MEM-CAL cover itself. **Caution:** *Use of unapproved removal or installation methods may damage the MEM-CAL or socket.*

30 Verify that the numbers on the old PCM and new PCM match up (or that the numbers of the old and new MEM-CALs match up, depending on what component(s) you're replacing) as described in the procedure for removing and installing the PROM **(see illustration 3.8a).**

31 To install the MEM-CAL in the MEM-CAL socket, press only on the ends of the MEM-CAL.

32 The small notches in the MEM-CAL must be aligned with the small notches in the MEM-CAL socket. Press on the ends of the MEM-CAL until the retaining clips snap into the ends of the MEM-CAL. Do not press on the middle of the MEM-CAL - press only on the ends.

33 The remainder of the installation is similar to that for the PROM.

34 Once the new MEM-CAL is installed in the old PCM (or the old MEM-CAL is installed in the new PCM), check your installation to verify that it has been installed properly by doing the following test:

a) Turn the ignition switch ON.
b) Enter the diagnostics mode at the ALDL (see Section 2).
c) Allow Code 12 to flash four times to verify that no other codes are present. This indicates that the MEM-CAL is installed properly and the PCM is functioning properly.

35 If trouble codes 41, 42, 43 or 51 occur, or if the SERVICE ENGINE SOON light is on constantly but is flashing no codes, the MEM-CAL is either not fully seated or is defective. If it's not fully seated, press firmly on the ends of the MEM-CAL. If it is necessary to remove the MEM-CAL, follow the above Steps again.

4 Information sensors

Note: *See component location illustrations in Section 2 for additional information on the location of the following information sensors.*

Engine coolant temperature sensor

Refer to illustrations 4.3a, 4.3b and 4.4

General description

1 The coolant sensor is a thermistor (a resistor which varies the value of its voltage output in accordance with temperature changes). The change in the resistance values will directly affect the voltage signal from the coolant sensor. As the sensor temperature DECREASES, the resistance values will INCREASE. As the sensor temperature INCREASES, the resistance values will DECREASE. A failure in the coolant sensor circuit should set either a Code 14 or a Code 15. These codes indicate a failure in the coolant temperature circuit, so in most cases the appropriate solution to the problem will be either repair of a wire or replacement of the sensor.

Check

2 To check the sensor, check the resistance value of the coolant temperature sensor while it is completely cold (50 to 80-degrees F = 5,600 to 2,400 ohms). Next, start the engine and warm it up until it reaches operating temperature. The resistance should be lower (180 to 200-degrees F = 300 to 200 ohms). **Note:** *Access to the coolant temperature sensor makes it difficult to position electrical probes on the terminals. If necessary, remove the sensor and perform the tests in a pan of heated water to simulate the conditions.*

Replacement

3 To remove the sensor, release the locking tab, unplug the electrical connector **(see illustrations)**, then carefully unscrew the sensor. **Caution:** *Handle the coolant sensor with care. Damage to this sensor will affect the operation of the entire fuel injection system.*

4 Before installing the new sensor, wrap the threads with Teflon sealing tape to prevent leakage and thread corrosion **(see illustration)**.

5 Installation is the reverse of removal.

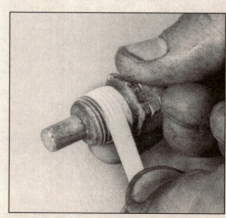
4.4 To prevent leakage, wrap the threads of the coolant temperature sensor with Teflon tape before installing it

4.8 Apply vacuum to the MAP sensor and observe the voltage. Position the positive (+) probe on the signal wire (terminal B) and the negative probe (-) on the ground (terminal A). The voltage should DECREASE

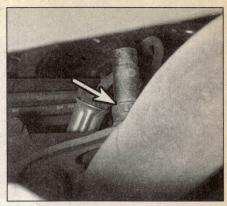

4.9 Be sure the engine is completely cooled down before attempting to remove the oxygen sensor from the exhaust manifold

Manifold Absolute Pressure (MAP) sensor (3.1L engine)

Refer to illustration 4.8

General description

6   The Manifold Absolute Pressure (MAP) sensor monitors the intake manifold pressure changes resulting from changes in engine load and speed and converts the information into a voltage output. The ECM uses the MAP sensor to control fuel delivery and ignition timing.
7   A failure in the MAP sensor circuit should set a Code 33 or a Code 34.

Check

8   With the ignition key on, connect a voltmeter between terminals A and B **(see illustration)**. The reading should be about 4.0 to 5.5 volts. Next, connect a hand vacuum pump to the MAP sensor and apply 15 in-Hg vacuum. The reading should now decrease to about 2.5 to 3.8 volts. The decrease should be about 1.5 volts. If the sensor does not perform as described, replace it. If the sensor is ok, check the vacuum hose to it.

Oxygen sensor

Refer to illustration 4.9

General description and check

9   The oxygen sensor, which is located in the exhaust manifold **(see illustration)**, monitors the oxygen content of the exhaust gas stream. The oxygen content in the exhaust reacts with the oxygen sensor to produce a voltage output which varies from 0.1-volt (high oxygen, lean mixture) to 0.9-volts (low oxygen, rich mixture). The ECM/PCM constantly monitors this variable voltage output to determine the ratio of oxygen to fuel in the mixture. The ECM/PCM alters the air/fuel mixture ratio by controlling the pulse width (open time) of the fuel injectors. A mixture ratio of 14.7 parts air to 1 part fuel is the ideal mixture ratio for minimizing exhaust emissions, thus allowing the catalytic converter to operate at maximum efficiency. It is this ratio of 14.7 to 1 which the ECM/PCM and the oxygen sensor attempt to maintain at all times.
10   The oxygen sensor produces no voltage when it is below its normal operating temperature of about 600-degrees F. During this initial period before warm-up, the ECM/PCM operates in open loop mode.
11   If the engine reaches normal operating temperature and/or has been running for two or more minutes, and if the oxygen sensor is producing a steady signal voltage between 0.35 and 0.55-volts, even though the TPS indicates that the engine is not at idle, the ECM/PCM will set a Code 13.
12   A delay of two minutes or more between engine start-up and normal operation of the sensor, followed by a low voltage signal or a short in the sensor circuit, will cause the ECM/PCM to set a Code 44. If a high voltage signal occurs, the ECM will set a Code 45.
13   When any of the above codes occur, the ECM/PCM operates in the open loop mode - that is, it controls fuel delivery in accordance with a programmed default value instead of feedback information from the oxygen sensor.
14   The proper operation of the oxygen sensor depends on four conditions:

a) **Electrical** - The low voltages generated by the sensor depend upon good, clean connections which should be checked whenever a malfunction of the sensor is suspected or indicated.
b) **Outside air supply** - The sensor is designed to allow air circulation to the internal portion of the sensor. Whenever the sensor is removed and installed or replaced, make sure the air passages are not restricted.
c) **Proper operating temperature** - The ECM/PCM will not react to the sensor signal until the sensor reaches approximately 600-degrees F. This factor must be taken into consideration when evaluating the performance of the sensor.
d) **Unleaded fuel** - The use of unleaded fuel is essential for proper operation of the sensor. Make sure the fuel you are using is of this type.

15   In addition to observing the above conditions, special care must be taken whenever the sensor is serviced.

a) The oxygen sensor has a permanently attached pigtail and electrical connector which should not be removed from the sensor. Damage or removal of the pigtail or electrical connector can adversely affect operation of the sensor.
b) Grease, dirt and other contaminants should be kept away from the electrical connector and the louvered end of the sensor.
c) Do not use cleaning solvents of any kind on the oxygen sensor.
d) Do not drop or roughly handle the sensor.
e) The silicone boot must be installed in the correct position to prevent the boot from being melted and to allow the sensor to operate properly.

Replacement

Note: *Because it is installed in the exhaust manifold or pipe, which contracts when cool, the oxygen sensor may be very difficult to loosen when the engine is cold. Rather than risk damage to the sensor (assuming you are planning to reuse it in another manifold or pipe), start and run the engine for a minute or two, then shut it off. Be careful not to burn yourself during the following procedure.*
16   Disconnect the cable from the negative terminal of the battery.
17   Raise the vehicle and place it securely on jackstands.
18   Carefully unsnap the electrical connector from the sensor.
19   Carefully unscrew the sensor from the exhaust manifold. **Caution:** *Excessive force may damage the threads.*
20   Anti-seize compound must be used on the threads of the sensor to facilitate future

Chapter 6 Emissions control systems

4.25a On 3.1L engines, the voltage should range 0.6 to 4.4 volts. Position the positive (+) probe on the B terminal and the negative probe (-) on the A terminal ground

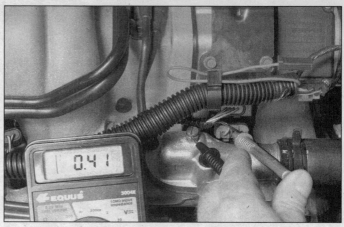

4.25b On 3800 engines, the voltage should range 0.4 to 4.2 volts. Position the positive (+) probe on the B terminal and the negative probe (-) on the C terminal ground

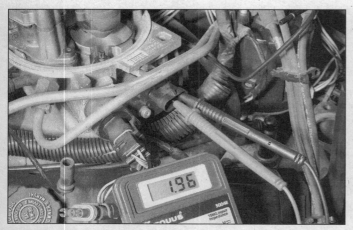

4.27a On 3.1L engines, the resistance should range 1,100 to 5,600 ohms. Position the positive (+) probe on the B terminal and the negative probe (-) on the A terminal (ground)

4.27b On 3800 engines, the resistance should range 1,400 to 4,700 ohms. Position the positive (+) probe on the B terminal and the negative probe (-) on the C terminal (ground) (this sensor reads out of range)

removal. The threads of new sensors will already be coated with this compound, but if an old sensor is removed and reinstalled, recoat the threads.
21 Install the sensor and tighten it securely.
22 Reconnect the electrical connector of the pigtail lead to the main engine wiring harness.
23 Lower the vehicle and reconnect the cable to the negative terminal of the battery.

Throttle Position Sensor (TPS)

General description

Refer to illustrations 4.25a, 4.25b, 4.27a and 4.27b

24 The Throttle Position Sensor (TPS) is located on the end of the throttle shaft on the throttle body. By monitoring the output voltage from the TPS, the ECM/PCM can determine fuel delivery based on throttle valve angle (driver demand). A broken or loose TPS can cause intermittent bursts of fuel from the injector and an unstable idle because the ECM/PCM thinks the throttle is moving.

Check

25 To check the TPS, turn the ignition switch to ON (engine not running) and install the probes of the volt-ohmmeter into the ground wire and signal wire on the backside of the electrical connector **(see illustrations)**.
26 The sensor should read 0.25 to 0.98-volts at idle. Have an assistant depress the accelerator pedal to simulate full throttle and the sensor should increase voltage to 4.0 to 4.8-volts. If the TPS voltage readings are incorrect, replace it with a new unit.
27 Also, check the TPS sensor resistance. First disconnect the electrical connector and install the probes **(see illustrations)** onto the terminals. Slowly move the throttle valve and observe a distinct change in the resistance values as the sensor travels from idle to full throttle.
28 A problem in any of the TPS circuits will set a Code 21 or 22. Once a trouble code is set, the ECM/PCM will use an artificial default value for TPS and some vehicle performance will return.

Replacement

29 Should the TPS require replacement, the complete procedure is contained in Chapter 4, Section 12 (3.1L engine) or Section 13 (3800 engine).

Park/Neutral switch

Refer to illustrations 4.31a and 4.31b

General description

30 The Park/Neutral (P/N) switch, located on the transaxle indicates to the ECM/PCM when the transmission is in Park or Neutral. This information is used for Transmission Converter Clutch (TCC), Exhaust Gas Recirculation (EGR) (3.1L engine) and Idle Air Control (IAC) valve operation. **Caution:** *The vehicle should not be driven with the Park/Neutral switch disconnected because idle quality will be adversely affected and a false Code 24 (failure in the Vehicle Speed Sensor circuit) may be set.*
31 The switch is closed to ground in Park or Neutral and open in Drive ranges **(see illustrations)**.

Chapter 6 Emissions control systems

4.31a Location of the Park/Neutral switch and electrical connector on the 3.1L engine

4.31b Location of the Park/Neutral switch and electrical connector on the 3800 engine

4.43a The Vehicle Speed Sensor (VSS) on models with the 3.1L engine is located on the right side of the transaxle housing

4.43b The Vehicle Speed Sensor (VSS) on models with the 3800 engine is located in the transaxle housing near the right driveaxle

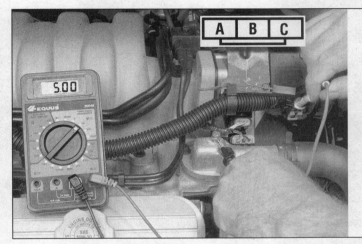

4.49 Check voltage on terminal A (yellow wire) with the positive probe of the VOM. Signal voltage should be between 4 and 6-volts

Adjustment

32 To adjust the Park/Neutral switch, Move the gear selector to the Park position.
33 Loosen the housing bolts and the housing back should pivot, providing proper switch adjustment.

Replacement

34 To replace the switch, place the gear selector in NEUTRAL.
35 Unplug the electrical connectors.
36 Remove the nut that retains the select lever and lift the lever from the transaxle.
37 Carefully mark the position with paint of the switch in relation to the transaxle.
38 Remove the bolts and lift the switch from the transaxle.
39 Position the notched grommet in correct alignment with the rod on the transaxle.
40 Push down on the top of the switch to engage the select rod and the switch.
41 Move the gear selector to PARK and adjust the switch.
42 Plug in the electrical connectors.

Vehicle Speed Sensor (VSS)

Refer to illustrations 4.43a and 4.43b

General description

43 The Vehicle Speed Sensor (VSS) is located in the housing directly above the right side driveaxle **(see illustrations)**. It sends a pulsing voltage signal to the ECM/PCM, which the ECM/PCM converts to miles per hour. The VSS is part of the Transmission Converter Clutch (TCC) system.

Check

44 To check the speed sensor, remove the I/P electrical connector in the wiring harness near the sensor. The brown wire (-437 on 3.1L engines) or yellow wire (GD14 on 3800 engines) should have 10 volts or more available.

Replacement

45 To replace the VSS, detach the sensor retaining screw and bracket and unplug the sensor. Trace the pigtail of the VSS to its electrical connector with the main wiring harness and unplug it.
46 Installation is the reverse of removal.

Mass Air Flow Sensor (MAF) (3800 engine)

Refer to illustrations 4.49 and 4.54

General description

47 The Mass Air Flow Sensor (MAF) is located on the throttle body directly following the air intake duct. The MAF sensor measures the amount of air entering the engine. The PCM uses this information to control fuel delivery. A large volume of air indicates acceleration, while a small volume of air indicates deceleration or idle.

Check

48 To check the MAF sensor, start the engine and allow it to idle for more than one minute and check for a stored trouble code 34 (see Section 3).
49 Turn the ignition OFF and connect the VOM to the MAF electrical connector. Turn the ignition key ON (engine NOT running). Use a voltmeter and check voltage at terminal A on the MAF sensor electrical connector **(see illustration)**. The meter should read between 4-volts and 6-volts. If the voltage reading is incorrect, check for a short in the

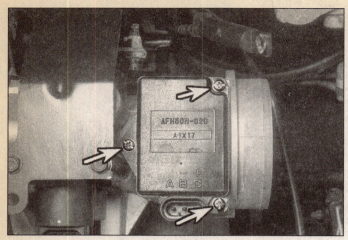

4.54 Remove the screws (arrows) from the MAF sensor and lift the sensor from the throttle body unit

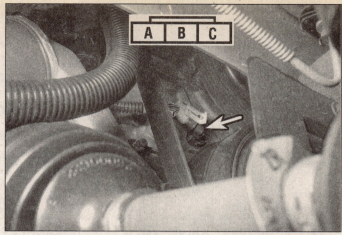

4.57 The camshaft sensor (arrow) is located on the timing chain cover (view from below)

Terminal A wire (yellow).

50   If the voltage reading is correct, connect a test light between terminal B and C (The ignition key should still be ON). The test light should be ON.

51   If the test light remains OFF, check the wiring harness for a short in the terminal C wire (pink/black wire).

52   Replace the MAF sensor with a new part if faulty.

Replacement

53   To replace the MAF sensor, disconnect the electrical connector.

54   Remove the two bolts **(see illustration)** that attach the MAF sensor to the throttle body unit and lift the sensor from the engine compartment.

55   Installation is the reverse of removal.

Camshaft sensor (3800 engine)

Refer to illustration 4.57

General description

56   The camshaft sensor is located in the timing chain cover behind the water pump near the camshaft sprocket. As the camshaft turns, a magnet on the camshaft activates the Hall-effect switch in the sensor. When the Hall-effect switch is activated, it grounds the signal line to the PCM causing the applied voltage to drop low. This creates an "interrupted signal". If the cam signal is not received by the PCM, a code 41 will set.

Check

57   To check the cam sensor circuit, disconnect the cam sensor electrical connector **(see illustration)**. Turn the ignition key to ON but do not start the engine.

58   Measure the voltage between pins A and B. It should be 5 to 7-volts.

59   Measure the voltage between pins B and C. It should be 8 to 11-volts. If A to B voltage is low, check the circuit for an open

or short. Also, check the ignition module.

60   If B to C voltage is low, check the circuit that directs the voltage for the corresponding pin designations.

Replacement

62   Disconnect the negative terminal from the battery.

63   Remove the serpentine drivebelt (see Chapter 1, Section 21).

64   Disconnect the electrical connector from the cam sensor.

65   Remove the bolt from the camshaft sensor and lift the sensor from the timing cover.

66   Installation is the reverse of removal. Be sure to torque the camshaft sensor bolt to the torque listed in this Chapter's Specifications.

Crankshaft sensor (3800 engine)

Refer to illustration 4.67

General description

67   The crankshaft sensor is located on an aluminum mounting bracket and bolted to the front, left side of the engine timing chain cover. It is partially behind the vibration damper **(see illustration)**. A four-way electrical connector plugs into the sensor connecting it to the ignition module. The dual crank sensor is equipped with two Hall-effect switches with one shared magnet mounted between them. The magnet and each Hall-effect switch are separated by an air gap. The vibration damper is equipped with specially designed blades or "interrupter rings" to create the ON OFF ON pulses necessary for the timing sequence. Although there is not an ignition timing procedure for the DIS system, the position of the crank sensor is very important. The sensor must not contact the rotating interrupter rings or the sensor will be damaged and the engine will shut down. Replacement of the crankshaft sensor is described in Chapter 5, Section 7.

Intake Air Temperature (IAT) Sensor (3800 engine)

Refer to illustrations 4.70 and 4.71

General description

68   The Intake Air Temperature sensor is located inside the air duct directly after the air filter housing. This sensor acts as a resistor which changes value according to the temperature of the air entering the engine. Low temperatures produce a high resistance value (for example, at -40-degrees F the resistance is 100,000 ohms) while high temperatures produce low resistance values (at 266-degrees F the resistance is 70 ohms. The PCM supplies approximately 5-volts (signal voltage) to the IAT sensor. The voltage will change according to the temperature of the incoming air. The voltage will be high when the air temperature is cold and low when the air temperature is warm.

Check

69   To check the IAT sensor, disconnect the two prong electrical connector and turn the ignition key ON but do not start the engine.

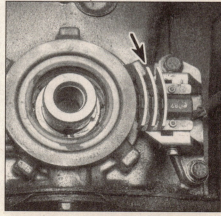

4.67 A typical crankshaft sensor (arrow) – vibration damper removed for clarity

Chapter 6 Emissions control systems

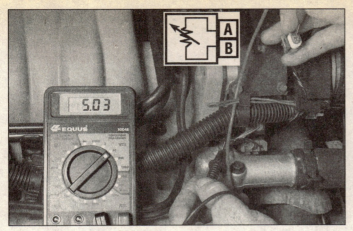

4.70 Probe terminal A (tan wire) to read the signal voltage. Note: Terminal B (black/white wire) can be used as a ground also

4.71 Example: The air temperature is approximately 70-degrees F (ambient) and the ohmmeter reads 2,920 ohms (2K scale)

70   Measure the voltage (signal voltage) on terminal A **(see illustration)**. The VOM should read approximately 5-volts.
71   Measure the resistance across the sensor terminals **(see illustration)**. The resistance should be HIGH when the air temperature is LOW. Next, start the engine and let it idle (cold). Wait awhile and let the engine reach operating temperature. Turn the ignition OFF, disconnect the IAT sensor and measure the resistance across the terminals. The resistance should be LOW when the air temperature is HIGH. If the sensor does not exhibit this change in resistance, replace it with a new part.

Knock sensor
General description
72   The knock sensor is located in the engine block directly behind the oil filter. The sensor plays a major role in the Electronic Spark Control (ESC) system. The knock sensor detects abnormal vibration in the engine. The sensor produces an AC output voltage which increases with the severity of the knock. The signal is fed into the PCM and the timing is retarded up to 10 degrees to compensate for the severe "rattling" or "heavy" detonation. The ESC system is explained in detail in Section 6 of this Chapter.

5   Electronic Spark Timing (EST)

3.1L engine
General description
1   To provide improved engine performance, fuel economy and control of exhaust emissions, the Electronic Control Module (ECM)/ Powertrain Control Module (PCM) controls distributor spark advance (ignition timing) with the Electronic Spark Timing (EST) system.
2   The EST system consists of the distributor, HEI module, an ECM and the connecting wires. The four terminals for the EST are lettered on the module. The distributor four-terminal electrical connector is lettered left-to-right, A-B-C-D **(see illustration 9.1 in Chapter 5)**. These circuits perform the following functions:

a) **Terminal A - EST**. This circuit triggers the HEI module. The ECM doesn't know what the actual timing is, but it does know when it gets the reference signal. It advances or retards the spark from that point. If the base timing is set incorrectly, the entire spark curve will be incorrect.
b) **Terminal B - Distributor Reference High**. This provides the ECM with rpm and crankshaft position information.
c) **Terminal C - Bypass**. At about 400 rpm the ECM applies 5-volts to this circuit to switch spark timing control from the HEI module to the ECM. An open or grounded bypass circuit will set a Code 42 and the engine will run at base timing, plus a small amount of advance built into the HEI module.
d) **Terminal D - Reference Ground Low**. This wire is grounded in the distributor and insures that the ground circuit has no voltage drop which could affect performance. If it is open, it may cause poor performance.

3   The ECM receives a reference pulse from the distributor, which indicates both engine rpm and crankshaft position. The ECM then determines the proper spark advance for the engine operating conditions and sends an EST pulse to the distributor.

Check
4   The ECM will set spark timing at a specified value when the diagnostic "Test" terminal in the ALDL electrical connector is grounded. To check for EST operation, the timing should be checked at 1500 rpm with the terminal ungrounded. Then ground the test terminal (see Section 2). If the timing changes at 1500 rpm, the EST is operating. A fault in the EST system will usually set Trouble Code 42.

Setting base timing
Note: *Always consult the VECI label for the exact timing procedure for your vehicle.*
5   To set the initial base timing, locate and disconnect the set timing electrical connector (the location and wire color of this connector is on the VECI label).
6   Set the timing as specified on the VECI label. This will cause a Code 42 to be stored in the ECM memory. Be sure to clear the memory after setting the timing (see Section 2).
7   For further information regarding the testing and component replacement procedures for either the HEI/EST distributor or the distributorless (DIS) ignition systems, refer to Chapter 5.

3800 engine
General description
8   The EST system consists of the module, the PCM and the connecting wires. The terminals for the EST are lettered on the module from A through P. The crankshaft sensor four-terminal electrical connector is lettered left-to-right, A-B-C-D. These circuits perform multiple functions.

6   Electronic Spark Control (ESC) system (1990 through 1992) or Knock Sensor (KS) system (1993 on)

Note: *The name of the Electronic Spark Control (ESC) system was changed in 1993. The system still operates in the exact same manner, but, from 1993 on, the system is referred to as the Knock Sensor (KS) system.*

3.1L engine
Refer to illustration 6.3

General description
1   Irregular octane levels in modern gasoline can cause detonation in an engine. Detonation is sometimes referred to as "spark knock."

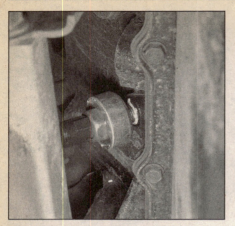

6.3 The knock sensor is located in the lower section of the engine block toward the rear

7.7a Apply vacuum to the EGR side of the EVRV and with the ignition ON and the ALDL not grounded, observe any movement on the EGR valve . . .

7.7b . . . by simply placing your finger directly onto the diaphragm inside the EGR valve housing (there should be NO movement)

2   The Electronic Spark Control (ESC) system is designed to retard spark timing up to 20-degrees to reduce spark knock in the engine. This allows the engine to use maximum spark advance to improve driveability and fuel economy.
3   The ESC knock sensor **(see illustration)** sends a voltage signal of 8 to 10-volts to the ECM when no spark knock is occurring and the ECM provides normal advance. When the knock sensor detects abnormal vibration (spark knock), the ESC module turns off the circuit to the ECM and the voltage at ECM terminal B7 drops to zero volts. The ECM then retards the EST distributor until spark knock is eliminated.
4   Failure of the ESC knock sensor signal or loss of ground at the ESC module will cause the signal to the ECM to remain high. This condition will result in the ECM controlling the EST as if no spark knock is occurring. Therefore, no retard will occur and spark knock may become severe under heavy engine load conditions. At this point, the ECM will set a Code 43.
5   Loss of the ESC signal to the ECM will cause the ECM to constantly retard EST. This will result in sluggish performance and cause the ECM to set a Code 43.

Check
6   With a tachometer connected, run the engine and tap the engine block with a wrench, near the knock sensor. The engine speed should decrease. If not, there is a problem with the knock sensor, ESC module or wiring between them.

Component replacement

ESC (knock) sensor
7   Detach the cable from the negative terminal of the battery.
8   Pull the insulator pipe off the sensor and disconnect the wiring harness electrical connector from the ESC sensor.
9   Remove the ESC sensor from the block.
10   Installation is the reverse of the removal procedure.

ESC module
11   Detach the cable from the negative terminal of the battery.
12   Remove the glove compartment assembly (see Chapter 11, Section 26).
13   Locate the module mounted to the ECM bracket behind the glove compartment.
14   Detach the electrical connector from the module.
15   Remove the module mounting bolts and remove the module.
16   Installation is the reverse of removal.

3800 engine
General description
17   The ESC system consists of the knock sensor and PCM. The MEM CAL equipped on this engine contains the functions which are similar to the parts on the ESC modules used on other models.

---

7   Exhaust Gas Recirculation (EGR) system (all 3.1L engines and 1993 and later 3800 engines)

Note: 1990 through 1992 3800 engines do not have an EGR system.

3.1L engines (vacuum-operated EGR valve)
Refer to illustrations 7.7a and 7.7b

General description
1   The EGR system is used to lower NOx (oxides of nitrogen) emission levels caused by high combustion temperatures. It does this by decreasing combustion temperatures. To reduce temperature, exhaust gasses are routed from the exhaust system back into the intake manifold during cruising-type driving conditions. These gasses reduce the combustion temperature and therefore the NOx levels produced in combustion.
2   The EGR system on 3.1L engines consists of a negative or positive backpressure EGR valve, a ported manifold vacuum source tube and an Electronic Vacuum Regulator Valve (EVRV) solenoid which controls this vacuum source.

Check
3   Too much EGR flow tends to weaken combustion, causing the engine to run rough or stop. When EGR flow is excessive, the engine can stall after a cold start or at idle after deceleration, the vehicle can surge at cruising speeds or the idle may be rough. If the EGR valve remains constantly open, the engine may not idle at all.
4   Too little or no EGR flow allows combustion temperatures to get too high during acceleration and load conditions. This can cause spark knock (detonation), engine overheating or emission test failure.
5   The following checks will help you pinpoint problems in the EGR system. Where the procedure says to lift up on the EGR valve diaphragm, it's a good idea to wear a heat-resistant glove to prevent burns.

EGR valve
6   The EGR valve is controlled by a normally open solenoid which allows vacuum to pass when energized. The ECM energizes the solenoid to turn on the EGR and monitor EGR flow with the EGR temperature switch. The ECM controls the EGR when three conditions are present: engine coolant temperature is above 113-degrees F, the TPS is at part throttle and the MAP sensor is in its mid-range. Code 32 will detect a faulty vacuum solenoid, temperature switch or vacuum supply.
7   The accompanying photos **(see illustrations)** outline the testing procedures for the EGR valve.

EGR control system
8   If a Code 32 is displayed by the SERVICE ENGINE SOON or CHECK ENGINE

7.29 The linear EGR valve (arrow) used on 1993 and later 3800 engines is attached with two nuts

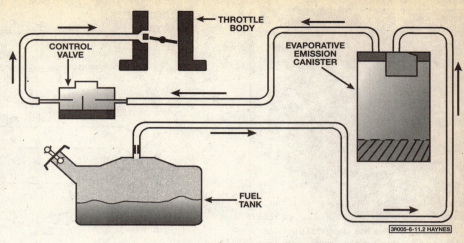

8.2 Details of a typical EECS system

light, there are several possibilities for EGR failure. Engine coolant temperature sensor, TPS, MAP sensor, BARO sensor, TCC system and the engine rpm govern the parameters the EGR system uses for distinguishing the correct ON time.

9   The EGR vacuum control solenoid (EVRV) uses an ECM controlled pulse width to modulate the EGR solenoid. The valve is normally open and the vacuum source is a ported signal. The ECM will turn the EGR ON and OFF (the "duty cycle") by grounding circuit 435 (gray wire). The duty cycle should be zero percent (no EGR) when in Park or Neutral, when the TPS input is below the specified value or when Wide Open Throttle (WOT) is indicated.

Component replacement

EGR valve

10   When buying a new EGR valve, make sure that you have the right EGR valve. Use the stamped code located on the top of the EGR valve.
11   Detach the cable from the negative terminal of the battery.
12   Remove the air cleaner housing assembly (see Chapter 4, Section 8).
13   Detach the vacuum line from the EGR valve.
14   Remove the EGR valve mounting bolts.
15   Remove the EGR valve and gasket from the manifold. Discard the gasket.
16   With a wire wheel, buff the exhaust deposits from the EGR valve mounting surface on the manifold and, if you plan to reuse the same valve, the mounting surface of the valve itself. Look for exhaust deposits in the outlet. Remove deposit build-up with a screwdriver. **Caution:** *Never wash the valve in solvents or degreaser - both agents will permanently damage the diaphragm. Sandblasting is also not recommended because it will affect the operation of the valve.*
17   If the EGR passage contains an excessive build-up of deposits, clean it out with a wire wheel. Make sure that all loose particles are completely moved to prevent them from clogging the EGR valve or from entering the engine.
18   Installation is the reverse of removal.

Hoses
19   When replacing hoses, use hose identified with the word Fluoroelastomer. Use the VECI label (on the fan shroud) as a hose routing guide. For more information regarding hose inspection and service, see Chapter 1, Section 10.

Electronic Vacuum Regulator Valve (EVRV)
20   Detach the cable from the negative terminal of the battery.
21   Remove the air cleaner housing assembly (see Chapter 4, Section 8).
22   Unplug the electrical connector from the solenoid.
23   Clearly label, then detach, both vacuum hoses.
24   Remove the solenoid mounting screw and remove the solenoid.
25   Installation is the reverse of removal.

1993 and later 3800 engines (linear EGR valve)

General description
26   The linear EGR valve system lowers NOx (oxides of nitrogen) emission levels caused by high combustion temperatures. It does this by decreasing combustion temperatures. As on the vacuum-operated system, the linear system routes burned gasses from the exhaust system back into the intake manifold during ;most driving conditions. These gasses reduce the combustion temperature and therefore the NOx levels produced in
27   The linear EGR valve is designed to accurately supply exhaust gasses to the engine, independent of intake manifold vacuum. The valve has a pintle, like on vacuum-controlled valves, but the pintle is moved by an electric stepper motor that is controlled by the PCM (computer). The linear EGR valve normally operates (opens) whenever the engine is at normal operating temperature and above idle speed. The PCM determines when and how much to open the linear EGR valve pintle by monitoring the Engine Coolant Temperature (ECT) sensor, Throttle Position (TP) sensor and Mass Air Flow (MAF) sensor.

Replacement
Refer to illustration 7.29

28   Disconnect the electrical connector from the EGR valve.
29   Remove the two mounting nuts and remove the EGR valve from the intake manifold **(see illustration)**.
30   Remove the gasket and clean the mounting surfaces of the valve and adapter assembly. Remove all traces of gasket material, using a gasket scraper, if necessary.
31   Install a new gasket and the EGR valve, tightening the nuts securely.
32   Connect the electrical connector to the valve.

8   Evaporative Emission Control System (EECS)

Refer to illustrations 8.2 and 8.13

General description
1   This system is designed to trap and store fuel vapors that evaporate from the fuel tank, throttle body and intake manifold.
2   The Evaporative Emission Control System (EECS) consists of a charcoal-filled canister and the lines connecting the canister to the fuel tank, ported vacuum and intake manifold vacuum **(see illustration)**.
3   Fuel vapors are transferred from the fuel tank, throttle body and intake manifold to a canister where they are stored when the engine is not operating. When the engine is running, the fuel vapors are purged from the canister by intake air flow and consumed in the normal combustion process.

6-18   Chapter 6 Emissions control systems

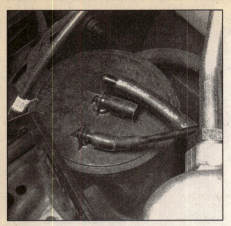

8.13 To remove the charcoal canister, label and detach the vacuum lines, then remove the canister clamp bolt and lift the canister out

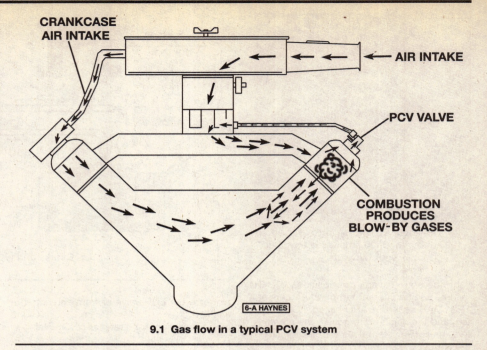

9.1 Gas flow in a typical PCV system

Check

4 Poor idle, stalling and poor driveability can be caused by an inoperative purge valve, a damaged canister, split or cracked hoses or hoses connected to the wrong tubes.
5 Evidence of fuel loss or fuel odor can be caused by fuel leaking from fuel lines or the TBI, a cracked or damaged canister, an inoperative bowl vent valve, an inoperative purge valve, disconnected, misrouted, kinked, deteriorated or damaged vapor or control hoses or an improperly seated air cleaner or air cleaner gasket.
6 Inspect each hose attached to the canister for kinks, leaks and breaks along its entire length. Repair or replace as necessary.
7 Inspect the canister. If it is cracked or damaged, replace it.
8 Look for fuel leaking from the bottom of the canister. If fuel is leaking, replace the canister and check the hoses and hose routing.
9 Apply a short length of hose to the lower tube of the purge valve assembly and attempt to blow through it. Little or no air should pass into the canister (a small amount of air will pass because the canister has a constant purge hole).
10 With a hand vacuum pump, apply vacuum through the control vacuum signal tube to the purge valve diaphragm.
11 If the diaphragm does not hold vacuum for at least 20 seconds, the diaphragm is leaking and the canister must be replaced.
12 If the diaphragm holds vacuum, again try to blow through the hose while vacuum is still being applied. An increased flow of air should be noted. If it isn't, replace the canister.

Component replacement

13 Clearly label, then detach, all vacuum lines from the canister **(see illustration)**.
14 Loosen the canister mounting clamp bolt and pull the canister out.
15 Installation is the reverse of removal.

9 Positive Crankcase Ventilation (PCV) system

Refer to illustration 9.1

1 The Positive Crankcase Ventilation (PCV) system reduces hydrocarbon emissions by scavenging crankcase vapors. It does this by circulating fresh air from the air cleaner through the crankcase, where it mixes with blow-by gases and is then rerouted through a PCV valve to the intake manifold **(see illustration)**.
2 The main components of the PCV system are the PCV valve, a fresh air filtered inlet and the vacuum hoses connecting these two components with the engine and the EECS system.
3 To maintain idle quality, the PCV valve restricts the flow when the intake manifold vacuum is high. If abnormal operating conditions arise, the system is designed to allow excessive amounts of blow-by gases to flow back through the crankcase vent tube into the air cleaner to be consumed by normal combustion.
4 Checking and replacement of the PCV valve and filter is covered in Chapter 1, Section 30.

10 Thermostatic air cleaner (THERMAC) (3.1L engine)

Refer to illustrations 10.2, 10.18, 10.27 and 10.28

General description

1 A heated air intake system is used to provide good driveability under varying climatic conditions. By having a uniform inlet air temperature, the fuel system can be calibrated to reduce exhaust emissions and to eliminate throttle valve icing.
2 The THERMAC air cleaner **(see illustration)** is operated by heated air and manifold vacuum. Air can enter the air cleaner from outside the engine compartment or from a heat stove built around the exhaust manifold. A temperature sensor located inside the air

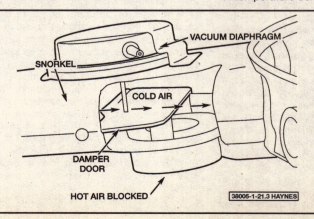

10.2 Typical THERMAC air cleaner shown with the snorkel passage (damper door) open

Chapter 6 Emissions control systems

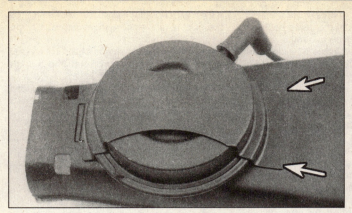

10.18 Once the air cleaner housing assembly is removed from the TBI unit, turn it upside down and detach the vacuum line from the temperature sensor, drill out the spot welds (arrows) with a 1/16-inch drill, then enlarge as necessary to remove the retaining strip

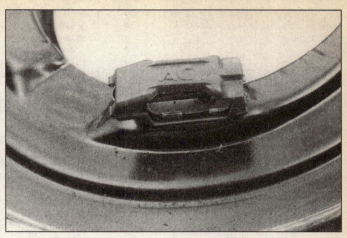

10.27 Note the position of the sensor in the air cleaner housing assembly

cleaner housing determines the operational mode of the THERMAC.

Check

3 If the engine hesitates during warm-up:
  a) The heat stove tube could be disconnected.
  b) The vacuum diaphragm motor could be inoperative, leaving the snorkel (mouth) of the air cleaner housing open to outside air.
  c) There may be no manifold vacuum.
  d) The damper door does not move.
  e) The air cleaner housing assembly-to-TBI adapter seal may be missing.
  f) The air cleaner housing assembly cover seal may be missing.
  g) The air cleaner housing assembly cover may be loose.
  h) The air cleaner housing assembly may be loose.

4 Lack of power and/or sluggish or spongy throttle response can be caused by:
  a) A stuck damper door that does not open to outside air.
  b) A temperature sensor that does not bleed off vacuum.

5 Inspect the system to be sure that all hoses and the heat stove tube are connected. Check for kinked, plugged or deteriorated hoses.

6 Check for the presence and condition of air cleaner to carburetor gasket seal.

7 With the air cleaner assembly installed, the damper door should be open to outside air.

8 Start the engine. Watch the damper door in the air cleaner snorkel. When the engine is first started, the damper door should move and close off outside air. As the air cleaner warms up, the damper door should open slowly to the outside air.

9 If the air cleaner fails to operate as described above, the door may not be moving at the right temperature. If the driveability problem is during warm-up, make the temperature sensor check below.

10 With the engine off, disconnect the vacuum hose at the vacuum diaphragm motor.

11 Apply at least seven inches of vacuum to the vacuum diaphragm motor. The damper door should completely block off outside air when vacuum is applied. If it doesn't, check to see if the linkage is hooked up correctly.

12 With vacuum still applied, trap vacuum in the vacuum diaphragm motor by bending the hose. The damper door should remain closed. If it doesn't, replace the vacuum diaphragm motor assembly. Failure of the vacuum diaphragm motor is more likely to be caused from binding linkage or a corroded snorkel than from a failed diaphragm. Check this first before replacing the diaphragm.

13 If the vacuum motor checks out okay, check the vacuum hoses and connections. If they're okay, replace the temperature sensor.

14 Start the test with the air cleaner temperature below 86-degrees F. If the engine has been run recently, remove the air cleaner cover and place the thermometer as close as possible to the sensor. Let the air cleaner cool until the thermometer reads below 86-degrees F for about five to ten minutes. Reinstall the air cleaner.

15 Start and idle the engine. The damper door should move to close off outside air immediately if the engine is cool enough. When the damper door starts to open the snorkel passage (in a few minutes), remove the air cleaner and read the thermometer. It must read about 131-degrees F.

16 If the damper door is not open to outside air at the indicated temperature, the temperature sensor is malfunctioning and must be replaced.

Component replacement

Vacuum diaphragm motor

17 Remove the air cleaner housing assembly (see Chapter 4, Section 8).

18 Detach the vacuum tube from the motor **(see illustration)**.

19 Drill out the two spot welds with a 1/16-inch drill, then enlarge as required to remove the retaining strap. Do not damage the

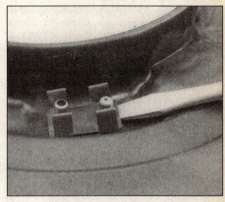

10.28 Pry off the sensor retaining clips with a small screwdriver

snorkel tube.

20 Bend the strap up and out of the way.

21 Lift up the motor, cocking it to one side to unhook the motor linkage at the control damper assembly.

22 To install a new motor, drill a 7/64-inch hole in the snorkel tube at the center of the vacuum motor retaining strap.

23 Install the vacuum motor linkage into the control damper assembly.

24 Use the motor retaining strap and sheet metal screw provided in the motor service package to secure the motor to the snorkel tube. Make sure that the screw does not interfere with the operation of the damper assembly. Shorten the screw if necessary.

25 Install the air cleaner housing assembly (see Chapter 4, Section 8). Be sure to attach the vacuum hose to the motor.

Sensor

26 Remove the air cleaner housing assembly (see Chapter 4, Section 8).

27 Note the position **(see illustration)** of the sensor in the air cleaner housing to facilitate reinstallation.

28 Pry up the tabs on the sensor retaining clip **(see illustration)**. Remove the clip and sensor from the air cleaner.

11 Transmission Converter Clutch (TCC)

General description

1   The Transmission Converter Clutch (TCC) uses a single solenoid-operated valve (3.1L engine) or two solenoids (3800 engine) in the automatic transmission to mechanically couple the engine flywheel to the output shaft of the transmission through the torque converter. This reduces the slippage losses in the converter, reducing emissions because engine rpm at any given speed is reduced. It also increases fuel economy. **Note:** *On 1992 models, the TCC system will engage at approximately 25 mph on vehicles equipped with front and rear air conditioning systems. On vehicles equipped with the front ONLY air conditioning system, the TCC will activate at two different speeds. With the air conditioning ON, the TCC system engages at 35 mph and with the air conditioning OFF, the TCC system engages at 25 mph. Each system uses a different style air conditioning compressor.*

2   For the converter clutch to operate properly, two conditions must be met:
   a) The engine must be warmed up before the clutch can apply. The engine coolant temperature sensor (see Section 4) tells the ECM/PCM when the engine is at operating temperature.
   b) The vehicle must be traveling at the necessary minimum speed to raise the pressure to the level necessary to apply the valve. If the hydraulic pressure is correct, the ECM/PCM signals the solenoid(s) to apply the converter clutch.

3   After the converter clutch applies, the ECM/PCM uses the information from the TPS to release the clutch when the car is accelerating or decelerating at a certain rate.

4   Another switch used in the TCC circuit is a brake switch which opens the power supply to the TCC solenoid when the brake is applied.

5   On the 3T40 transaxles (3.1L engines), the TCC system uses a 3rd gear switch to send a signal to the ECM telling it what gear the transmission is in. The ECM uses this information to vary the conditions under which the clutch applies or releases. However, the transmission does not have to be in high gear in order for the ECM to turn on the clutch. Transmissions using gear select switches can be identified by three wires coming out of the TCC electrical connector.

6   On 4T60-E transaxles (3800 engine), the apply solenoids control the amount of fluid that reaches the TCC system.

7   Also on 4T60-E transaxles (3800 engine), the TCC uses a Pulse Width Modulated (PWM) solenoid to vary the hydraulic pressure to make the TCC "shift" smoother and quieter.

Check

8   If the converter clutch is applied at all times, the engine will stall immediately, just like a manual transmission with the clutch applied.

9   If the converter clutch does not apply, fuel economy may be lower than expected. If the Vehicle Speed Sensor (VSS) (see Section 4) fails, the TCC will not apply.

10   A TCC-equipped transmission has different operating characteristics than an automatic transmission without TCC. If you detect a "chuggle" or "surge" condition, perform the following check.

11   Install a tachometer.

12   Drive the vehicle until normal operating temperature is reached, then maintain a 50 to 55 mph speed.

13   Lightly touch the brake pedal and check it for a slight bumpy sensation, indicating that the TCC is releasing. A slight increase in rpm should also be noted.

14   Release the brake and check for reapplication of the converter clutch and a slight decrease in engine rpm.

15   If the TCC fails to perform satisfactorily during this test, take your vehicle to a dealer to have the TCC serviced.

12 Catalytic converter

General description

1   The catalytic converter is an emission control device added to the exhaust system to reduce pollutants from the exhaust gas stream. A single-bed converter design is used in combination with a three-way (reduction) catalyst. The catalytic coating on the three-way catalyst contains platinum and rhodium, which lowers the levels of oxides of nitrogen (NOx) as well as hydrocarbons (HC) and carbon monoxide (CO).

Check

2   The test equipment for a catalytic converter is expensive and highly sophisticated. If you suspect that the converter on your vehicle is malfunctioning, take it to a dealer or authorized emissions inspection facility for diagnosis and repair.

3   Whenever the vehicle is raised for servicing of underbody components, check the converter for leaks, corrosion and other damage. If damage is discovered, the converter should be replaced.

Replacement

4   Because the converter part of the exhaust system, converter replacement requires removal of the exhaust pipe assembly (see Chapter 4, Section 14). Take the vehicle, or the exhaust system, to a dealer or a muffler shop.

Chapter 7
Automatic transaxle

Contents

	Section		Section
Automatic transaxle fluid and filter change	See Chapter 1	Neutral start switch - replacement and adjustment	4
Automatic transaxle fluid level check	See Chapter 1	Neutral start (starter safety) switch check	See Chapter 1
Automatic transaxle - removal and installation	9	Park/lock cable - removal and installation	6
Diagnosis - general	2	Shift cable - replacement and adjustment	5
Differential oil seals - replacement	7	Throttle valve (TV) cable - replacement and adjustment	3
General information	1	Transaxle mount - check and replacement	8

Specifications

General
Fluid type and capacity .... See Chapter 1

Torque specifications
Ft-lbs

3T40 (three-speed)
- Starter safety switch-to-case bolts 22
- Transaxle-to-engine bolts 55
- Torque converter-to-driveplate bolts 35
- Torque converter cover bolts 5
- Transaxle mount bolts 32
- Transaxle mount nuts 35
- TV cable-to-transaxle case bolt 6

4T60E (four-speed)
- Starter safety switch-to-case bolts 20
- Transaxle-to-engine bolts 55
- Torque converter-to-driveplate bolts .... 46 (in two steps)
- Torque converter cover bolts 5
- Transaxle-to-engine mount bolts 55
- Transaxle mount nut/bolts
  - Frame bracket-to-frame bolt 38
  - Frame nut 30
  - Mount-to-transaxle bolt 40
  - Transaxle bracket to mount nut 30
- TV cable-to-transaxle case bolt 6

Chapter 7 Automatic transaxle

1.1 An underside view of a automatic transaxle and its related components

1 Transaxle cooler lines
2 Transaxle fluid pan

1 General information

Refer to illustration 1.1

The vehicles covered by this manual are equipped with a three or four-speed Hydra-Matic automatic transaxles. Two models of Hydra-Matic automatic transaxles are used on these vehicles: 3T40 (three-speed) and 4T60E (four-speed) **(see illustration)**.

Due to the complexity of the clutches and the hydraulic control system, and because of the special tools and expertise required to perform an automatic transaxle overhaul, it should not be undertaken by the home mechanic. Therefore, the procedures in this Chapter are limited to general diagnosis, routine maintenance, adjustment and transaxle removal and installation.

If the transaxle requires major repair work, it should be left to a dealer service department or an automotive or transmission repair shop. You can, however, remove and install the transaxle yourself and save the expense, even if the repair work is done by a transmission shop.

Replacement and adjustment procedures the home mechanic can perform include those involving the throttle valve (TV) cable and the shift linkage.

Caution: *Never tow a disabled vehicle with an automatic transaxle at speeds greater than 35 mph or for distances over 50 miles if the front wheels are on the ground.*
Note: *On models equipped with the Delco Loc II audio system, be sure the lockout feature is turned off before performing any procedure which requires disconnecting the battery.*

2 Diagnosis - general

Note: *Automatic transaxle malfunctions may be caused by five general conditions: poor engine performance, improper adjustments, hydraulic malfunctions, mechanical malfunctions or malfunctions in the computer or its signal network. Diagnosis of these problems should always begin with a check of the easily repaired items: fluid level and condition (see Chapter 1, Section 4), shift linkage adjustment and throttle linkage adjustment. Next, perform a road test to determine if the problem has been corrected or if more diagnosis is necessary. If the problem persists after the preliminary tests and corrections are completed, additional diagnosis should be done by a dealer service department or transmission repair shop. Refer to the Troubleshooting section at the front of this manual for transaxle problem diagnosis.*

Preliminary checks

1 Drive the vehicle to warm the transaxle to normal operating temperature.
2 Check the fluid level as described in Chapter 1, Section 4:
   a) If the fluid level is unusually low, add enough fluid to bring the level within the designated area of the dipstick, then check for external leaks.
   b) If the fluid level is abnormally high, drain off the excess, then check the drained fluid for contamination by coolant. The presence of engine coolant in the automatic transmission fluid indicates that a failure has occurred in the internal radiator walls that separate the coolant from the transmission fluid.
   c) If the fluid is foaming, drain it and refill the transaxle, then check for coolant in the fluid or a high fluid level.
3 Check the engine idle speed. **Note:** *If the engine is malfunctioning, do not proceed with the preliminary checks until it has been repaired and runs normally.*
4 Check the throttle valve cable for freedom of movement. Adjust it if necessary (see Section 3). **Note:** *The throttle valve cable may*

Chapter 7 Automatic transaxle

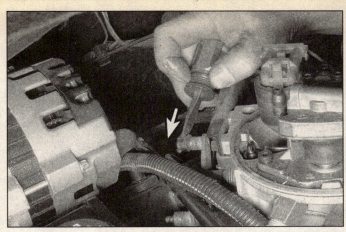

3.1 Use a screwdriver to push the TV cable forward, then lift it up to detach it from the throttle lever

3.4 Use a socket and extension to remove the TV cable bolt

function properly when the engine is shut off and cold, but it may malfunction once the engine is hot. Check it cold and at normal engine operating temperature.

5 Inspect the shift cable (see Section 5). Make sure that it's properly adjusted and that the linkage operates smoothly.

Fluid leak diagnosis

6 Most fluid leaks are easy to locate visually. Repair usually consists of replacing a seal or gasket. If a leak is difficult to find, the following procedure may help.

7 Identify the fluid. Make sure it's transmission fluid and not engine oil or brake fluid (automatic transmission fluid is a deep red color).

8 Try to pinpoint the source of the leak. Drive the vehicle several miles, then park it over a large sheet of cardboard. After a minute or two, you should be able to locate the leak by determining the source of the fluid dripping onto the cardboard.

9 Make a careful visual inspection of the suspected component and the area immediately around it. Pay particular attention to gasket mating surfaces. A mirror is often helpful for finding leaks in areas that are hard to see.

10 If the leak still cannot be found, clean the suspected area thoroughly with a degreaser or solvent, then dry it.

11 Drive the vehicle for several miles at normal operating temperature and varying speeds. After driving the vehicle, visually inspect the suspected component again.

12 Once the leak has been located, the cause must be determined before it can be properly repaired. If a gasket is replaced but the sealing flange is bent, the new gasket will not stop the leak. The bent flange must be straightened.

13 Before attempting to repair a leak, check to make sure that the following conditions are corrected or they may cause another leak. **Note:** *Some of the following conditions cannot be fixed without highly specialized tools and expertise. Such prob-*

lems must be referred to a transmission shop or a dealer service department.

Gasket leaks

14 Check the pan periodically. Make sure the bolts are tight, no bolts are missing, the gasket is in good condition and the pan is flat (dents in the pan may indicate damage to the valve body inside).

15 If the pan gasket is leaking, the fluid level or the fluid pressure may be too high, the vent may be plugged, the pan bolts may be too tight, the pan sealing flange may be warped, the sealing surface of the transaxle housing may be damaged, the gasket may be damaged or the transaxle casting may be cracked or porous. If sealant instead of gasket material has been used to form a seal between the pan and the transaxle housing, it may be the wrong sealant.

Seal leaks

16 If a transaxle seal is leaking, the fluid level or pressure may be too high, the vent may be plugged, the seal bore may be damaged, the seal itself may be damaged or improperly installed, the surface of the shaft protruding through the seal may be damaged or a loose bearing may be causing excessive shaft movement.

17 Make sure the dipstick tube seal is in good condition and the tube is properly seated. Periodically check the area around the speedometer gear or sensor for leakage. If transmission fluid is evident, check the O-ring for damage. Also inspect the side gear shaft oil seals for leakage.

Case leaks

18 If the case itself appears to be leaking, the casting is porous and will have to be repaired or replaced.

19 Make sure the oil cooler hose fittings are tight and in good condition.

Fluid comes out vent pipe or fill tube

20 If this condition occurs, the transaxle is overfilled, there is coolant in the fluid, the

case is porous, the dipstick is incorrect, the vent is plugged or the drain back holes are plugged.

3 Throttle valve (TV) cable - replacement and adjustment

Refer to illustrations 3.1, 3.4 and 3.5

Replacement

1 Disconnect the TV cable from the throttle lever by pushing the connector forward with a screwdriver to detach it, then lifting up and off the lever pin **(see illustration)**.

2 Disconnect the TV cable housing from the bracket by compressing the tangs and pushing the housing back through the bracket.

3 Disconnect any clips or straps retaining the cable to the transaxle.

4 Remove the bolt retaining the cable to the transaxle **(see illustration)**.

5 Pull up on the cover until the end of the cable can be seen, then disconnect it from the transaxle TV link **(see illustration)**. Remove the cable from the vehicle.

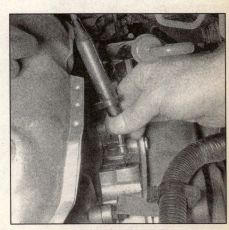

3.5 Pull the TV cable out of the transaxle and slide the cable link off the pin

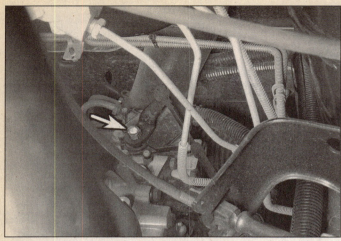

4.5 Remove the nut (arrow) and detach the shift lever so the switch can be removed

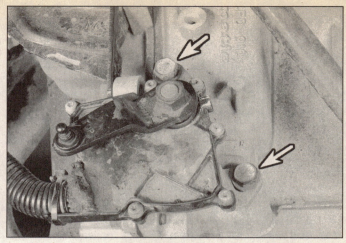

4.6 Neutral safety switch retaining bolts (arrows)

6 To install the cable, connect it to the transaxle TV link then push the cover securely over the cable and install the bolt. Route the cable to the top of the engine, push the housing through the bracket until it clicks into place, place the connector over the throttle lever pin and pull back to lock it.

Adjustment

7 The engine MUST NOT be running during this adjustment.
8 Depress the re-adjust tab and push the slider through the fitting (away from the throttle lever) as far as it will go.
9 Release the re-adjust tab.
10 Manually turn the throttle lever to the "wide open throttle" position until the re-adjust tab makes an audible click, then release the throttle lever. The cable is now adjusted. *Note: Don't use excessive force at the throttle lever to adjust the TV cable. If great effort is required to adjust the cable, disconnect the cable at the transaxle end and check for free operation. If it's still difficult, replace the cable. If it's now free, suspect a bent TV link in the transaxle or a problem with the throttle lever.*

4 Neutral start switch - replacement and adjustment

Replacement

Refer to illustrations 4.5, 4.6 and 4.9

1 Disconnect the negative cable from the battery.
2 Shift the transaxle into Neutral.
3 Raise the hood. Disconnect the electrical and vacuum connectors from the cruise control servo (if equipped). Remove the servo to provide access to the Neutral Start switch.
4 Trace the wire harness from the Neutral Start switch and unplug it.
5 Remove the nut and detach the shift lever from the transaxle **(see illustration)**.

4.9 The outer alignment notch must line up with the inner one on the shift shaft

6 Remove the bolts and detach the switch **(see illustration)**.
7 To install a new switch, line up the flats on the shift shaft with the flats in the switch and lower the switch onto the shaft.
8 Install the bolts. If the switch is new and the shaft hasn't been moved, tighten the bolts. If the switch requires adjustment, leave the bolts loose and follow the adjustment procedure below. The remainder of installation is the reverse of removal.

Adjustment

9 Line up the outer notch on the switch with the inner notch on the shift shaft and tighten the bolts **(see illustration)**. Some later models use an alignment hole for adjustment. On these models, rotate the switch and line up the adjustment hole in the switch with the hole in the shift lever, insert a 3/32-in drill bit and tighten the switch bolts.
10 Connect the negative battery cable and verify that the engine will start only in Neutral or Park.

5.2a Use a screwdriver to detach the cable from the transaxle shift lever

5 Shift cable - replacement and adjustment

Note: On Delco Loc II equipped audio systems, be sure the lockout feature is turned off before disconnecting the battery.

1 Disconnect the negative cable from the battery.

Replacement

Refer to illustrations 5.2a, 5.2b and 5.2c

2 Working in the engine compartment, remove the cruise control servo unit (if equipped), then disconnect the cable from the transaxle lever and bracket **(see illustrations)**.
3 Working in the passenger compartment, remove the left sound insulator located under the dash.
4 Use a large screwdriver to detach the cable from the bracket at the base of the steering column.
5 Dislodge the grommet in the firewall and withdraw the cable from the vehicle.
6 Installation is the reverse of removal.

Chapter 7 Automatic transaxle   7-5

5.2b Pry the locking tab up to release the cable from the bracket, then . . .

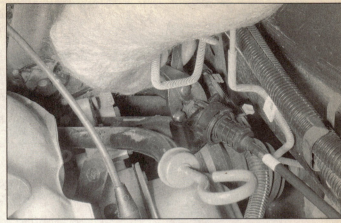

5.2c . . . use pliers to detach the cable and push it back through the bracket

After installation, adjust the cable as described below.

Adjustment

7 Place the shift lever and the transaxle lever in Neutral, then push the locking tab on the shift cable to automatically adjust the cable **(see illustration 5.2b)**.
8 Reconnect the negative battery cable.

---

6 Park/lock cable - removal and installation

Removal

1 Disconnect the negative cable from the battery.
2 Place the shift lever in Park and the ignition switch in the Run position.
3 Remove the steering column cover located under the dash.
4 Insert a screwdriver blade into the slot in the ignition switch inhibitor, depress the cable latch and detach the cable.
5 Push the cable connector lock button (located at the lower end of the steering column) to the Up position and detach the cable from the park lock lever pin. Depress the two cable connector latches and remove the cable from the shift control base.
6 Remove the cable clips.

Installation

7 Make sure the cable lock button is in the Up position and the shift lever is in Park. Snap the cable connector into the shift control base.
8 With the ignition key in the Run position (this is very important), snap the cable into the inhibitor housing.
9 Turn the ignition key to the Lock position.
10 Snap the end of the cable onto the shifter park/lock pin.
11 Push the nose of the cable connector forward to remove the slack.
12 With no load on the connector nose,

8.3a Typical front transaxle mount details

snap the cable connector lock button on.
13 Check the operation of the park/lock cable as follows.

  a) With the shift lever in Park and the key in Lock, make sure the shift lever cannot be moved to another position and the key can be removed.
  b) With the key in Run and the shift lever in Neutral, make sure the key cannot be turned to Lock.

14 If it operates as described above, the park/lock cable system is properly adjusted. Proceed to Step 16.
15 If the park/lock system doesn't operate as described, return the cable connector lock to the up position and repeat the adjustment procedure. Push the cable connector down and recheck the operation.
16 If the key cannot be removed in the Park position, snap the lock button to the up position and move the nose of the cable connector to the rear until the key can be removed from the ignition switch.
17 Install the cable in the retaining clips.

---

7 Differential oil seals - replacement

1 Raise the vehicle and support it securely on jackstands.

2 Remove the driveaxle(s) (see Chapter 8, Section 2).
3 Use a hammer and chisel to pry up the outer lip of the seal to dislodge it so it can be pried out of the housing. The manufacturer recommends using a slide hammer to remove the seal.
4 Compare the new seal to the old one to make sure they're the same.
5 Coat the lips of the new seal with transmission fluid.
6 Place the new seal in position and tap it into the bore with a hammer and a large socket or a piece of pipe that's the same diameter as the outside edge of the seal.
7 Reinstall the various components in the reverse order of removal.

---

8 Transaxle mount - check and replacement

Refer to illustrations 8.3a and 8.3b

1 Insert a large screwdriver or pry bar into the space between the transaxle bracket and the mount and try to pry the transaxle up slightly.
2 The transaxle bracket should not move away from the insulator much at all.
3 To replace the mount, remove the nuts attaching the insulator to the crossmember

8.3b Rear transaxle mount

and the nuts attaching the insulator to the transaxle **(see illustrations)**.

4 Raise the transaxle slightly with a jack and remove the insulator, noting which holes are used in the support for proper alignment during installation.

5 Installation is the reverse of the removal procedure. Be sure to tighten the nuts/bolts securely.

9 Automatic transaxle - removal and installation

Note: *The automatic transaxle can't be removed with the engine in the vehicle. The transaxle can only be separated from the engine after the engine/transaxle/subframe unit has first been removed from the vehicle. This is a dangerous and complicated procedure requiring a vehicle lift and is best left to a dealer or properly equipped shop.*

Removal

1 Remove the engine/transaxle/subframe unit from the vehicle (see Chapter 2C, Section 5).
2 Remove the torque converter cover.
3 Mark the torque converter-to-driveplate relationship with white paint so they can be installed in the same position.
4 Remove the torque converter-to-driveplate bolts. Turn the crankshaft pulley bolt for access to each bolt.
5 Remove the bolts securing the transaxle to the engine.
6 Remove the transaxle mount nuts and bolts.
7 Move the transaxle back to disengage it from the engine block dowel pins and make sure the torque converter is detached from the driveplate. Secure the torque converter to the transaxle so it will not fall out during removal.
8 Remove the transaxle from the engine.

Installation

9 Prior to installation, make sure that the torque converter hub is securely engaged in the pump. Lubricate the torque converter hub with multi-purpose grease.
10 Move the transaxle into position. Be sure to keep it level so the torque converter does not slide out.
11 Turn the torque converter to line up the bolt holes with the holes in the driveplate. The white paint mark on the torque converter and the driveplate made in Step 3 must line up.
12 Move the transaxle forward carefully until the dowel pins and the torque converter are engaged.
13 Install the transaxle housing-to-engine bolts. Tighten them securely.
14 Install the torque converter-to-driveplate bolts. Tighten the bolts to the torque specified at the beginning of this Chapter.
15 Install the torque converter cover.
16 Once the engine/transaxle/subframe unit has been reinstalled (see Chapter 2C, Section 5), connect the fluid cooler lines and adjust the shift linkage (see Section 5).
17 Fill the transaxle (see Chapter 1), run the vehicle and check for fluid leaks.

Chapter 8 Driveaxles

Contents

	Section		Section
Driveaxle boot check	See Chapter 1	Driveaxles - removal and installation	2
Driveaxle boot replacement and constant velocity (CV) joint overhaul	3	Driveaxles - general information	1

Specifications

CV joint boot dimensions (collapsed)
1990	5-1/16 inches
1991 on	4-29/32 inches

Torque specifications
Driveaxle hub nut	184
Wheel lug nuts	See Chapter 1

1 Driveaxles - general information

Refer to illustration 1.2

Power is transmitted from the transaxle to the front wheels by two driveaxles, which consist of splined solid axles with constant velocity (CV) joints at each end. Each driveaxle has two types of CV joints; inner and outer. The outer joint is a double-offset design using ball bearings with an inner and outer race to allow angular movement **(see illustration on next page)**. The inner CV joint is a tri-pot design, with a spider bearing assembly and tri-pot housing to allow angular movement and permit the driveaxle to slide in and out.

The CV joints are protected by rubber boots, which are retained by clamps so the joints are protected from water and dirt. The boots should be inspected periodically (see Chapter 1, Section 13). Damaged CV joint boots must be replaced immediately or the joints can be damaged. Boot replacement involves removing the driveaxles (see Section 2). It's a good idea to disassemble, clean, inspect and repack the CV joint whenever replacing a CV joint boot to make sure the joint isn't contaminated with moisture or dirt, which would cause premature failure of the CV joint.

The most common symptom of worn or damaged CV joints, besides lubricant leaks, are a clicking noise in turns, a clunk when accelerating from a coasting condition or vibration at highway speeds.

Note: *On models equipped with the Delco Loc II audio system, be sure the lockout feature is turned off before performing any procedure which requires disconnecting the battery.*

1.2 Two types of CV joints are used on each driveaxle

A   Inner CV joint   B   Outer CV joint

2.2 A screwdriver or prybar (arrow) inserted through the caliper and into a disc cooling vane will hold the hub stationary while loosening the hub nut

2.5 Swing the steering away from the transaxle and detach the driveaxle – support the driveaxle in the level position when disconnecting the inner end and removing from the vehicle

Chapter 8 Driveaxles

2.6 Pry the inner end of the driveaxle out of the transaxle with a large screwdriver or prybar (arrow)

2.8 A large punch or screwdriver, positioned in the groove on the CV joint housing, can be used to seat the CV joint in the transaxle

2 Driveaxles - removal and installation

Refer to illustrations 2.2, 2.5, 2.6 and 2.8

Removal

1 Loosen the wheel lug nuts, raise the front of the vehicle and support it securely on jackstands. Apply the parking brake and block the rear wheels to keep the vehicle from rolling off the jackstands. Remove the front wheel.
2 Remove the driveaxle hub nut. To prevent the hub from turning, insert a screwdriver through the caliper and into a disc cooling vane, then remove the nut **(see illustration)**.
3 Remove the brake caliper (don't disconnect the brake hose) and disc and support the caliper out of the way with a piece of wire (see Chapter 9, Sections 3 and 4).
4 Separate the control arm from the steering knuckle (see Chapter 10, Section 5).
5 Push the driveaxle out of the hub, then support the outer end of the driveaxle with a piece of wire to prevent damage to the inner CV joint **(see illustration)**. If the driveaxle sticks in the hub, push it from the hub using a two-jaw puller.
6 Use a large screwdriver or prybar to pry the inner joint out of the transaxle **(see illustration)**.
7 Support the CV joints and carefully remove the driveaxle from the vehicle.

Installation

8 Lubricate the differential seal with multi-purpose grease, raise the driveaxle into position while supporting the CV joints and insert the splined end of the inner CV joint into the differential side gear. Seat the shaft in the side gear by positioning the end of a screwdriver in the groove in the CV joint and tapping it into position with a hammer **(see illustration)**.
9 Grasp the inner CV joint housing (not the driveaxle) and pull out to make sure the axle has seated securely.
10 Apply a light coat of multi-purpose grease to the outer CV joint splines, pull out on the strut/steering knuckle assembly and install the stub axle in the hub.
11 Insert the control arm balljoint stud into the steering knuckle and tighten the nut. Be sure to use a new cotter pin (refer to Chapter 10, Section 5).
12 Install the brake disc and caliper (see Chapter 9, Sections 3 and 4 if necessary).
13 Install the hub nut. Lock the disc so it can't turn, using a screwdriver or punch inserted through the caliper into a disc cooling vane, and tighten the hub nut to the torque listed in this Chapter's Specifications.
14 Install the wheel and lower the vehicle. Tighten the lug nuts to the torque listed in the Chapter 1 Specifications.
15 Install the wheel cover.

3 Driveaxle boot replacement and Constant Velocity (CV) joint overhaul

Note: *If the CV joints exhibit wear indicating the need for an overhaul (usually due to torn boots), explore all options before beginning the job. Complete rebuilt driveaxles are available on an exchange basis, which eliminates a lot of time and work. Whatever is decided, check on the cost and availability of parts before disassembling the vehicle.*

1 Remove the driveaxle (see Section 2).
2 Place the driveaxle in a vise lined with rags to avoid damage to the shaft.

Inner CV joint

Refer to illustrations 3.3a through 3.3s

3 Refer to the accompanying illustrations

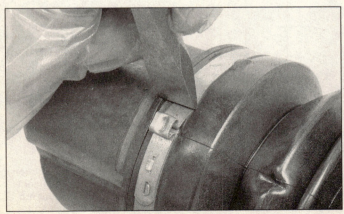

3.3a Cut off the boot seal retaining clamps, using wire cutters or a chisel and hammer

3.3b Slide the housing off the spider assembly

Chapter 8 Driveaxles

3.3c Slide the boot towards the center of the driveaxle

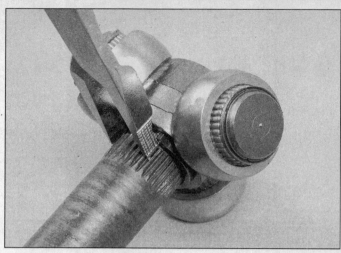

3.3d Spread the spider assembly stop-ring apart and slide it towards the center of the shaft

3.3e Slide the spider assembly back to expose the front retaining ring and remove it

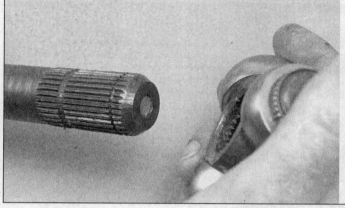

3.3f Slide the spider off the driveaxle, holding the bearings in place - it's a good idea to use tape or a cloth wrapped around the spider bearing assembly to retain them

and perform the inner CV joint overhaul and boot installation procedure (see illustrations 3.3a through 3.3s).

and perform the outer CV joint overhaul and boot installation procedure (see illustrations 3.4a through 3.4s).

Outer CV joint

Refer to illustrations 3.4a through 3.4s

4   Refer to the accompanying illustrations

3.3g Slide the stop-ring and boot off the axle

3.3h Clean all the old grease out of the housing and spider assembly, then remove each bearing, one at a time

3.3l Carefully disassemble each section of the spider assembly, clean the needle bearings with solvent and inspect the rollers, spider cross, bearings and housing for scoring, pitting and other signs of abnormal wear

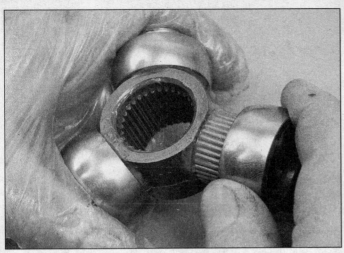
3.3j Apply a coat of CV joint grease to the inner bearing surfaces to hold the needle bearings in place and slide the bearing over them

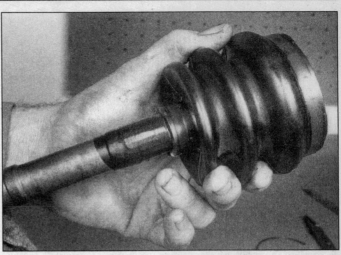
3.3k Wrap the driveaxle splines with tape to avoid damaging the boot, then slide the boot onto the axle

3.3l Slide the spider stop-ring onto the axleshaft, past the groove in which it seats

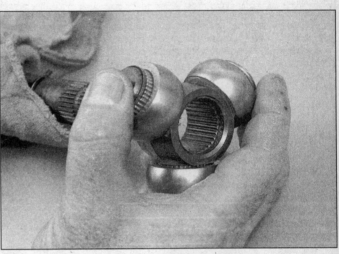
3.3m Install the spider bearing with the recess in the counterbore facing the end of the driveaxle

3.3n Use a screwdriver to install the outer snap-ring, then slide the spider assembly against it and install the stop-ring in the groove

3.3o Pack the housing with half of the grease furnished with the new boot and place the remainder in the boot

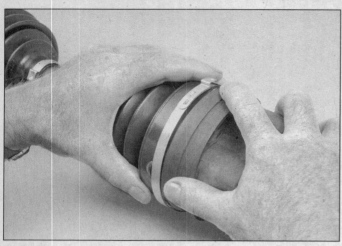

3.3p With the retaining clamp in place (but not tightened), install the tri-pot housing

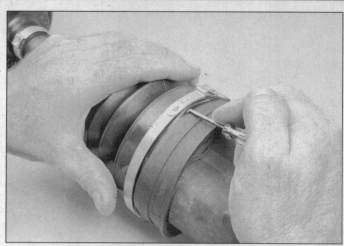

3.3q Seat the boot in the housing and axle seal grooves - a small screwdriver can make the job easier (make sure the boot isn't dimpled, stretched or out of shape)

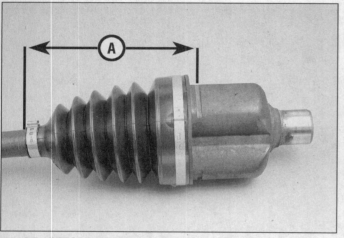

3.3r BEFORE tightening the clamps, adjust the collapsed dimension of the joint (A) (from the small end of the boot to the groove in the housing) to the measurement in the Specifications Section at the beginning of this Chapter

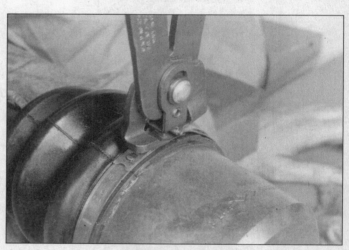

3.3s Secure the boot clamps with special pliers (available at auto parts stores)

3.4a Cut off the band retaining the boot to the shaft, then slide the boot toward the center of the shaft

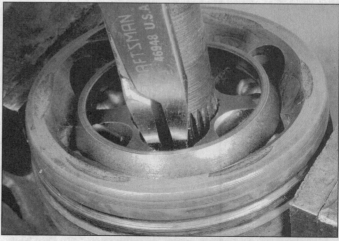

3.4b Remove the snap-ring, slide the joint assembly off and remove the old boot

3.4c Press down on the inner race far enough to allow a ball bearing to be removed - if it's difficult to tilt, gently tap the cage and inner race with a brass punch and hammer

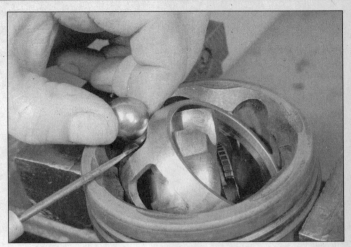

3.4d Pry the balls out of the cage, one at a time, with a blunt screwdriver or wooden tool

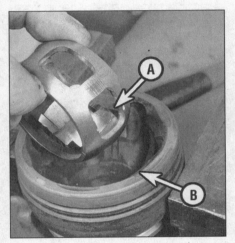

3.4e Tilt the inner race and cage 90-degrees, then align the windows in the cage (A) with the lands of the housing (B) and rotate the inner race up and out of the outer race

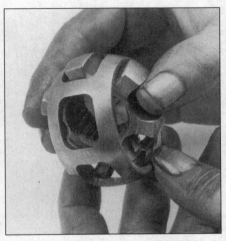

3.4f Align the inner race lands with the cage window and rotate the inner race out of the cage

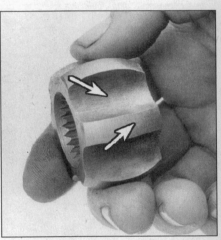

3.4g After cleaning the components with solvent, check the inner race lands and grooves for pitting and score marks

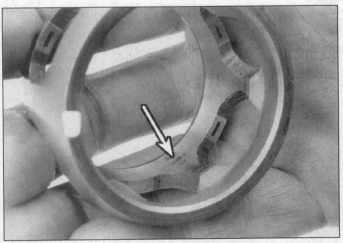

3.4h Check the cage for cracks, pitting and scoring marks - shiny spots are normal and don't affect operation

3.4i With the race and cage tilted at 90-degrees, lower the assembly into the housing

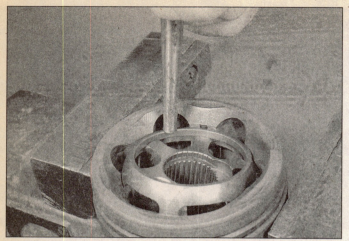

3.4j Rotate the assembly by gently tapping with a hammer and brass punch, then . . .

3.4k . . . press the balls into the cage windows, repeating until all of the balls are installed

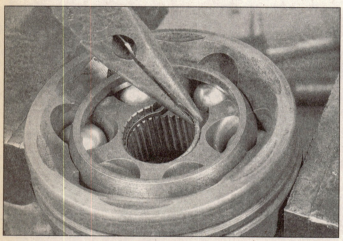

3.4l Use needle-nose pliers to lower the snap-ring into the groove, then . . .

3.4m . . . seat it into the groove with snap-ring pliers

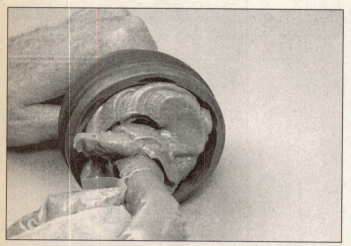

3.4n Pack the CV joint assembly with lubricant through the inner splined hole, then . . .

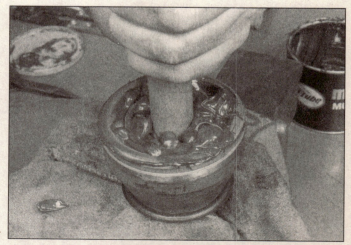

3.4o . . . insert a wooden dowel (with a diameter slightly less than that of the axle) through the splined hole and push down - the dowel will force the grease into the joint - repeat until the bearing is completely packed

3.4p Install the boot on the driveaxle and apply grease to the inside of the axle boot until . . .

3.4q . . . the level is up to the end of axle

3.4r Position the CV joint assembly on the drivebelt, aligning the splines, then use a soft-face hammer to drive the joint onto the driveaxle until the snap-ring is seated in the groove

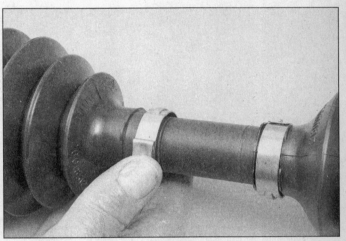

3.4s Seat the inner end of the boot in the groove and install the retaining clamp, then do the same on the other end of the boot - tighten boot clamps with the special tool (see illustration 3.3s)

Notes

Chapter 9 Brakes

Contents

	Section		Section
Brake check...	See Chapter 1	General information..	1
Brake fluid level check...	See Chapter 1	Master cylinder - removal, overhaul and installation.....	7
Brake hoses and lines - inspection and replacement.....	8	Parking brake - adjustment...	10
Brake light switch - removal, installation and adjustment.....	13	Parking brake cables - removal and installation...........	11
Brake disc - inspection, removal and installation.....	4	Power brake booster - inspection, removal and installation.....	12
Brake system bleeding...	9	Rear brake shoes - inspection and replacement.........	5
Disc brake caliper - removal, overhaul and installation.....	3	Rear wheel cylinder - removal, overhaul and installation.....	6
Disc brake pads - replacement.................................	2		

Specifications

General
Brake fluid type... See Chapter 1

Disc brakes
Disc thickness
    Standard... 1.043 inches
    Minimum thickness after refinishing...................... 0.972 inch
    Discard thickness*.. 0.957 inch
Disc runout... 0.004 inch
Disc thickness variation (parallelism)............................ 0.0005 inch
Caliper-to-bracket stop clearance................................. 0.005 to 0.012 inch

*Refer to the marks cast into the drum. They supersede information printed here.

Drum brakes
Drum diameter
    Standard... 8.863 inches
    Service limit.. 8.877 inches
    Discard diameter.. 8.909 inches
Out-of-round (maximum).. 0.006 inch

*Refer to the marks cast into the drum. They supersede information printed here.

Torque specifications   Ft-lbs
Caliper mounting bolts... 38
Brake hose-to-caliper bolt.. 33
Master cylinder-to-booster.. 20
Booster-to-pedal bracket... 25
Wheel lug nuts... See Chapter 1

Chapter 9 Brakes

2.5 A large C-clamp can be used to compress the piston into the caliper for removal

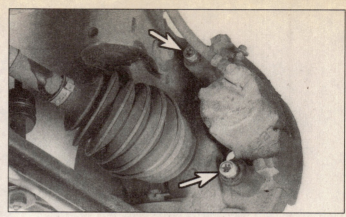

2.6a A No. 50 Torx bit must be used to remove the two caliper mounting bolts (arrows) on 1990 and 1991 models - don't attempt to loosen them with a hex-head wrench, because the bolt heads could be damaged

1 General information

Conventional (non-ABS) brake system

The vehicles covered in this manual are equipped with hydraulically operated front and rear brake systems. The front brakes are disc type and the rear are drum type. Both the front and rear are self adjusting and automatically compensate for pad and shoe wear.

Hydraulic system

The hydraulic system consists of two separate circuits. The master cylinder has separate reservoirs for the two circuits and in the event of a leak or failure in one hydraulic circuit, the other circuit will remain operative. A visual warning of circuit failure or air in the system is given by a warning light activated on the dash.

Proportioner valves

The proportioner valves are designed to provide better front to rear braking balance with heavy brake application. These valves allow more pressure to be applied to the front brakes (under certain braking operations) due to the fact that the rear of the vehicle is lighter and does not require as much braking force.

Power brake booster

The power brake booster, utilizing engine manifold vacuum and atmospheric pressure to provide assistance to the hydraulically operated brakes, is mounted on the firewall in the engine compartment.

Anti-lock Brake System (ABS)

This system is available as an option. It is designed to reduce lost traction while braking. The system is similar to the non-ABS system except for the controller (computer) and related wiring, speed sensors and the hydraulic pump which replaces the master cylinder and power brake booster.

Anti-lock braking occurs only when a wheel is about to lock up (lose traction). Input signals from the wheel speed sensors to the computer are used to determine when a wheel is about to lose traction during braking. Hydraulic pressure will be reduced for the wheel about to loose traction. Diagnosis of this system is beyond the scope of the home mechanic. **Note:** *The ABS system is equipped with a self diagnosis system similar to the engine codes. However, it is necessary to use a scan tool linked into the ALDL for access to these codes. Any failures in the ABS system should be repaired by a dealer service department or other repair shop.*

Parking brake

The parking brake operates the rear brakes only, through cable actuation. It's activated by a pedal mounted on the left side kick panel.

Service

Note: *On models equipped with the Delco Loc II audio system, be sure the lockout feature is turned off before performing any procedure which requires disconnecting the battery.*

After completing any operation involving disassembly of any part of the brake system, always test drive the vehicle to check for proper braking performance before resuming normal driving. When testing the brakes, perform the tests on a clean, dry flat surface. Conditions other than these can lead to inaccurate test results.

Test the brakes at various speeds with both light and heavy pedal pressure. The vehicle should stop evenly without pulling to one side or the other. Avoid locking the brakes because this slides the tires and diminishes braking efficiency and control of the vehicle.

Tires, vehicle load and front-end alignment are factors which also affect braking performance.

2 Disc brake pads - replacement

Refer to illustrations 2.5 and 2.6a through 2.6k

Warning: *Disc brake pads must be replaced on both front wheels at the same time - never replace the pads on only one wheel. Also, the dust created by the brake system may contain asbestos, which is harmful to your health. Never blow it out with compressed air and do not inhale any of it. An approved filtering mask should be worn whenever servicing the brake system. Do not, under any circumstances, use petroleum-based solvents to clean brake parts. Use brake cleaner or denatured alcohol only.*

Note: *When servicing the disc brakes, use only high quality, nationally recognized brand-name pads.*

1 Remove the cover from the brake fluid reservoir, siphon off about two ounces of the fluid into a container and discard it.
2 Loosen the wheel lug nuts, raise the vehicle and support it securely on jackstands.
3 Remove the front wheel, then reinstall three lug nuts (flat side toward the disc) to hold the disc in place. Work on one brake assembly at a time, using the assembled brake for reference if necessary.
4 Inspect the disc carefully as outlined in Section 4. If machining is necessary, follow the information in that Section to remove the disc, at which time the pads can be removed from the calipers as well.
5 Push the piston back into its bore. If necessary, a C-clamp can be used, but a flat bar will usually do the job **(see illustration)**. As the piston is depressed to the bottom of the caliper bore, the fluid in the master cylinder will rise. Make sure that it does not overflow. If necessary, siphon off more of the fluid as directed in Step 1.
6 Refer to the accompanying illustrations and perform the procedure **(see illustrations 2.6a through 2.6k)**.

2.6b The caliper mounting bolts (arrows) on 1992 and later models are regular hex-head type

2.6c Remove the inner pad by snapping it out of the piston

2.6d Unclip the outer pad from the caliper

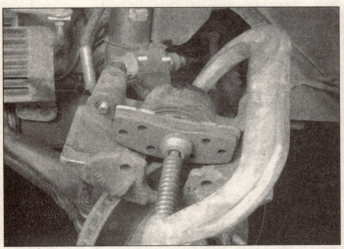
2.6e Now, use the C-clamp to completely bottom the piston in the caliper bore

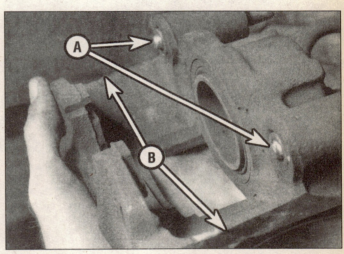
2.6f Inspect the caliper bolts and bushings (A) for damage and the contact surfaces (B) for corrosion

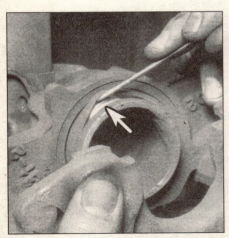
2.6g Carefully peel back the edge of the piston boot and check for corrosion and leaking fluid

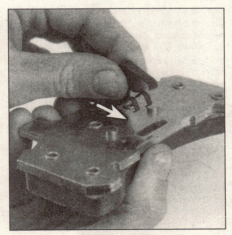
2.6h Snap the inner pad retainer spring into the new pad in the direction shown (arrow) - if the new pad already has a retainer spring, disregard this step

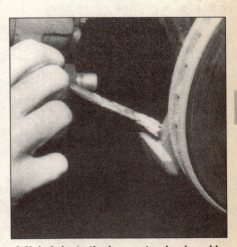
2.6i Lubricate the lower steering knuckle contact surface lightly with white lithium base grease

Chapter 9 Brakes

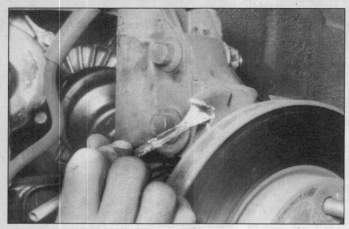

2.6j Apply a light coat of white lithium base grease to the upper steering knuckle-to-caliper contact surface

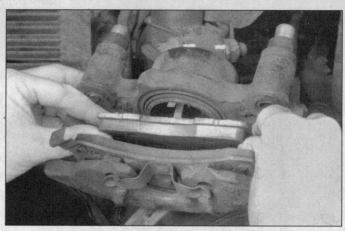

2.6k Place the pads in position and snap the inner pad into place in the piston

3.9 With the caliper padded to catch the piston, use compressed air to force the piston out of its bore - make sure your hands or fingers are not between the piston and caliper!

3.10 Carefully pry the dust boot out of the housing, taking care not to scratch the bore surface

7   When reinstalling the caliper, be sure to tighten the mounting bolts to the torque listed in this Chapter's Specifications. After the job has been completed, firmly depress the brake pedal a few times to bring the pads into contact with the disc.

3 Disc brake caliper - removal, overhaul and installation

Refer to illustrations 3.9, 3.10, 3.11, 3.15, 3.16, 3.17 and 3.18

Warning 1: *This procedure should not be undertaken on vehicles equipped with ABS (Anti-lock Brake System), since special tools are needed to properly bleed the brakes. Take the vehicle to a dealer service department or other repair shop that has the proper tools.*

Warning 2: *Dust created by the brake system may contain asbestos, which is harmful to your health. Never blow it out with compressed air and do not inhale any of it. An approved filtering mask should be worn whenever servicing the brake system. Do not,* under any circumstances, use petroleum-based solvents to clean brake parts. Use brake cleaner or denatured alcohol only.

Note: *If an overhaul is indicated (usually because of fluid leakage) explore all options before beginning the job. New and factory rebuilt calipers are available on an exchange basis, which makes this job quite easy. If it is decided to rebuild the calipers, make sure that a rebuild kit is available before proceeding.*

Removal

1   Remove the cover from the brake fluid reservoir, siphon off two/thirds of the fluid into a container and discard it.
2   Loosen the wheel lug nuts, raise the front of the vehicle and support it securely on jackstands. Remove the front wheel.
3   Reinstall two lug nuts with the flat side against the disc, to hold the disc in place.
4   Bottom the piston in the caliper bore **(see illustration 2.5)**.
    Remove the brake hose inlet fitting bolt and disconnect the fitting. Have a rag handy to catch the spilling fluid and wrap a plastic bag tightly around the end of the hose to pre- vent fluid loss and contamination.
6   Remove the two mounting bolts and lift the caliper from the vehicle **(see illustration 2.6a or 2.6b)**.

Overhaul

7   Refer to Section 2 and remove the brake pads from the caliper.
8   Clean the exterior of the caliper with brake cleaner or denatured alcohol. Never use gasoline, kerosene or any petroleum-based cleaning solvents. Place the caliper on a clean workbench.
9   Position a wooden block or numerous shop rags in the caliper as a cushion, then use compressed air to remove the piston from the caliper **(see illustration)**. Use only enough air pressure to ease the piston out of the bore. If the piston is blown out, even with the cushion in place, it may be damaged.
Warning: *Never place your fingers in front of the piston in an attempt to catch or protect it when applying compressed air, as serious injury could occur.*
10 Carefully pry the dust boot out of the caliper bore **(see illustration)**.
11 Using a wood or plastic tool, remove the

Chapter 9 Brakes

3.11 To avoid damage to the caliper bore or seal groove, it is best to remove the seal with a plastic or wooden tool

3.15 Position the seal in the caliper bore, making sure it is not twisted

3.16 Install the new dust boot in the piston groove with the folds toward the open end of the piston

3.17 Install the piston squarely into the caliper bore

3.18 Use a seal driver to seat the boot in the caliper housing counterbore - if a seal driver isn't available, carefully tap around the circumference of the boot with a drift until it is seated

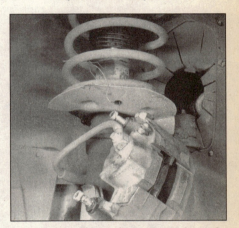

4.2 Suspend the caliper with a piece of wire whenever it is necessary to reposition it (don't let it hang by the brake hose!)

piston seal from the groove in the caliper bore **(see illustration)**. Metal tools may cause bore damage.

12 Remove the caliper bleeder valve, then remove and discard the sleeves and bushings from the caliper ears. Discard all rubber parts.

13 Clean the remaining parts with brake cleaner or denatured alcohol then blow them dry with compressed air.

14 Carefully examine the piston for nicks and burrs and loss of plating. If surface defects are present, parts must be replaced. Check the caliper bore in a similar way. Light polishing with crocus cloth is permissible to remove light corrosion and stains. Discard the mounting bolts if they are corroded or damaged.

15 When assembling, lubricate the piston bores and seal with clean brake fluid. Position the seal in the caliper bore groove **(see illustration)**.

16 Lubricate the piston with clean brake fluid, then install a new boot in the piston groove with the fold toward the open end of the piston **(see illustration)**.

17 Insert the piston squarely into the caliper bore, then apply force to bottom the piston in

the bore **(see illustration)**.

18 Position the dust boot in the caliper counterbore, then use a drift to drive it into position **(see illustration)**. Make sure that the boot is evenly installed below the caliper face.

19 Install the bleeder valve.

20 Install new bushings in the mounting bolt holes and fill the area between the bushings with silicone grease, supplied in the rebuild kit. Push the sleeves into the mounting bolt holes.

Installation

21 Inspect the mounting bolts for excessive corrosion.

22 Place the caliper in position over the disc and mounting bracket, install the bolts and tighten to the specified torque.

23 On 1990 models, check to make sure the clearance between the caliper and the bracket stops is between 0.005 and 0.012 inch.

24 Install the inlet fitting bolt, using new copper washers, then tighten the retaining bolt to the torque listed in this Chapter's

Specifications. It will be necessary to bleed the brakes (see Section 9).

25 Install the wheels and lower the vehicle.

26 After the job has been completed, firmly depress the brake pedal a few times to bring the pads into contact with the disc.

4 Brake disc - inspection, removal and installation

Refer to illustrations 4.2, 4.3, 4.4a, 4.4b, 4.5a and 4.5b

Inspection

1 Loosen the wheel lug nuts, raise the vehicle and support it securely on jackstands. Remove the wheel and install three lug nuts (flat side against the disc) to hold the disc in place.

2 Remove the brake caliper as outlined in Section 3. It is not necessary to disconnect the brake hose. After removing the caliper bolts, suspend the caliper out of the way with a piece of wire **(see illustration)**.

4.3 The brake pads on this vehicle were obviously neglected, as they wore down completely and cut deep grooves into the disc - wear this severe will require replacement of the disc

4.4a Check for runout with a dial indicator, mounted with the indicator needle about 1/2-inch from the outside edge

4.4b If you elect not to have the discs machined, at the very least be sure to break the glaze on the surface with emery cloth

4.5a The minimum wear (or discard) thickness (arrow) is cast into the inside of the disc

3 Visually inspect the disc surface for scoring or damage. Light scratches and shallow grooves are normal after use and may not always be detrimental to brake operation, but deep scoring - over 0.015 inch (0.38 mm) - requires disc removal and refinishing by an automotive machine shop. Be sure to check both sides of the disc (see illustration). If pulsating has been noticed during application of the brakes, suspect disc runout.

4 To check disc runout, place a dial indicator at a point about 1/2-inch from the outer edge of the disc (see illustration). Set the indicator to zero and turn the disc. The indicator reading should not exceed the specified allowable runout limit. If it does, the disc should be refinished by an automotive machine shop. **Note:** *It is recommended that the discs be resurfaced regardless of the dial indicator reading, as this will impart a smooth finish and ensure a perfectly flat surface, eliminating any brake pedal pulsation or other undesirable symptoms related to questionable discs. At the very least, if you elect not to have the discs resurfaced, remove the glazing from the surface with medium-grit emery cloth using a swirling motion* (see illustration).

5 It is absolutely critical that the disc not be machined to a thickness under the specified minimum allowable disc refinish thickness. The minimum wear (or discard) thickness is cast into the inside of the disc (see illustration). This should not be confused with the minimum refinish thickness. The disc thickness can be checked with a micrometer (see illustration).

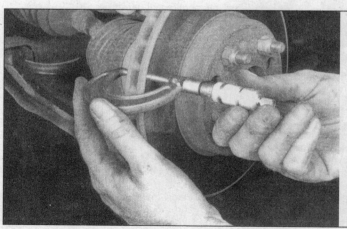

4.5b Measure the thickness of the disc with a micrometer

Chapter 9 Brakes

5.4a If the brake drum will not come off easily, it may be necessary to remove the lanced cutout with a hammer and chisel (on some models it must be drilled out using a 7/16-inch drill bit), then turn the adjuster screw to move the brake shoes away from the drum

5.4b Before removing anything, clean the brake assembly with brake cleaner or denatured alcohol - DO NOT use compressed air to blow the dust from the brake assembly!

5.4c 1992 and later models are equipped with a single horseshoe type retractor spring (arrow)

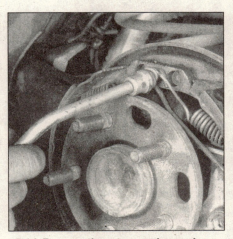

5.4d Remove the return springs using a brake spring tool (1990 and 1991 models)

5.4e Use a small pair of needle nose pliers to remove the actuator spring on 1992 and later models

Removal

6 Remove the lug nuts which were put on to hold the disc in place and remove the disc from the hub.

Installation

7 Place the disc in position over the threaded studs.
8 Install the caliper and brake pad assembly over the disc and position it on the steering knuckle (refer to Section 3 for the caliper installation procedure, if necessary). Tighten the caliper bolts to the torque listed in this Chapter's Specifications.
9 Install the wheel, then lower the vehicle to the ground. Depress the brake pedal a few times to bring the brake pads into contact with the disc. Bleeding of the system will not be necessary unless the fluid hose was disconnected from the caliper. Check the operation of the brakes carefully before placing the vehicle into normal service.

5 Rear brake shoes - inspection and replacement

Refer to illustrations 5.4a through 5.4ae and 5.5

Warning: Drum brake shoes must be replaced on both wheels at the same time - never replace the shoes on only one wheel. Also, the dust created by the brake system may contain asbestos, which is harmful to your health. Never blow it out with compressed air and do not inhale any of it. An approved filtering mask should be worn whenever servicing the brake system. Do not, under any circumstances, use petroleum-based solvents to clean brake parts. Use brake cleaner or denatured alcohol only.

Caution: Whenever the brake shoes are replaced, the retractor and hold-down springs should also be replaced. Due to the continuous heating/cooling cycle that the springs are subjected to, they lose their tension over a period of time and may allow the shoes to drag on the drum and wear at a much faster rate than normal. When replacing the rear brake shoes, use only high quality nationally recognized brand name parts.

1 Loosen the wheel lug nuts, raise the vehicle and support it securely on jackstands.
2 Release the parking brake.
3 Remove the wheel. **Note:** All four rear shoes must be replaced at the same time, but to avoid mixing up parts, work on only one brake assembly at a time.
4 Refer to the accompanying illustrations and perform the brake shoe inspection and, if necessary, the replacement procedure **(see illustrations 5.4a through 5.4ae)**. **Note:** If the brake drum cannot be easily pulled off the axle and shoe assembly, make sure that the parking brake is completely released, then squirt some penetrating oil around the center hub area. Allow the oil to soak in and try to

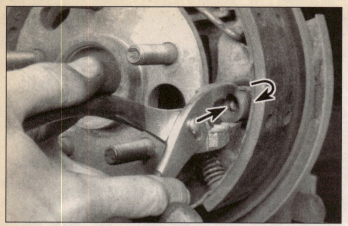

5.4f Remove the hold down springs and pins by pushing in with pliers and turning (arrows) (1990 and 1991 models)

5.4g Wedge a flat bladed screwdriver under the spring and pry it out of the leading brake shoe, then remove the shoe (1992 models)

5.4h Lift up on the actuator lever (1990 and 1991 models) and remove the actuating link from the anchor pin pivot along with the actuator lever and return spring (arrows)

5.4i Remove the parking brake strut (all models) - on 1992 models, also remove the adjusting screw assembly, noting the direction in which it is installed

5.4j With the shoe assembly spread to clear the hub flange, lift it from the backing plate (1990 and 1991 models)

pull the drum off. If the drum still cannot be pulled off, the brake shoes will have to be retracted. This is accomplished by first removing the lanced cutout in the brake drum with a hammer and chisel **(see illustration 5.4a)**. With the cutout removed, pull the lever off the adjusting screw wheel with one small screwdriver while turning the adjusting wheel with another small screwdriver, moving the shoes away from the drum. The drum may now be pulled off.

5.4k On 1992 models, lift the retractor spring from the trailing brake shoe and swing the shoe out from the hub area to gain access to the parking brake cable

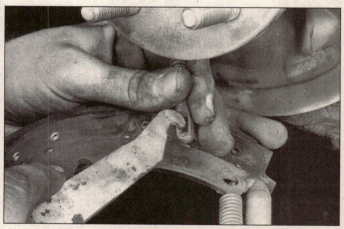

5.4l Disconnect the parking brake lever from the cable and remove the shoe assembly (1990 and 1991 models)

Chapter 9 Brakes

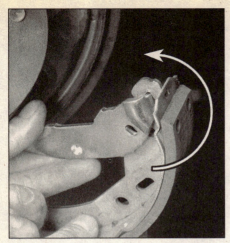

5.4m On 1992 models, rotate the brake shoe to release the parking brake lever from the shoe

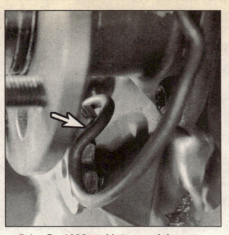

5.4n On 1992 and later models, use a screwdriver to pry the retractor spring over the alignment peg (arrow) - to install the new shoes on these models, perform steps 5.4r and 5.4s, then reverse the removal steps and proceed to step 5.4ac

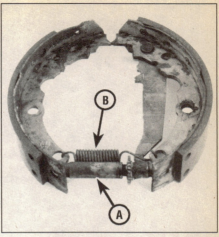

5.4o Remove he adjusting screw (A) and spring (B) from the shoe assembly, making sure to note the direction in which they are installed (1990 and 1991 models)

5.4p On 1990 and 1991 models, remove the parking brake lever by prying off the C-clip

5.4q Install the parking brake lever on the new brake shoe and press the C-clip into place with needle-nose pliers (1990 and 1991 models)

9.4r Lubricate the contact surfaces of the backing plate with white lithium base grease (all models)

5.4s Lubricate the adjuster screw with white lithium base grease prior to installation (all models)

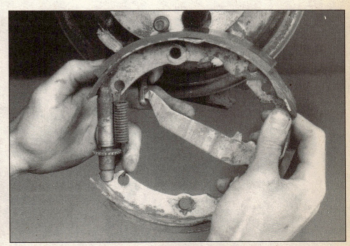

5.4t Connect the parking brake lever to the cable (1990 and 1991 models)

5.4u Spread the brake assembly apart sufficiently to clear the hub flange and raise it into position (1990 and 1991 models)

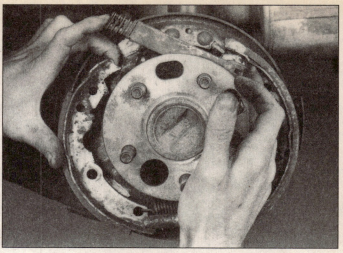
5.4v Install the parking brake strut and spring (1990 and 1991 models)

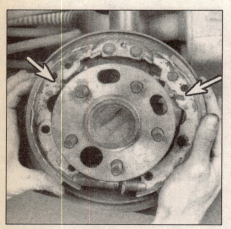
5.4w Make sure the parking brake strut is positioned in the shoes properly (arrows)

5.4x Install the hold-down pin and spring on the primary brake shoe (1990 and 1991 models)

5.4y Install the actuator link and lever to the secondary brake shoe (1990 and 1991 models)

5.4z Install the actuator lever return spring (1990 and 1991 models)

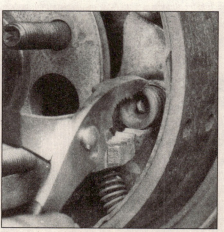
5.4aa Install the hold-down pin and spring on the secondary brake shoe (1990 and 1991 models)

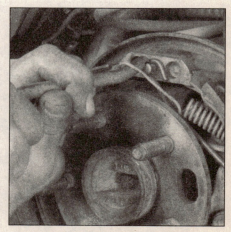
5.4ab Install the return springs (1990 and 1991 models)

Chapter 9 Brakes

5.4ac Center the brake shoe assembly so the drum will slide over it (all models)

5.4ad Adjust the star wheel so the drum fits over the shoes with a slight drag (all models)

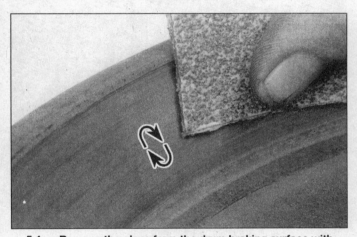

5.4ae Remove the glaze from the drum braking surface with emery cloth, working in a circular motion

5.5 The drum has a maximum permissible diameter cast into it (arrow) which is a wear dimension, not a refinish dimension

5 Before reinstalling the drum it should be checked for cracks, score marks, deep scratches and hard spots, which will appear as small discolored areas. If the hard spots cannot be removed with emery cloth or if any of the other conditions listed above exist, the drum must be taken to an automotive machine shop to have it turned. **Note:** *It is recommended that the drums be resurfaced regardless of the surface appearance, as this will impart a smooth finish and ensure a perfectly round drum, eliminating any brake pedal pulsation or other undesirable symptoms related to questionable drums. At the very least, if you elect not to have the drums resurfaced, remove the glazing from the surface with medium-grit emery cloth using a swirling motion. If the drum will not "clean up" before the maximum service limit is reached in the machining operation, the drum will have to be replaced with a new one. The maximum wear diameter is cast into the each brake drum* **(see illustration)**.

6 Install the brake drum on the axle flange.

7 Mount the wheel, install the lug nuts, then lower the vehicle. Tighten the lug nuts to the torque listed in this Chapter's Specifications.

8 Make a number of forward and reverse stops to adjust the brakes until a satisfactory pedal action is obtained.

6 Rear wheel cylinder - removal, overhaul and installation

Refer to illustrations 6.4, 6.6 and 6.12
Warning: *This procedure should not be undertaken on vehicles equipped with ABS (Anti-lock Brake System), since special tools are needed to properly bleed the brakes. Take the vehicle to a dealer service department or other repair shop that has the proper tools.*
Note: *If an overhaul is indicated (usually because of fluid leakage or sticky operation) explore all options before beginning the job. New wheel cylinders are available, which makes this job quite easy. If it is decided to rebuild the wheel cylinder, make sure that a rebuild kit is available before proceeding.*

Removal

1 Raise the rear of the vehicle and support it securely on jackstands.

2 Remove the brake shoe assembly (see Section 5).

3 Carefully clean all dirt and foreign material from around the wheel cylinder.

4 Disconnect the brake fluid inlet line **(see illustration)**. Do not pull the brake line away from the wheel cylinder.

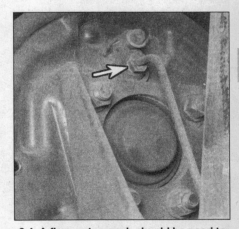

6.4 A flare nut wrench should be used to disconnect the brake line (arrow)

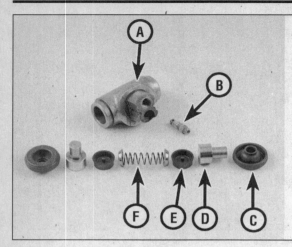

6.6 Wheel cylinder components - exploded view

A   Wheel cylinder body
B   Bleeder screw
C   Boot
D   Piston
E   Seal
F   Spring

6.12 A wood block (arrow) should be used to hold he wheel cylinder in position

7.2 Unplug the fluid level sensor connector

7.4 Disconnect the brake lines using a flare-nut wrench, if available

5   Remove the wheel cylinder from the brake backing plate and place it on a clean workbench. Immediately plug the brake line to prevent fluid loss and contamination.

Overhaul

6   Remove the bleeder valve, seals, pistons, boots and spring assembly from the cylinder body **(see illustration)**.
7   Clean the wheel cylinder with brake fluid, denatured alcohol or brake system cleaner. **Warning:** *Do not, under any circumstances, use petroleum-based solvents to clean brake parts.*
8   Use compressed air to remove excess fluid from the wheel cylinder and to blow out the passages.
9   Check the cylinder bore for corrosion and scoring. Crocus cloth may be used to remove light corrosion and stains, but the cylinder must be replaced with a new one if the defects cannot be removed easily, or if the bore is scored.
10 Lubricate the new seals with brake fluid.
11 Assemble the brake cylinder, making sure the boots are properly seated.

Installation

12 Place the wheel cylinder in position and use a wooden block wedged between the axle flange and the cylinder body to hold it in place **(see illustration)**.
13 Install the wheel cylinder retainer over the wheel cylinder using a 1-1/8 inch 12 point socket to press it into place.
14 Connect the brake line and install the brake shoe assembly.

7   Master cylinder - removal, overhaul and installation

Warning: *This procedure should not be undertaken on vehicles equipped with ABS (Anti-lock Brake System), since special tools are needed to properly bleed the brakes. Take the vehicle to a dealer service department or other repair shop that has the proper tools.*
Note: *Before deciding to overhaul the master cylinder, check on the availability and cost of a new or factory-rebuilt unit and the availability of a rebuild kit..*

Removal

Refer to illustrations 7.2 and 7.4

1   Detach the cable from the negative battery terminal.
2   Unplug the electrical connector from the fluid level sensor switch **(see illustration)**.
3   Place rags under the line fittings and prepare caps or plastic bags to cover the ends of the lines once they're disconnected. **Caution:** *Brake fluid will damage paint. Cover all painted parts and be careful not to spill fluid during this procedure.*
4   Loosen the fittings at the ends of the brake lines where they enter the master cylinder **(see illustration)**. To prevent rounding off the flats on the fittings, use a flare-nut wrench, which wraps around the hex.
5   Pull the brake lines away from the master cylinder and plug the ends to prevent contamination.
6   Remove the two mounting nuts and detach the master cylinder from the vehicle.
7   Remove the reservoir cover and reservoir diaphragm, then discard any remaining fluid.

Overhaul

Refer to illustrations 7.11, 7.12, 7.16, 7.18, 7.19a, 7.19b, 7.19c, 7.19d, 7.19e, 7.19f, 7.20 and 7.32

8   Mount the master cylinder in a vise. Be sure to line the vise jaws with rags or blocks of wood to prevent damage to the cylinder body.

Chapter 9 Brakes

7.11 Press down on the piston and remove the primary piston lock ring

7.12 Remove the primary piston assembly

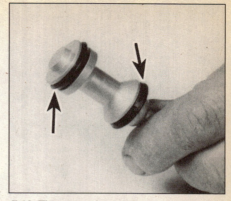

7.16 The secondary piston seals must be installed with the lips facing out as shown

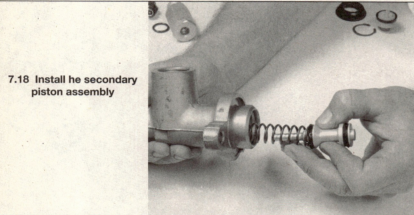

7.18 Install he secondary piston assembly

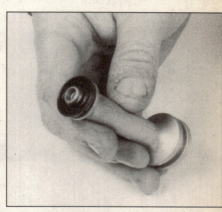

7.19a The primary piston seal must be installed with the lip facing away from the piston

9 Drive out the roll pins with a 1/8-inch punch. Pull straight up on the reservoir assembly and separate it from the master cylinder body. Remove and discard the two O-rings.

10 Remove the proportioner valve caps **(see illustration 7.9)**. Using needle-nose pliers, carefully remove the O-rings, the springs, the proportioner valve pistons and seals. Make sure you don't scratch or otherwise damage the piston stems or the bores. Set each proportioner valve assembly aside.

11 Remove the primary piston lock ring by depressing the piston and prying the ring out with a screwdriver **(see illustration)**.

12 Remove the primary piston assembly from the bore **(see illustration)**.

13 Remove the secondary piston assembly from the bore. It may be necessary to remove the master cylinder from the vise and invert it, carefully tapping it against a block of wood to expel the piston.

14 Clean the master cylinder body, the primary and secondary piston assemblies, the proportioner valve assemblies and the reservoir in brake cleaner or denatured alcohol and dry them off with unlubricated compressed air or a clean (lint-free) shop rag. **Warning:** *DO NOT, under any circumstances, use petroleum-based solvents to clean brake parts.*

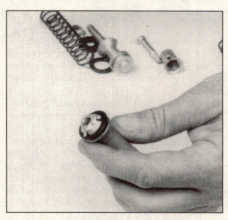

7.19b Install the seal guard over the seal

15 Inspect the master cylinder piston bore for corrosion and score marks. If any corrosion or damage in the bore is evident, replace the master cylinder body - don't use abrasives to try to clean it up.

16 Remove the old seals from the secondary piston assembly and install the new seals with the cup lips facing out **(see illustration)**.

17 Attach the spring retainer to the secondary piston assembly.

18 Lubricate the cylinder bore with clean brake fluid and install the spring and secondary piston assembly **(see illustration)**.

19 Disassemble the primary piston assembly, noting the locations of the parts, then lubricate the new seals with clean brake fluid and install them on the piston **(see illustrations)**.

7.19c Place the primary piston spring in position

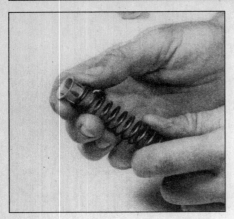

7.19d Insert the spring retainer into the spring

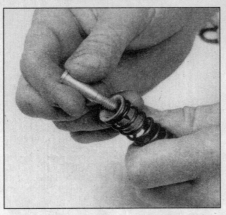

7.19e Insert the spring retaining bolt through the retainer and spring and thread it into the piston

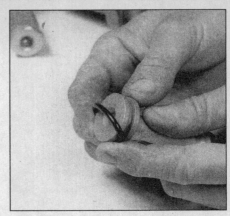

7.19f Lubricate the O-ring with clean brake fluid, then install it on the piston

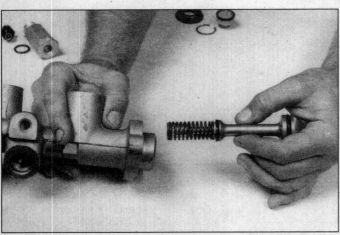

7.20 Insert the primary piston assembly into the bore

7.32 Install the reservoir diaphragm in the cover

20 Install the primary piston assembly in the cylinder bore **(see illustration)**, depress it and install the lock ring.

21 Inspect the proportioner valves for corrosion and score marks. Replace them if necessary.

22 Lubricate the new O-rings and proportioner valve seals with the silicone grease supplied with the rebuild kit. Also lubricate the stem of the proportioner valve pistons.

23 Install the new seals on the proportioner valve pistons with the seal lips facing toward the cap assembly.

24 Install the proportioner valve pistons and seals in the master cylinder body.

25 Install the springs in the master cylinder body.

26 Install the new O-rings in their respective grooves in the proportioner valve cap assemblies.

27 Install the proportioner valve caps in the master cylinder and tighten them to the torque listed in this Chapter's Specifications.

28 Inspect the reservoir for cracks and distortion. If any damage is evident, replace it.

29 Lubricate the new reservoir O-rings with clean brake fluid and press them into their respective grooves in the master cylinder body. Make sure they're properly seated.

30 Lubricate the reservoir fittings with clean brake fluid and install the reservoir on the master cylinder body by pressing it straight down.

31 Drive in new reservoir retaining (roll) pins. Make sure you don't damage the reservoir or master cylinder body.

32 Inspect the reservoir diaphragm and cover for cracks and deformation. Replace any damaged parts with new ones and attach the diaphragm to the cover **(see illustration)**.

Note: *Whenever the master cylinder is removed, the complete hydraulic system must be bled. The time required to bleed the system can be reduced if the master cylinder is filled with fluid and bench bled (refer to Steps 33 through 36) before it's installed on the vehicle.*

33 Insert threaded plugs of the correct size into the brake line outlet holes and fill the reservoirs with brake fluid. The master cylinder should be supported so brake fluid won't spill during the bench bleeding procedure.

34 Loosen one plug at a time and push the piston assembly into the bore to force air from the master cylinder. To prevent air from being drawn back in, the appropriate plug must be replaced before allowing the piston to return to its original position.

35 Stroke the piston three or four times for each outlet to ensure that all air has been expelled.

36 Since high pressure isn't involved in the bench bleeding procedure, there is an alternative to the removal and replacement of the plugs with each stroke of the piston assembly. Before pushing in on the piston assembly, remove one of the plugs completely. Before releasing the piston, however, instead of replacing the plug, simply put your finger tightly over the hole to keep air from being drawn back into the master cylinder. Wait several seconds for the brake fluid to be drawn from the reservoir into the piston bore, then repeat the procedure. When you push down on the piston it'll force your finger off the hole, allowing the air inside to be expelled. When only brake fluid is being ejected from the hole, replace the plug and go on to the other port.

37 Refill the master cylinder reservoirs and install the diaphragm and cover assembly.

Note: *The reservoirs should only be filled to the top of the reservoir divider to prevent overflowing when the cover is installed.*

Chapter 9 Brakes

8.2a Use a small screwdriver to pry the retaining clip from the notch in the brake line end

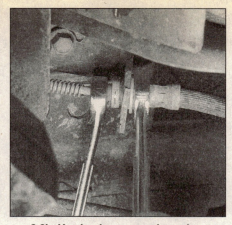

8.2b Use back-up wrenches when loosening brake lines

Installation

38 Carefully install the master cylinder by reversing the removal steps, fill the reservoir with the recommended fluid (see Chapter 1, Section 4), then bleed the brakes at the master cylinder and at each wheel (see Section 9).

8 Brake hoses and lines - inspection and replacement

Refer to illustration 8.2a and 8.2b

Inspection

1 About every six months, with the vehicle raised and placed securely on jackstands, the flexible hoses which connect the steel brake lines with the front and rear brake assemblies should be inspected for cracks, chafing of the outer cover, leaks, blisters and other damage. These are important and vulnerable parts of the brake system and inspection should be complete. A light and mirror will prove helpful for a thorough check. If a hose exhibits any of the above conditions, replace it with a new one.

Replacement

Warning: *This procedure should not be undertaken on vehicles equipped with ABS (Anti-lock Brake System), since special tools are needed to properly bleed the brakes. Take the vehicle to a dealer service department or other repair shop that has the proper tools.*

Front brake hose

2 Using a back-up wrench, disconnect the brake line from the hose fitting, being careful not to bend the frame bracket or brake line **(see illustrations)**.
3 Use pliers to remove the U-clip from the female fitting at the bracket, then remove the hose from the bracket.
4 At the caliper end of the hose, remove the bolt from the fitting block, then remove the hose and the copper gaskets on either side of the fitting block.
5 When installing the hose, always use new copper gaskets on either side of the fitting block and lubricate all bolt threads with clean brake fluid before installation.
6 With the fitting flange engaged with the caliper locating ledge, attach the hose to the caliper.
7 Without twisting the hose, install the female fitting in the hose bracket. It will fit the bracket in only one position.
8 Install the U-clip retaining the female fitting to the frame bracket.
9 Using a back-up wrench, attach the brake line to the hose fitting.
10 When the brake hose installation is complete, there should be no kinks in the hose. Make sure the hose does not contact any part of the suspension. Check this by turning the wheels to the extreme left and right positions. If the hose makes contact, remove the hose and correct the installation as necessary.

Rear brake hose

11 Using a back-up wrench, disconnect the hose at both ends, being careful not to bend the bracket or steel lines **(see illustration 8.2b)**.
12 Remove the two U-clips with pliers and separate the female fittings from the brackets.
13 Unbolt the hose retaining clip and remove the hose.
14 Without twisting the hose, install the female ends of the hose in the frame brackets. It will fit the bracket in only one position.
15 Install the U-clips retaining the female end to the bracket.
16 Using a back-up wrench, attach the steel line fittings to the female fittings. Again, be careful not to bend the bracket or steel line.
17 Check that the hose installation did not loosen the frame bracket. Tighten the bracket if necessary.
18 Fill the master cylinder reservoir and bleed the system (see Section 9).

Metal brake lines

19 When replacing brake lines, it is important that the proper replacements be purchased. Do not use copper pipe for any brake system connections. Purchase proper brake line from a dealer parts department, auto parts store or brake specialist.
20 Prefabricated brake line, with the tube ends already flared and connectors installed, is available at auto parts stores or dealers. These lines are also bent to the proper shapes if necessary.
21 If prefabricated lengths are not available, obtain the recommended steel tubing and fittings to match the line to be replaced. Determine the correct length by measuring the old brake line (a piece of string can usually be used for this) and cut the new tubing to length, allowing about 1/2-inch extra for flaring the ends.
22 Install the fitting onto the cut tubing and flare the ends of the line with an ISO flaring tool.
23 If necessary, carefully bend the line to the proper shape. A tube bender is recommended for this. **Caution:** *Do not crimp or damage the line.*
24 When installing the new line make sure it is well supported in the brackets and has plenty of clearance between moving or hot components.
25 After installation, check the master cylinder fluid level and add fluid as necessary. Bleed the brake system as outlined in the next Section and test the brakes carefully before placing the vehicle into normal service.

9 Brake system bleeding

Refer to illustration 9.8
Warning 1: *This procedure should not be undertaken on vehicles equipped with ABS (Anti-lock Brake System), since special tools are needed to properly bleed the brakes. Take the vehicle to a dealer service department or other repair shop that has the proper tools.*
Warning 2: *Wear eye protection whenever bleeding the brake system. If the fluid comes in contact with your eyes, immediately rinse them with water and seek a physician's advice.*
Note: *Bleeding the hydraulic system is necessary to remove any air that manages to find its way into the system when it has been opened during removal and installation of a hose, line, caliper or master cylinder.*

1 It will probably be necessary to bleed the system at all four brakes if air has entered the system due to low fluid level, or if the brake lines have been disconnected at the master cylinder.
2 If a brake line was disconnected only at a wheel, then only that caliper or wheel cylinder must be bled.
3 If a brake line is disconnected at a fitting located between the master cylinder and any

9.8 When bleeding the brakes, a hose is connected to the bleeder valve at the caliper (or wheel cylinder) and then submerged in brake fluid. Air will be seen as bubbles in the container or in the tube. All air must be expelled before continuing to the next wheel

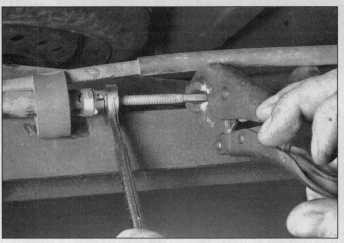

10.4 With a pair of locking pliers clamped to the end of the threaded rod to prevent it from turning, turn the adjusting nut until the right wheel can just be turned in a rear direction but not in a forward direction

of the brakes, that part of the system served by the disconnected line must be bled.
4   Remove any residual vacuum from the brake power booster by applying the brake several times with the engine OFF.
5   Remove the master cylinder reservoir cover and fill the reservoir with brake fluid. Reinstall the cover. **Note:** *Check the fluid level often during the bleeding operation and add fluid as necessary to prevent the fluid level from falling low enough to allow air bubbles into the master cylinder.*
6   Have an assistant on hand, as well as a supply of new brake fluid, an empty clear plastic container, a length of 3/16-inch plastic, rubber or vinyl tubing to fit over the bleeder valve and a wrench to open and close the bleeder valve.
7   Beginning at the right rear wheel, loosen the bleeder valve slightly, then tighten it to a point where it is snug but can still be loosened quickly and easily.
8   Place one end of the tubing over the bleeder valve and submerge the other end in brake fluid in the container **(see illustration)**.
9   Have the assistant pump the brakes slowly a few times to get pressure in the system, then hold the pedal firmly depressed.
10 While the pedal is held depressed, open the bleeder valve just enough to allow a flow of fluid to leave the valve. Watch for air bubbles to exit the submerged end of the tube. When the fluid flow slows after a couple of seconds, close the valve and have your assistant release the pedal.
11 Repeat Steps 9 and 10 until no more air is seen leaving the tube, then tighten the bleeder valve and proceed to the left rear wheel, the right front wheel and the left front wheel, in that order, and perform the same procedure. Be sure to check the fluid in the master cylinder reservoir frequently.
12 Never use old brake fluid. It contains moisture which will deteriorate the brake system components.

13 Refill the master cylinder with fluid at the end of the operation.
14 Check the operation of the brakes. The pedal should feel solid when depressed, with no sponginess. If necessary, repeat the entire process. **Warning:** *Do not operate the vehicle if you are in doubt about the effectiveness of the brake system.*

10 Parking brake - adjustment

Refer to illustration 10.4

1   Depress the parking brake pedal exactly three ratchet clicks.
2   Raise the vehicle and support it securely on jackstands.
3   Before adjusting, make sure the equalizer nut groove is lubricated liberally with multi-purpose grease.
4   Tighten the adjusting nut **(see illustration)** until the right rear wheel can just be turned rearward with two hands, but locks when forward motion is attempted.
5   Release the parking brake lever and check to make sure the rear wheels turn freely in both directions with no drag.
6   Lower the vehicle.

11.5 First remove the clip (A) from the parking brake lever and then depress the tangs (B) and slide the parking brake cable assembly through the bracket

11 Parking brake cables - removal and installation

Refer to illustrations 11.5 and 11.10

Front cable

1   Raise the rear of the vehicle and support it securely on jackstands.
2   Loosen the equalizer nut (adjusting nut) and separate the front cable from the connector and the equalizer.
3   Remove the clips at the frame and maneuver the cable out of the wire bracket.
4   Remove the left side under-dash panel to gain access to the parking brake pedal mechanism.
5   Remove the clip from the cable end at the parking brake lever **(see illustration)** and push the cable and casing assembly through the control assembly while simultaneously squeezing the tangs.
6   Slide the cable casing out of the equalizer and disconnect the cable from the cable joiner.
7   Installation is the reverse of the removal procedure. Refer to Section 10 for the cable adjustment procedure.

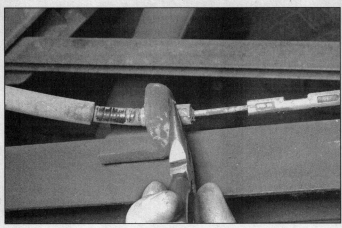

11.10 Depress the tangs and separate the cable casing from the mounting bracket

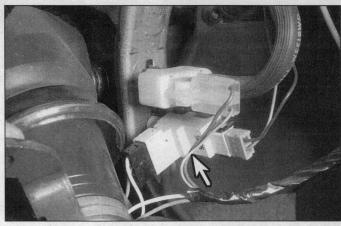

13.3 The brake light switch (arrow) is located to the right of the steering column at the end of the mounting bracket

Rear cable

8 Raise the vehicle and support it securely on jackstands.
9 Loosen the equalizer (adjusting) nut and disconnect the rear cable from the equalizer.
10 Disconnect the cable casing at the frame mounting bracket by depressing the tangs on the retainer with a pair of pliers (see illustration).
11 Remove the brake drum and brake shoes as described in Section 5. Disconnect the parking brake cable from the parking brake lever.
12 Using a pair of pliers, depress the tangs on the cable casing retainer and push the cable and casing out through the backing plate.
13 Installation is the reverse of the removal procedure. Be sure to adjust the cable as described in Section 10.

12 Power brake booster - inspection, removal and installation

Note: *This procedure applies to non-ABS-equipped models only.*

1 The power brake booster unit requires no special maintenance apart from periodic inspection of the vacuum hose and the case. Early models have an in-line filter which should be inspected periodically and replaced if clogged or damaged.
2 Dismantling of the power unit requires special tools and is not ordinarily performed by the home mechanic. If a problem develops, it is recommended that a new or factory rebuilt unit be used.
3 Remove the nuts attaching the master cylinder to the booster and carefully pull the master cylinder forward until it clears the mounting studs. Use caution so as not to bend or kink the brake lines.
4 Disconnect the vacuum hose where it attaches to the power brake booster.
5 From the passenger compartment, disconnect the power brake pushrod from the top of the brake pedal.
6 Also from this location, remove the nuts attaching the booster to the firewall.
7 Carefully lift the booster unit away from the firewall and out of the engine compartment.
8 To install, place the booster into position and tighten the retaining nuts. Connect the brake pedal.
9 Install the master cylinder and vacuum hose.
10 Carefully test the operation of the brakes before placing the vehicle in normal operation.

13 Brake light switch - removal, installation and adjustment

Refer to illustration 13.3

Removal

1 The brake light switch, or stop light switch as it is sometimes called, is located on a bracket at the top of the brake pedal. The switch activates the brake lights at the rear of the vehicle whenever the pedal is depressed.
2 Remove the under-dash cover and disconnect the wiring to the courtesy lamp in this panel.
3 Locate the switch at the top of the brake pedal (see illustration). If equipped with cruise control, there will be another switch very similar in appearance. The brake light switch is the one towards the end of the bracket.
4 Disconnect the negative battery cable and secure it out of the way so that it cannot come back into contact with the battery post.
5 Disconnect the wiring connectors at the brake light switch.
6 Depress the brake pedal and pull the switch out of its clip. The switch appears to be threaded, but it is designed to be pushed into and out of the clip and not turned.

Installation and adjustment

7 With the brake pedal depressed, push the new switch into the clip. Note that audible clicks will be heard as this is done.
8 Pull the brake pedal fully rearward against the pedal stop until the clicking sounds can no longer be heard. This action will automatically move the switch the proper amount and no further adjustment will be required. **Caution:** *Do not apply excessive force during this adjustment procedure, as power booster damage may result.*
9 Connect the wiring at the switch and the battery. With an assistant, check that the rear brake lights are functioning properly.

Notes

Chapter 10
Suspension and steering systems

Contents

	Section
Balljoint - check and replacement	8
Chassis lubrication	See Chapter 1
Control arm - removal and installation	5
Electronic Level Control (ELC) - general information	14
Front hub and wheel bearing assembly - removal and installation	6
Front stabilizer bar and bushings - removal and installation	4
General information	1
Power steering fluid level check	See Chapter 1
Power steering pump - removal and installation	19
Power steering system - bleeding	20
Rear axle assembly - removal and installation	13
Rear coil springs - removal, inspection and installation	11
Rear hub and wheel bearing assembly - removal and installation	12
Rear shock absorbers - removal, inspection and installation	9
Steering gear boots - replacement	18
Steering gear - removal and installation	17
Steering knuckle and hub - removal and installation	7
Steering system - general information	15
Steering wheel - removal and installation	21
Strut assembly - removal and installation	2
Strut - replacement	3
Suspension and steering check	See Chapter 1
Tie-rod ends - removal and installation	16
Tire and tire pressure checks	See Chapter 1
Tire rotation	See Chapter 1
Track bar assembly - removal and installation	10
Wheel alignment - general information	23
Wheels and tires - general information	22

Specifications

Torque specifications
Ft-lbs (unless otherwise indicated)

Front suspension
Control arm pivot bolt nuts	61
Balljoint pinch bolt nut	33
Stabilizer bar	
Bushing clamp nuts	33
Reinforcement plate bolts	40
Front hub and wheel bearing assembly bolts	70
Strut assembly	
Damper shaft nut	65
Strut-to-body nuts	18
Strut-to-steering knuckle bolt nuts	140
Balljoint kit bolts	Specifications included with instructions

Rear suspension
Rear hub and wheel bearing assembly bolts	45
Shock absorber	
Upper mounting nuts	16
Lower mounting bolt nut	44
Trailing arm-to-bracket bolt	83
Track arm	
Upper bolt and nut	
1990 through 1993	35
1994 on	48
Lower bolt and nut	
1990 through 1993	44
1994 on	63

Steering system
Steering gear mounting bolt nuts	70
Tie-rod end-to-steering arm nut	30 (52 maximum to install cotter pin)
Intermediate shaft pinch-bolt	35
Steering wheel hub nut	30

Chapter 10 Suspension and steering systems

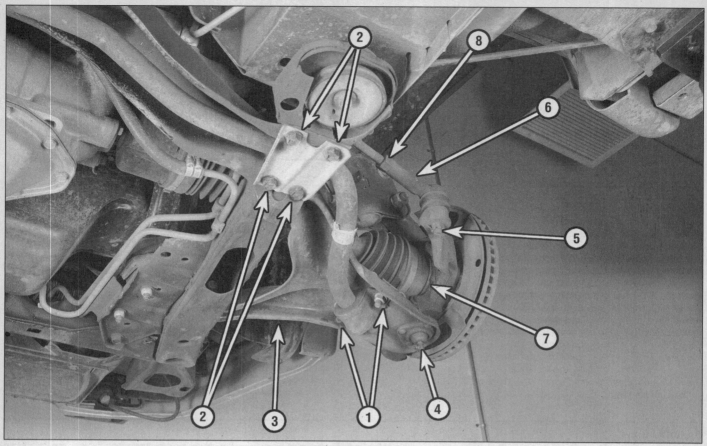

1.1a Front suspension components

1. Stabilizer bar clamp bushing nuts
2. Stabilizer bar reinforcement plate nuts
3. Lower control arm
4. Balljoint
5. Outer tie-rod balljoint stud nut
6. Outer tie-rod
7. Steering knuckle
8. Outer tie-rod jam nut

1 General information

Refer to illustrations 1.1a, 1.1b and 1.2

Front suspension is by MacPherson struts. The steering knuckle is located by a control arm and both front control arms are connected by a stabilizer bar **(see illustrations)**.

The rear suspension features a beam-type axle with coil springs, located by trailing arms and a track bar **(see illustration)**. Damping is handled by vertically-mounted shock absorbers located between the axle and the chassis.

The rack-and-pinion steering gear is located behind the engine and actuates the steering arms which are integral with the steering knuckles. Power assist is standard and the steering column is designed to collapse in the event of an accident. **Note:** *These vehicles use a combination of standard and metric fasteners on the various suspension and steering components, so it would be a good idea to have both types of tools available when beginning work.*

Frequently, when working on the suspension or steering system components, you may come across fasteners which seem impossible to loosen. These fasteners on the underside of the vehicle are continually subjected to water, road grime, mud, etc., and can become rusted or frozen, making them extremely difficult to remove. In order to unscrew these stubborn fasteners without damaging them (or other components), be sure to use lots of penetrating oil and allow it to soak in for a while. Using a wire brush to clean exposed threads will also ease removal of the nut or bolt and prevent damage to the threads. Sometimes a sharp blow with a hammer and punch will break the bond between a nut and bolt threads, but care must be taken to prevent the punch from slipping off the fastener and ruining the threads. Heating the stuck fastener and surrounding area with a torch sometimes helps too, but isn't recommended because of the obvious dangers associated with fire. Long breaker bars and extension, or "cheater", pipes will increase leverage, but never use an extension pipe on a ratchet - the ratcheting mechanism could be damaged. Sometimes tightening the nut or bolt first will help to break it loose. Fasteners that require drastic measures to remove should always be replaced with new ones.

Since most of the procedures dealt with in this Chapter involve jacking up the vehicle and working underneath it, a good pair of jackstands will be needed. A hydraulic floor jack is the preferred type of jack to lift the vehicle, and it can also be used to support certain components during various operations. **Warning:** *Never, under any circumstances, rely on a jack to support the vehicle while working on it. Whenever any of the suspension or steering fasteners are loosened or removed they must be inspected and, if necessary, replaced with new ones of the same part number or of original equipment quality and design. Torque specifications must be followed for proper reassembly and component retention. Never attempt to heat or straighten any suspension or steering components. Instead, replace any bent or damaged part with a new one.* **Note:** *On models equipped with the Delco Loc II audio system, be sure the lockout feature is turned off before performing any procedure which requires disconnecting the battery.*

Chapter 10 Suspension and steering systems

1.1b Front suspension components

1. Steering knuckle
2. Outer tie-rod
3. Outer tie-rod jam nut
4. Coil spring
5. Strut-to-steering knuckle bolts/nuts
6. Strut assembly
7. Lower control arm rear pivot bolt/nut
8. Lower control arm front pivot bolt/nut

1.2 Rear suspension components

1. Lower track bar-to-axle beam bolt
2. Shock absorber
3. Coil spring
4. Track bar
5. Upper track bar-to-body bracket bolt
6. Axle beam assembly

2.3 Scribe or paint an alignment mark between the strut and the steering knuckle to ensure proper reassembly, then remove the two large strut-to-steering knuckle nuts and bolts (arrows)

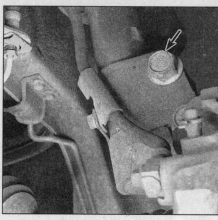

2.5 Remove the brake hose bracket bolt (arrow) and detach the brake hose bracket from the strut assembly

2.6 Remove the three upper strut mounting nuts (arrows) from the strut tower and remove the strut from the vehicle - DON'T loosen the center nut!

2 Strut assembly - removal and installation

Refer to illustration 2.3, 2.5 and 2.6

1 Loosen the front wheel lug nuts.
2 Raise the vehicle and support it securely on jackstands. Remove the front wheels.
3 Mark the relationship of the strut to the steering knuckle **(see illustration)**.
4 Remove the strut-to-steering knuckle nuts, bolts and washer plate.
5 Disconnect the brake hose bracket from the strut **(see illustration)**.
6 Remove the upper mounting nuts **(see illustration)**, disengage the strut from the steering knuckle and detach it from the vehicle.
7 Inspect the strut and coil spring assembly for leaking fluid, dents, damage and corrosion. If the strut is damaged, see Section 3.
8 To install the strut, place it in position with the studs extending up through the shock tower. Install the nuts and tighten them to the torque listed in this Chapter's Specifications.
9 Attach the strut to the steering knuckle, then insert the strut-to-steering knuckle bolts and washer plate.
10 Align the marks you made on the knuckle and strut. Install the nuts on the strut-to-steering knuckle bolts and tighten them to the torque listed in this Chapter's Specifications.
11 Attach the brake hose bracket to the strut.
12 Install the wheels and lower the vehicle. Tighten the lug nuts to the torque listed in the Chapter 1 Specifications.

3 Strut - replacement

Refer to illustrations 3.4, 3.5, 3.6a, 3.6b and 3.6c

Note: *You'll need a spring compressor for*

3.4 A spring compressor is essential for disassembling the strut and coil spring assembly

3.5 After the spring has been compressed, remove the damper shaft nut

this procedure. Spring compressors are available on a daily rental basis at most auto parts stores or equipment yards.

1 If the struts or coil springs exhibit the telltale signs of wear (leaking fluid, loss of damping capability, chipped, sagging or cracked coil springs) explore all options before beginning any work. The struts are not serviceable and must be replaced if a problem develops. However, strut assemblies complete with springs may be available on an exchange basis, which eliminates much time and work. Whichever route you choose to take, check on the cost and availability of parts before disassembling your vehicle. **Warning:** *Disassembling a strut assembly is a potentially dangerous undertaking and utmost attention must be directed to the job at hand, or serious bodily injury may result. Use only a high-quality spring compressor and carefully follow the manufacturer's instructions furnished with the tool. After removing the coil spring from the strut assembly, set it aside in a safe, isolated area (a steel cabinet is preferred).*

2 Remove the strut and spring assembly (see Section 2).
3 Mount the strut assembly in a vise. Line the vise jaws with wood or rags to prevent damage to the unit and don't tighten the vise excessively.
4 Install the spring compressor in accordance with the manufacturer's instructions **(see illustration)**. Compress the spring until you can wiggle the mount assembly and spring seat.
5 To loosen the damper shaft nut, hold the shaft with a box-end wrench while loosening the shaft nut with another box-end wrench **(see illustration)**.
6 Disassemble the strut assembly **(see illustrations)** and lay out the parts in exactly the same order as shown **(see illustration)**.
Warning: *When removing the compressed spring, lift it off very carefully and set it in a*

Chapter 10 Suspension and steering systems

3.6a Remove the bearing cap . . .

3.6b . . . and the upper spring seat and insulator from the damper shaft

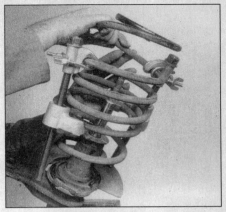

3.6c Remove the compressed spring assembly - use extreme caution whenever handling the spring

Installation

7 Assemble the shaft bushings and clamps on the bar, place the bar in position, install the reinforcement plate bolts and the bushing clamp nuts and tighten all fasteners to the torque listed in this Chapter's Specifications.
8 Install the wheels and lower the vehicle. Tighten the lug nuts to the torque listed in the Chapter 1 Specifications.

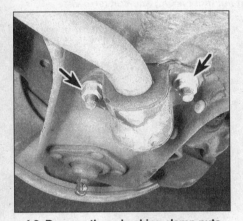

4.3 Remove these bushing clamp nuts (arrows) on both control arms

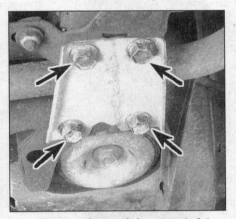

4.4 Remove these reinforcement plate bolts (arrows) and remove the stabilizer bar

safe place, such as a steel cabinet. Keep the ends of the spring away from your body.
7 Inspect all rubber parts for damage, cracking and hardness and replace as necessary.
8 Reassembly is the reverse of disassembly. Be careful not to damage the damper shaft or the strut will leak. When installing the spring, be sure the spring ends mesh with the spring anti-rotation stops provided by the spring upper and lower seats.

4 Front stabilizer bar and bushings - removal and installation

Removal

Refer to illustrations 4.3 and 4.4
1 Loosen the lug nuts on both front wheels, raise the vehicle and support it securely on jackstands. Remove the front wheels.
2 Remove the steering shaft pinch bolt.
3 Remove the nuts from the stabilizer bar-to-control arm bushing clamps (see illustration).
4 Remove the bolts from the stabilizer bar

reinforcement plate (see illustration).
5 Remove the stabilizer bar, clamps and bushings.
6 Inspect the bushings for wear and damage and replace them if necessary. To remove them, pry the bushing clamp off with a screwdriver and pull the bushings off the bar. To ease installation, spray the inside and outside of the bushings with a silicone-based lubricant. Do not use petroleum-based lubricants on any rubber suspension part!

5 Control arm - removal and installation

Removal

Refer to illustrations 5.3, 5.4, 5.5a and 5.5b
1 Loosen the wheel lug nuts, raise the front of the vehicle and support it securely on jackstands. Apply the parking brake and block the rear wheels to keep the vehicle from rolling off the jackstands. Remove the wheel.
2 If only one control arm is being removed, disconnect only that end of the stabilizer bar. If both control arms are being removed, disconnect both ends (see Section 4).
3 Remove the balljoint stud-to-steering knuckle nut and pinch bolt (see illustration).

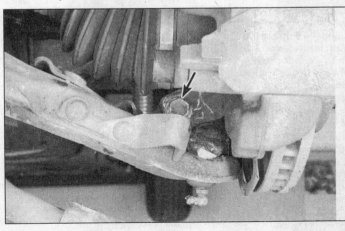

5.3 Before separating the lower control arm from the steering knuckle, remove the pinch bolt nut (not visible in this illustration) and pinch bolt (arrow)

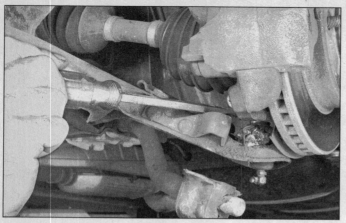

5.4 To separate the lower control arm from the steering knuckle, pry them apart with a pry bar or large screwdriver

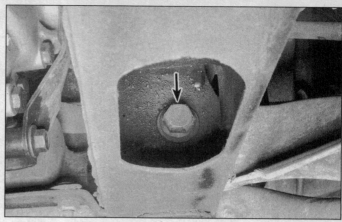

5.5a To detach the lower control arm from the subframe, remove the front pivot nut (not shown) and bolt (arrow) . . .

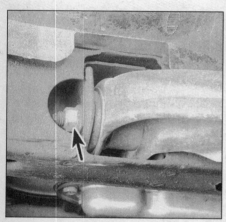

5.5b . . . and the rear pivot nut (arrow) and bolt (threaded portion visible in this illustration)

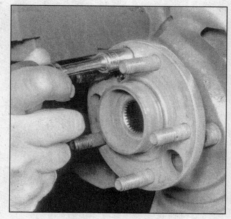

6.6 You'll need a No. 55 Torx bit to remove the hub bolts - DO NOT use an Allen wrench or the bolts will be damaged

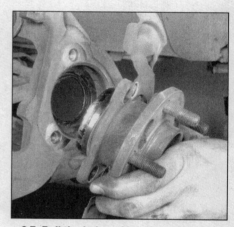

6.7 Pull the hub and bearing assembly and the disc shield out of the steering knuckle

4   Position a large pry bar or screwdriver between the control arm and the steering knuckle, then pop the balljoint stud loose from the knuckle **(see illustration)**. **Caution:** *When removing the balljoint from the knuckle, be careful not to overextend the inner CV joint or it may be damaged.*

5   Remove the two control arm pivot bolts and detach the control arm **(see illustrations)**.

6   The control arm bushings are replaceable, but special tools and expertise are necessary to do the job. Carefully inspect the bushings for hardening, excessive wear and cracks. If they appear to be worn or deteriorated, take the control arm to a dealer service department or repair shop.

Installation

7   Position the control arm in the suspension support and install the pivot bolts. Don't tighten the bolts completely yet.

8   Insert the balljoint stud into the steering knuckle boss, install a NEW pinch bolt and nut and tighten the nut to the torque listed in this Chapter's Specifications.

9   Place a floor jack under the balljoint and raise it until the vehicle just raises off the jackstand. Tighten the control arm pivot bolts to the torque listed in this Chapter's Specifications.

10   Reattach the stabilizer bar (see Section 4).

11   Install the wheel and lower the vehicle. Tighten the lug nuts to the torque specified in Chapter 1.

12   Drive the vehicle to a dealer service department or an alignment shop to have the front wheel alignment checked and, if necessary, adjusted.

6   Front hub and wheel bearing assembly - removal and installation

Refer to illustrations 6.6, 6.7, 6.8 and 6.9

Note: *The front hub and wheel bearing assembly is sealed-for-life and must be replaced as a unit.*

1   Loosen the wheel lug nuts, raise the front of the vehicle and support it securely on jackstands. Don't place the jackstands under the control arms. Apply the parking brake and block the rear wheels to keep the vehicle from rolling off the jackstands. Remove the wheel.

2   Disconnect the stabilizer bar from the control arm (see Section 4).

3   Remove the balljoint-to-steering knuckle nut and separate the control arm from the knuckle (see Section 5).

4   Remove the caliper from the steering knuckle and hang it out of the way with a piece of wire (see Chapter 9, Section 3).

5   Pull the disc off the hub and remove the driveaxle (see Chapter 8, Section 2).

6   Using a no. 55 Torx bit, remove the three hub retaining bolts through the opening in the flange **(see illustration)**.

7   Wiggle the hub and bearing assembly back-and-forth and pull it out of the steering knuckle, along with the disc shield **(see illustration)**.

8   If the hub and bearing assembly is being replaced with a new one, it's a good idea to replace the dust seal in the back of the steering knuckle. Pry it out of the knuckle with a screwdriver **(see illustration)**.

9   Drive the new dust seal into the knuckle

Chapter 10 Suspension and steering systems

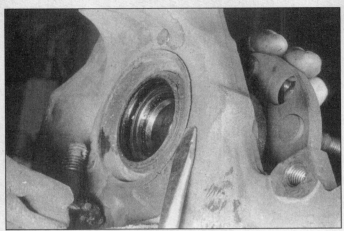

6.8 Pry the seal out of the knuckle with a screwdriver

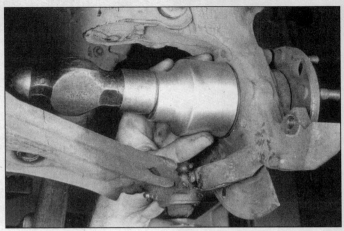

6.9 Using a large socket, drive the new seal into place

8.3a Check for movement between the balljoint and steering knuckle (arrow) when prying up

8.3b With the pry bar positioned between the steering knuckle boss and the balljoint, pry down and check for play in the balljoint - if there is any play, replace the balljoint

with a large socket or a seal driver and a hammer (see illustration). Try not to cock the seal in the bore.
10   Install a new O-ring around the rear of the bearing and push it up against the bearing flange.
11   Clean the mating surfaces on the steering knuckle, bearing flange and knuckle bore. Lubricate the outside diameter of the bearing and the seal lips with high-temperature grease and insert the hub and bearing into the steering knuckle. Position the disc shield and install the three bolts, tightening them to the torque listed in this Chapter's Specifications.
12   Install the driveaxle (see Chapter 8, Section 2).
13   Attach the control arm to the steering knuckle (see Section 5).
14   Reconnect the stabilizer bar to the control arm (see Section 4).
15   Install the brake disc and caliper (see Chapter 9, Section 3).
16   Install the hub nut and tighten it securely. Prevent the axle from turning by inserting a screwdriver through the caliper and into a cooling vane in the brake disc (see Chapter 8, Section 2).
17   Install the wheel, lower the vehicle and tighten the lug nuts to the torque listed in the Chapter 1 Specifications.
18   Tighten the hub nut to the torque listed in the Chapter 8 Specifications.

7 Steering knuckle and hub - removal and installation

Removal

1   Loosen the wheel lug nuts, raise the front of the vehicle and support it securely on jackstands. Apply the parking brake and block the rear wheels to keep the vehicle from rolling off the jackstands. Remove the wheel.
2   Remove the hub nut. Insert a screwdriver through the caliper and into a disc cooling vane to prevent the driveaxle from turning (see Chapter 8, Section 2).
3   Remove the caliper and suspend it out of the way with a piece of wire (see Chapter 9, Section 3). Lift the disc off the hub.
4   Mark the position of the two strut-to-knuckle nuts and remove them (see illustration 2.3). Don't drive out the bolts at this time.
5   Separate the control arm balljoint from the steering knuckle (see Section 5).
6   Push the driveaxle out of the hub (see Chapter 8, Section 2). Hang the driveaxle with a piece of wire to prevent damage to the inner CV joint.
7   Support the knuckle and drive out the two strut-to-knuckle bolts with a soft-face hammer. Remove the steering knuckle assembly from the strut.

Installation

8   Position the knuckle in the strut and insert the two bolts and nuts, but don't tighten them at this time.
9   Install the driveaxle in the hub.
10   Connect the control arm to the steering knuckle (see Section 5) and tighten the pinch bolt nut to the torque listed in this Chapter's Specifications.
11   Align the strut-to-knuckle nuts with the previously applied marks and tighten them to the torque listed in this Chapter's Specifications.

9.2 Place a floor jack under the rear axle before removing a shock absorber - otherwise, when you disconnect the lower shock mounting bolt, the axle - which is under spring pressure from the coil springs - could fly down, out of control

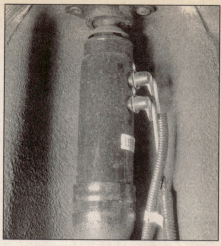

9.3 On vehicles with ELC, the left shock (shown) is connected to the compressor and to the right shock (the right shock has only one connector, for the crossover line) - to detach an air line, simply squeeze the spring clip and pull off the connector; to reattach a line, lubricate the O-ring seals with petroleum jelly, then push the connector all the way onto the fitting until it pops into place

12 Install the brake disc and caliper (see Chapter 9, Sections 4 and 3, respectively).
13 Tighten the hub nut securely.
14 Install the wheel, lower the vehicle and tighten the lug nuts to the torque listed in the Chapter 1 Specifications.
15 Tighten the hub nut to the torque listed in the Chapter 8 Specifications.

8 Balljoint - check and replacement

Check

Refer to illustrations 8.3a and 8.3b

1 Raise the front of the vehicle and support it securely on jackstands. Apply the parking brake and block the rear wheels to keep the vehicle from rolling off the jackstands.
2 Visually inspect the rubber seal for damage, deterioration and leaking grease. If any of these conditions are noticed, the balljoint should be replaced.
3 Place a large pry bar under the balljoint and attempt to push the balljoint up. Next, position the pry bar between the steering knuckle and control arm and pry down **(see illustrations)**. If any movement is seen or felt during either of these checks, a worn-out balljoint is indicated.
4 Have an assistant grasp the tire at the top and bottom and move the top of the tire in-and-out.
5 Separate the control arm from the steering knuckle (see Section 5). Using your fingers (don't use pliers), try to twist the stud in the socket. If the stud turns, replace the balljoint.

Replacement

6 Loosen the wheel lug nuts, raise the front of the vehicle and support it securely on jackstands. Apply the parking brake and block the rear wheels to keep the vehicle from rolling off the jackstands. Remove the wheel.
7 Separate the control arm from the steering knuckle (see Section 5). Temporarily insert the balljoint stud back into the steering knuckle (loosely). This will ease balljoint removal after Step 9 has been performed, as well as hold the assembly stationary while drilling out the rivets.
8 Using a 1/8-inch drill bit, drill a pilot hole into the center of each balljoint-to-control arm rivet. Be careful not to damage the CV joint boot in the process.
9 Using a 1/2-inch drill bit, drill the head off each rivet. Work slowly and carefully to avoid deforming the holes in the control arm.
10 Loosen (but don't remove) the stabilizer bar-to-control arm bushing clamp nuts. Pull the control arm and balljoint down to remove the balljoint stud from the steering knuckle, then dislodge the balljoint from the control arm.
11 Position the new balljoint on the control arm and install the bolts (supplied in the balljoint kit) from the top of the control arm. Tighten the bolts to the torque specified in the new balljoint instruction sheet.
12 Insert the balljoint into the steering knuckle, install the nut, tighten it to the torque listed in this Chapter's Specifications and install a new cotter pin. It may be necessary to tighten the nut some to align the cotter pin hole with an opening in the nut, which is acceptable. Never loosen the nut to allow cotter pin insertion.
13 Tighten the stabilizer bar-to-control arm bushing clamp nuts to the torque listed in this Chapter's Specifications.
14 Install the wheel, lower the vehicle and tighten the lug nuts to the torque listed in the Chapter 1 Specifications. It's a good idea to take the vehicle to a dealer service department or alignment shop to have the front end alignment checked and, if necessary, adjusted.

9 Rear shock absorbers - removal, inspection and installation

Removal

Refer to illustrations 9.2, 9.3, 9.4a, 9.4b, 9.4c, 9.4d, 9.4e, 9.4f and 9.5

1 Loosen the rear wheel lug nuts, raise the rear of the vehicle and support it securely on jackstands.
2 Support the axle with a jack **(see illustration)** and remove the rear wheels.
3 If the vehicle has Electronic Level Control (ELC), the air lines must be disconnected before the shocks can be removed. The air lines use spring clip connections with molded sealing shoulders in the retainer and on the end of the air line with double O-ring seals. Before you release a connector, clean it and the immediate area thoroughly, then squeeze the spring clip **(see illustration)** and detach the fitting from the shock (the left shock has an inlet line and a crossover line; the right shock has only the crossover line).
4 Remove the upper shock mounting nuts **(see illustrations)**.

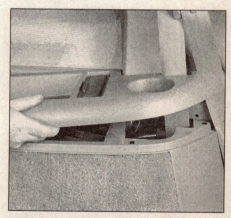

9.4a To get at the upper shock absorber mounting nuts, pry this tray loose with your hand . . .

Chapter 10 Suspension and steering systems

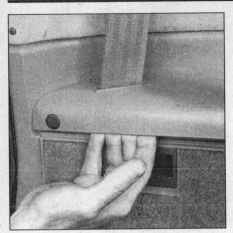

9.4b ... run your hand along the edge of the tray and pop loose each fastener ...

9.4c ... tip the tray up and unplug the electrical connector for the courtesy light, and remove the tray ...

5 Remove the lower shock mounting nuts and bolts **(see illustration)**.
6 Detach the shock absorbers.

Inspection

7 Holding each shock upright, grasp it at each end and pump it in and out several times. The action should be smooth, with no binding or dead spots. Check for fluid leakage. Replace a shock if it's leaking, or if the pumping action is rough. Always replace the shocks in pairs.

Installation

8 Hold each shock in position and install the bolts. Tighten the upper nuts to the torque listed in this Chapter's Specifications. Lower the vehicle and tighten the lower bolts to the torque listed in this Chapter's Specifications.

9 If the vehicle has ELC, reattach the air line(s). Lubricate the O-rings with petroleum jelly or a suitable equivalent, and push the air line and connector all the way into the fitting. After inflating the system to 100 psi, check for leaks with a soapy water solution, a stethoscope or an electronic leak detector.

10 Track bar assembly - removal and installation

Refer to illustrations 10.3a and 10.3b

1 Loosen the rear wheel lug nuts, raise the vehicle, place it securely on jacks and remove the rear wheels.
2 Raise the rear axle to normal ride height with a jack.
3 Support the track bar while you remove the track bar-to-axle pivot bolt and the track bar-to-frame pivot bolt **(see illustrations)**.

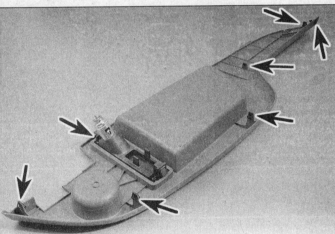

9.4d This view of the tray, removed and upside down, shows all the pop fasteners

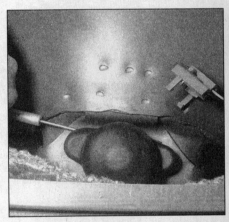

9.4e Pry loose this rubber protective cap from the top of the shock ...

9.4f ... and remove these three nuts (arrows)

9.5 Lower shock mounting nut and bolt (arrows)

10.3a To remove the track bar, remove the track bar-to-axle nut (not visible in this illustration) and bolt (arrow) ...

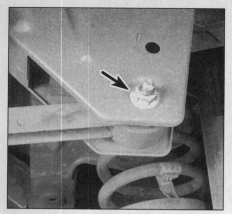

10.3b ... then remove the track bar-to-frame nut (arrow) and bolt and remove the bar

12.3 Using a No. 55 Torx bit, remove the four hub and bearing assembly bolts

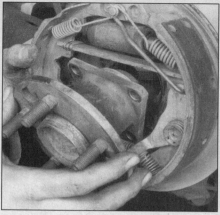

12.4a Angle the hub and bearing assembly out through the brake assembly

12.4b Temporarily reinstall two bolts (arrows) to retain the brake assembly to the trailing arm, rather then let it hang by the hydraulic line

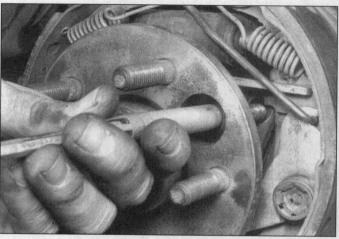

12.5 A magnet is useful for installing the bolts

4   Remove the track bar. Inspect the bushing for cracks and deterioration. A dealer service department or automotive machine shop can install new bushings in the track bar, if necessary.
5   Installation is the reverse of removal. Be sure to tighten the upper and lower bolts to the torque listed in this Chapter's Specifications.

11 Rear coil springs - removal, inspection and installation

Removal

1   Loosen the rear wheel lug nuts, raise the rear of the vehicle and support it securely on jackstands.
2   Support the axle with a jack (see illustration 9.2) and remove the rear wheels.
3   Detach the shock absorbers (see Section 9).
4   Carefully lower the jack, supporting the axle until all compression is relieved and remove the coil springs.

Inspection

5   Inspect the insulators at the upper and lower ends of each coil spring. If an insulator is cracked or dried out, replace it.
6   Check the springs for breakage, nicks and cracks. Replace a spring if you notice any of these conditions.

Installation

7   To install the coil springs, make sure they're seated properly, then carefully raise the axle back into position with the jack.
8   Install the shock absorbers (see Section 9).

12 Rear hub and wheel bearing assembly - removal and installation

Refer to illustrations 12.3, 12.4a, 12.4b and 12.5
Note: The rear hub and wheel bearing assembly is sealed-for-life and must be replaced as a unit.

Removal

1   Loosen the wheel lug nuts, raise the vehicle and support it securely on jackstands. Remove the wheel.
2   Pull the brake drum from the hub. If difficulty is encountered, refer to Chapter 9 for the removal procedure.
3   Using a No. 55 Torx bit, remove the four hub-to-trailing arm bolts, accessible by turning the hub flange so that the circular cutout exposes each bolt (see illustration). Save the upper rear bolt for last, because there isn't much clearance between the bolt and the parking brake strut.
4   Remove the hub and bearing assembly from its seat, maneuvering it out through the brake assembly. Reinstall two bolts through the brake backing plate into the trailing arm to avoid hanging the brake assembly by the hydraulic line (see illustrations).

Installation

5   Position the hub and bearing assembly to the trailing arm and align the holes in the backing plate. Install the bolts, beginning with the upper rear bolt. A magnet is useful in

Chapter 10 Suspension and steering systems

13.3a On vehicles with Electronic Level Control (ELC), you'll have to disconnect the height sensor link from the ball and stud on the axle; first, pop loose the locking spring clip with a small screwdriver . . .

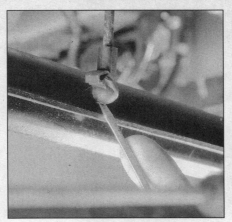

13.3b . . . then pry the link head off the ball stud until it pops loose

13.7 To detach the axle assembly from the frame, remove the trailing arm pivot nut (arrow) and bolt from each trailing arm (left trailing arm shown, other side similar)

guiding the bolts through the hub flange and into position **(see illustration)**. After all four bolts have been installed, tighten them to the torque listed in this Chapter's Specifications.
6 Install the brake drum and wheel. Lower the vehicle and tighten the wheel lug nuts to the torque listed in the Chapter 1 Specifications.

13 Rear axle assembly - removal and installation

Refer to illustrations 13.3a, 13.3b and 13.7

1 Loosen the rear wheel lug nuts, raise the vehicle, place it securely on jackstands and remove the rear wheels.
2 Support the axle at the center with a floor jack.
3 If the vehicle is equipped with Electronic Level Control (ELC), disconnect the height sensor link from the ball stud on the axle **(see illustrations)**.
4 Disconnect the rear parking brake cables from the middle cable at the connectors, detach the retaining brackets and remove the cable retainer clips from the brackets on the trailing arms (see Chapter 9, Section 11). Pull the rear cables through the holes in the trailing arms so they're clear of the axle assembly. Disconnect the brake lines from the hoses at the trailing arm pivot points (see Chapter 9, Section 8).
5 Disconnect the lower ends of the shock absorbers (see Section 9) and the track bar (see Section 10). Support the lower end of the track bar with a piece of sturdy wire.
6 Carefully lower the axle assembly until the coil springs are no longer under compression, then remove the coil springs and insulators.
7 Support the forward ends of the trailing arms with stands and remove the pivot nuts and bolts **(see illustration)**.
8 Remove the axle assembly.
9 Inspect the pivot bushings. If they're cracked or dried out, replace them.
10 Installation is the reverse of removal. Be sure to tighten the trailing arm pivot bolt nuts to the torque listed in this Chapter's Specifications.
11 Bleed the brakes after the job is complete (see Chapter 9, Section 9).

14 Electronic Level Control (ELC) - general information

Refer to illustrations 14.1a and 14.1b

Some vehicles are equipped with Electronic Level Control (ELC), which adjusts the rear trim height in response to changes in the vehicle load. The ELC system consists of an air compressor assembly **(see illustration)**, an air dryer, an exhaust solenoid, a compressor relay, a height sensor **(see illustration)**, air adjustable shocks (see illustration 9.3) and the tubing between various components.
When the ignition is turned on, and weight is added to the vehicle, the compressor is automatically activated; when the igni-

14.1a The compressor assembly (plastic cover removed)

1 Compressor head
2 Compressor air dryer
3 Compressor motor

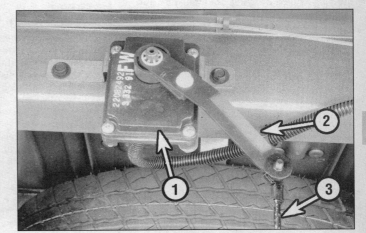

14.1b Height sensor assembly

1 Height sensor
2 Height sensor link arm
3 Height sensor link

16.2 Loosen the jam nut on the inner tie-rod

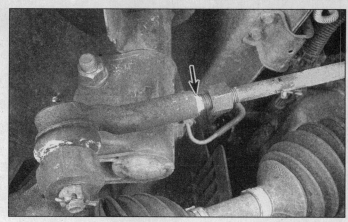

16.3 To ensure that the tie-rod end is screwed onto the inner tie-rod correctly, mark the threaded portion of the inner tie-rod with paint (arrow)

tion is turned off, and excess weight is removed, the exhaust solenoid, which is hooked up to the positive side of the battery, is activated, relieving system pressure. The solenoid, which is located in the compressor assembly, also acts as a relief valve, limiting maximum pressure to about 180 psi.

15 Steering system - general information

Warning: *Whenever any of the steering fasteners are removed, they must be inspected and, if necessary, replaced with new ones of the same part number or of original equipment quality and design. Torque specifications must be followed for proper reassembly and component retention. Never attempt to heat or straighten any suspension or steering components. Instead, replace any bent or damaged part with a new one.*

All vehicles covered by this manual have power rack-and-pinion steering systems. The components making up the system are the steering wheel, steering column, rack and pinion steering gear, tie-rods and tie-rod ends. The power steering system has a belt-driven pump to provide hydraulic pressure.

In the power steering system, the motion of turning the steering wheel is transferred through the column to the pinion shaft in the rack-and-pinion assembly. Teeth on the pinion shaft are meshed with teeth on the rack, so when the shaft is turned, the rack is moved left or right in the housing. A rotary control valve in the rack-and-pinion unit directs hydraulic fluid under pressure from the power steering pump to either side of the integral rack piston, which is connected to the rack, thereby reducing steering force. Depending on which side of the piston this hydraulic pressure is applied to, the rack will be forced either left or right, which moves the tie-rods, etc. If the power steering system loses hydraulic pressure it will still function manually, though with increased effort.

The steering column is a collapsible, energy-absorbing type, designed to compress in the event of a front end collision to minimize injury to the driver. The column also houses the ignition switch lock, key warning buzzer, turn signal controls, headlight dimmer control and windshield wiper controls. The ignition and steering wheel can both be locked while the vehicle is parked.

In addition to the standard steering column, optional tilt and key release versions are also offered. The tilt model can be set in five different positions, while with the key release model the ignition key is locked in the column until a lever is depressed to extract it.

Because disassembly of the steering column is more often performed to repair a switch or other electrical part than to correct a problem in the steering, the upper steering column disassembly and reassembly procedure is included in Chapter 12.

16 Tie-rod ends - removal and installation

Removal

Refer to illustrations 16.2, 16.3 and 16.4

1 Loosen the wheel lug nuts, raise the front of the vehicle and support it securely on jackstands. Apply the parking brake and block the rear wheels to keep the vehicle from rolling off the jackstands. Remove the wheel.
2 Loosen the tie-rod end jam nut **(see illustration)**.
3 Mark the relationship of the tie-rod end to the threaded portion of the tie-rod **(see illustration)**. This will ensure the toe-in setting is restored when reassembled.
4 Disconnect the tie-rod end from the steering knuckle arm with a puller **(see illustration)**.
5 Unscrew the tie-rod end from the tie-rod.

Installation

6 Thread the tie-rod end onto the tie-rod

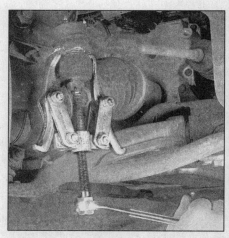

16.4 After removing the cotter pin from the castellated nut on the tie-rod end ball stud, and loosening - but not removing - the nut, install a two-jaw puller and separate the tie-rod end from the steering knuckle

to the marked position and connect the tie-rod end to the steering arm. Install the castellated nut and tighten it to the torque listed in this Chapter's Specifications. Install a new cotter pin.
7 Tighten the jam nut securely and install the wheel. Lower the vehicle and tighten the lug nuts to the torque listed in the Chapter 1 Specifications.
8 Have the front end alignment checked by a dealer service department or an alignment shop.

17 Steering gear - removal and installation

Removal

Refer to illustrations 17.3, 17.7, 17.8, 17.9, 17.10a and 17.10b

1 Disconnect the cable from the negative

Chapter 10 Suspension and steering systems

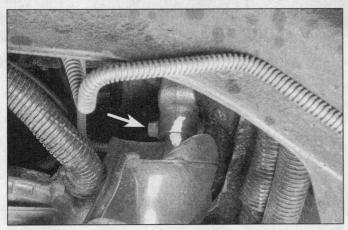

17.3 Scribe or paint an alignment mark on the coupler between the intermediate shaft and the steering gear stub shaft, then remove the coupler pinch bolt (arrow)

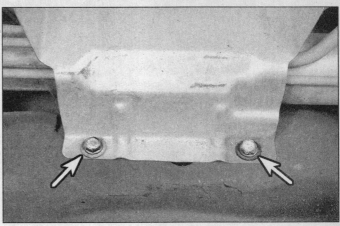

17.7 You'll have to remove these bolts (arrows) and the heat shield to get at the fluid pipe clip (engine/transaxle/steering gear subframe assembly removed from vehicle)

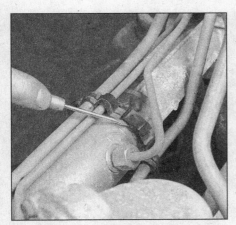

17.8 To remove the fluid pipe clip from the steering gear, put the point of a small screwdriver here and pop the clip open (engine/transaxle/steering gear subframe assembly removed from vehicle)

17.9 Use a flare nut wrench to break loose the threaded fittings from the steering gear box

17.10a The left (driver's side) steering gear mounting bolt and nut (arrow)

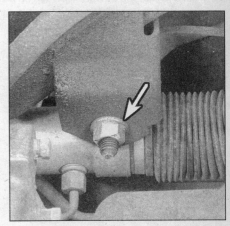

17.10b The right (passenger's side) steering gear mounting bolt and nut (arrow)

battery terminal.
2 Loosen the front wheel lug nuts, raise the front of the vehicle and support it securely on jackstands. Apply the parking brake and block the rear wheels to keep the vehicle from rolling off the jackstands. Remove both front wheels.
3 Roll back the boot at the bottom of the steering column to expose the flange and steering coupler assembly. Mark the coupler and steering column shaft **(see illustration)**.
4 Remove the coupler pinch bolt **(see illustration 17.3)**.
5 Separate the tie-rod ends from the steering arms (see Section 16).
6 Support the rear of the subframe assembly with a jack and remove the rear bolts; loosen the front bolts of the subframe and lower the rear approximately five inches (see Chapter 2, Part C, Section 5).
7 Remove the heat shield from the steering gear **(see illustration)**.
8 Remove the fluid pipe clip from the rack **(see illustration)**.

9 Place a drain pan or tray under the vehicle, positioned beneath the steering gear. Using a flare-nut wrench, disconnect the pressure and return lines **(see illustration)** from the steering gear. Plug the lines to prevent excessive fluid loss.
10 Remove the steering gear through-bolts and nuts **(see illustrations)**.
11 Lift the steering gear out of the mounts, then move it forward and remove the pinch bolt from the coupler.
12 Support the steering gear and carefully maneuver the entire assembly out through the left side wheel opening.

Installation

13 Installation is the reverse of removal.
14 Pass the steering gear assembly through the left wheel opening and place it in position in the mounts. Install the through-bolts and nuts, tightening them to the torque listed in this Chapter's Specifications.
15 Attach the pressure and return lines to the steering gear. Connect the fluid pipe clip and install the heat shield.
16 Raise the subframe into position and install the bolts. Tighten the frame bolts to the torque listed in the Chapter 2, Part C Specifications.
17 Connect the tie-rod ends to the steering arms and tighten the nuts to the torque listed in this Chapter's Specifications. Install new

10-14   Chapter 10   Suspension and steering systems

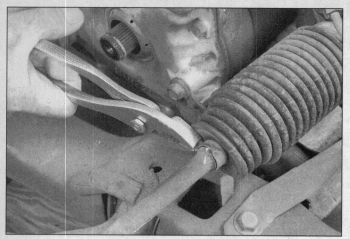

18.3 Remove the outer boot clamp

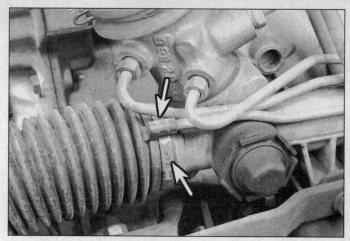

18.5 Disconnect the breather tube (upper arrow) and remove the inner boot clamp (lower arrow)

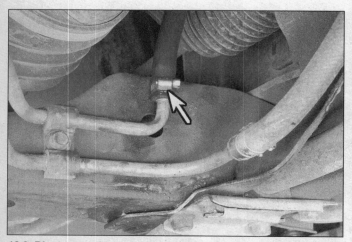

19.3 Disconnect the power steering pump return line here (arrow) and drain the power steering pump system

19.4 Using a flare nut wrench, disconnect the power steering pump pressure line at this fitting (arrow)

cotter pins.
18   Center the steering gear and have an assistant guide the coupler onto the steering column shaft, with the previously applied marks aligned. Install and tighten the pinch bolt securely.
19   Install the front wheels, lower the vehicle and tighten the lug nuts to the torque listed in the Chapter 1 Specifications.
20   Reconnect the negative battery cable.
21   Fill the power steering pump with the recommended fluid, bleed the system (see Section 20) and recheck the fluid level. Check for leaks.
22   Have the front end alignment checked by a dealer service department or an alignment shop.

18 Steering gear boots - replacement

Refer to illustrations 18.3 and 18.5

1   Remove the steering gear from the vehicle (see Section 17).
2   Remove the tie-rods ends from the steering gear (see Section 16).
3   Remove the jam nuts and outer boot clamps **(see illustration)**.
4   Cut off both inner boot clamps and discard them.
5   Mark the location of the breather tube, then remove the boots and the tube **(see illustration)**.
6   Install a new clamp on the inner end of the boot.
7   Apply multi-purpose grease to the tie-rod and the mounting groove on the steering gear.
8   Line up the breather tube with the marks made during removal and slide the boot onto the steering gear housing.
9   Make sure the boot isn't twisted, then tighten the clamp.
10   Install the outer clamps and tie-rod end jam nuts.
11   Install the tie-rod ends **(see illustration 14.4)** and tighten the jam nuts securely.
12   Install the steering gear assembly.

19 Power steering pump - removal and installation

Removal

Refer to illustrations 19.3, 19.4, 19.5a and 19.5b

1   Disconnect the cable from the negative battery terminal.
2   Remove the pump drivebelt (see Chapter 1).
3   Position a drain pan under the vehicle. Disconnect the return line **(see illustration)**. Remove as much fluid as possible with a suction pump, then remove the return line from the pump.
4   Using a flare-nut wrench and a back-up wrench, disconnect the pressure hose **(see illustration)** from the pump.
5   If you're planning to install a new pump, remove the pulley with a special puller and puller bolt, or an equivalent setup **(see illustration)** BEFORE you remove the pump mounting bolts. Re- move the pump mount-

Chapter 10 Suspension and steering systems

19.5a Using a pulley removal tool, remove the power steering pump pulley

19.5b If you're not removing the power steering pump pulley, you can get at the power steering pump mounting bolts by rotating the pulley to line up one of its holes with each bolt

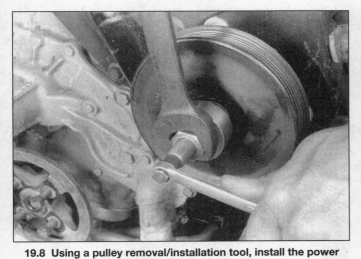

19.8 Using a pulley removal/installation tool, install the power steering pump pulley

21.3 Pry off the horn pad with a small screwdriver (models without airbag)

ing bolts and detach the pump from the engine, being careful not to spill the remaining fluid **(see illustrations)**.

Installation

Refer to illustration 19.8

6 Position the pump on the mounting bracket and install the bolts.
7 Connect the pressure and return lines to the pump.
8 If you're replacing the pump, install the pulley onto the new pump **(see illustration)**.
9 Fill the reservoir with the recommended fluid and bleed the system, following the procedure described in Section 20.

20 Power steering system - bleeding

1 Following any operation in which the power steering fluid lines have been disconnected, the power steering system must be bled to remove air and obtain proper steering performance.

2 With the front wheels turned all the way to the left, check the power steering fluid level and, if low, add fluid until it reaches the Cold mark on the dipstick.
3 Start the engine and allow it to run at fast idle. Recheck the fluid level and add more if necessary to reach the Cold mark on the dipstick.
4 Bleed the system by turning the wheels from side-to-side, without hitting the stops. This will work the air out of the system. Don't allow the reservoir to run out of fluid.
5 When the air is worked out of the system, return the wheels to the straight ahead position and leave the engine running for several minutes before shutting it off. Recheck the fluid level.
6 Road test the vehicle to be sure the steering system is functioning normally with no noise.
7 Recheck the fluid level to be sure it's up to the Hot mark on the dipstick while the engine is at normal operating temperature. Add fluid if necessary.

21 Steering wheel – removal and installation

Refer to illustrations 21.3, 21.4, 21.5 and 21.6
Warning: *Some models have airbags. Always disable the air bag system prior to working in the vicinity of the impact sensors, steering column or instrument panel to avoid the possibility of accidental deployment or the airbag, which could cause personal injury. Refer to Chapter 12 for additional airbag information.*

1 Place the wheels in the straight ahead position, place the ignition in Lock and remove the key. For models with airbags, refer to Chapter 12 and disable the air bag system.
2 Disconnect the negative cable from the battery. Wait at least 10 minutes before proceeding.
3 Remove the horn pad from the steering wheel **(see illustration)**. For models without air bags, use a small screwdriver to carefully

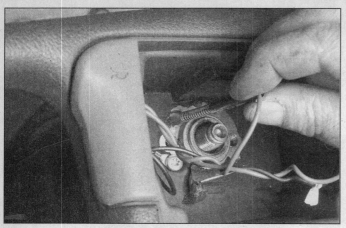

21.4 Remove the horn contact spring and lead

21.5 Make alignment marks (arrows) on the steering wheel and shaft

21.6 Use a steering wheel puller to separate the steering wheel from the shaft - DO NOT attempt to remove the wheel with a hammer!

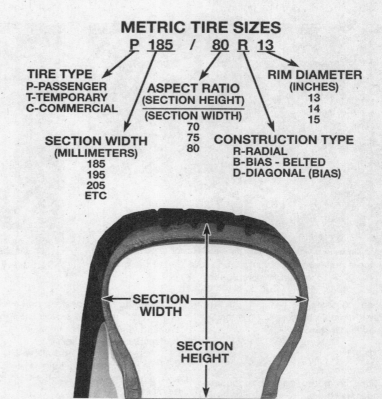

22.1 Metric size code

pry the horn pad up. For models with airbags, there are two screws on the back side of the steering wheel which need to be removed.

4   Disconnect the horn contact spring, horn wiring and air bag wiring **(see illustration)** and lift the horn pad or air bag module assembly away. **Warning:** *When carrying the airbag module, keep the driver's side (trim side) facing away from your body, and when you place it on a bench, have the driver's side facing up.*

5   Remove the steering wheel retaining nut, then mark the relationship of the steering shaft and hub to simplify installation and insure steering wheel alignment **(see illustration)**.

6   Use a puller to disconnect the steering wheel from the shaft **(see illustration)**.

7   Installation in the reverse of removal. Insure the steering wheel is positioned properly on the shaft by using the alignment marks made on disassembly and the retaining nut is tightened to the proper torque specification. Insure all electrical connectors are secure.

8   If equipped with airbags, refer to Chapter 12 to enable the airbag system.

22 Wheels and tires - general information

Refer to illustration 22.1

All vehicles covered by this manual are equipped with metric-size fiberglass or steel belted radial tires **(see illustration)**. The use of other size or type tires may affect the ride and handling of the vehicle. Don't mix different types of tires, such as radials and bias belted, on the same vehicle, since handling may be seriously affected. Tires should be replaced in pairs on the same axle, but if only one tire is being replaced, be sure it's the same size, structure and tread design as the other.

Because tire pressure affects handling and wear, the tire pressures should be checked at least once a month or before any

Chapter 10 Suspension and steering systems

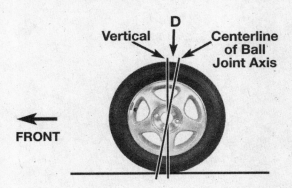

CAMBER ANGLE (FRONT VIEW)

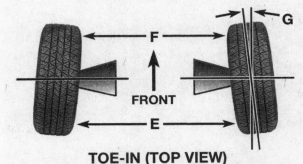

CASTER ANGLE (SIDE VIEW)

TOE-IN (TOP VIEW)

23.1 Typical front end alignment details

A minus B = C (degrees camber)
D = degrees caster
E minus F = toe-in (measured in inches)
G = toe-in (expressed in degrees)

extended trips (see Chapter 1).

Wheels must be replaced if they're bent, dented, leak air, have elongated bolt holes, are heavily rusted, out of vertical symmetry or if the lug nuts won't stay tight. Wheel repairs by welding or peening aren't recommended.

Tire and wheel balance is important to the overall handling, braking and performance of the vehicle. Unbalanced wheels can adversely affect handling and ride characteristics as well as tire life. Whenever a tire is installed on a wheel, the tire and wheel should be balanced by a shop with the proper equipment.

23 Wheel alignment - general information

Refer to illustration 23.1

A wheel alignment refers to the adjustments made to the wheels so they're in proper angular relationship to the suspension and the ground.

Wheels that are out of proper alignment not only affect steering control, but also increase tire wear. The adjustment most commonly required is the toe-in adjustment **(see illustration)**, but camber adjustment is also possible.

Getting the proper wheel alignment is a very exacting process, one in which complicated and expensive machines are necessary to perform the job properly. Because of this, you should have a technician with the proper equipment perform these tasks. We will, however, attempt to give you a basic idea of what's involved with wheel alignment so you can better understand the process and deal intelligently with the shop that does the work.

Toe-in is the turning in of the wheels. The purpose of a toe specification is to ensure parallel rolling of the wheels. In a vehicle with zero toe-in, the distance between the front edges of the wheels will be the same as the distance between the rear edges of the wheels. The actual amount of toe-in is normally only a fraction of an inch. Incorrect toe-in will cause the tires to wear improperly by making them scrub against the road surface. Toe adjustment is controlled by the tie-rod end position on the tie-rod.

Camber is the tilting of the wheels from the vertical when viewed from the front or rear of the vehicle. When the wheels tilt out at the top, the camber is said to be positive (+). When the wheels tilt in at the top the camber is negative (-). The amount of tilt is measured in degrees from the vertical and this measurement is called the camber angle. This angle affects the amount of tire tread which contacts the road and compensates for changes in the suspension geometry when the vehicle is cornering or travelling over an undulating surface. Caster is not adjustable. If the caster angle is not to specification, the suspension components must be checked for damage.

Notes

Chapter 11 Body

Contents

	Section		Section
Ashtray assembly (1990 through 1992) - removal and installation	24	Liftgate strut - replacement	17
Body repair - major damage	7	Lower extension housing (1993 on) - removal and installation	25
Body repair - minor damage	6	Maintenance - body	2
Door lock and handles - removal and installation	18	Maintenance - hinges and locks	5
Door trim panels - removal and installation	11	Maintenance - upholstery and carpets	3
Front door - removal and installation	14	Maintenance - vinyl trim	4
Front door window glass regulator - removal and installation	20	Outside mirror - removal and installation	13
Front door window glass - removal and installation	19	Seats - removal and installation	23
General information	1	Side trim panel - removal and installation	12
Glove box - removal and installation	27	Sliding door - removal and installation	15
Hood release latch and cable - removal and installation	10	Sound insulation panels - removal and installation	26
Hood - removal, installation and adjustment	9	Stereo, air conditioning and heater cover panel - removal and installation	28
Instrument cluster - removal and installation	See Chapter 12	Steering column cover - removal and installation	21
Instrument panel trim - removal and installation	22	Windshield and fixed glass - replacement	8
Liftgate - removal and installation	16		

Specifications

Torque specifications
Ft-lbs (unless otherwise indicated)

Front seat securing nuts	20
Front-door-to-hinge bolts	18
Front door hinge-to-pillar bolts	34
Outside mirror nuts	89 in-lbs
Roller bracket bolts	20
Seat belt retaining bolts	32

1 General information

Note: *On models equipped with the Delco Loc II audio system, be sure the lockout feature is turned off before performing any procedure which requires disconnecting the battery.*

These models are of unitized construction. The frame consists of a floorpan with front and rear frame side rails which support the body components, front and rear suspension systems and other mechanical components.

The exterior body panels of these models are made from various kinds of plastic, such as Sheet Molded Compound (SMC) or thermo-plastic olefin (TPO). All these materials are rustproof and can withstand minor impacts without damage.

Only general body maintenance practices and body panel repair procedures within the scope of the average home mechanic are included in this Chapter.

2 Maintenance - body

1 The condition of the body is very important, because the value of the vehicle is dependent on it. It's much more difficult to repair a neglected or damaged body than it is to repair mechanical components. The hidden areas of the body, such as the fender wells and the engine compartment, are equally important, although they obviously don't require as frequent attention as the rest of the body.

2 Once a year, or every 12,000 miles, it's a good idea to have the underside of the body steam cleaned. All traces of dirt and oil will be removed and the underside can then be inspected carefully for damaged brake lines, frayed electrical wiring, damaged cables and other problems.

3 At the same time, clean the engine and the engine compartment with a water soluble degreaser.

4 The body should be washed as needed. Wet the vehicle thoroughly to soften the dirt, then wash it down with a soft sponge and plenty of clean soapy water. If the surplus dirt isn't washed off very carefully, it will in time wear down the paint.

5 Spots of tar or asphalt coating thrown up from the road should be removed with a cloth soaked in solvent.

6 Once every six months, wax the body thoroughly.

3 Maintenance - upholstery and carpets

1 Every three months remove the carpets or mats and clean the interior of the vehicle (more frequently if necessary). Vacuum the upholstery and carpets to remove loose dirt and dust.

2 If the upholstery is soiled, apply upholstery cleaner with a damp sponge and wipe it off with a clean, dry cloth.

4 Maintenance - vinyl trim

Vinyl trim should not be cleaned with detergents, caustic soaps or petroleum-based cleaners. Plain soap and water or a mild vinyl cleaner is best for stains. Test a small area for color fastness. Bubbles under the vinyl can be eliminated by piercing them with a pin and then working the air out.

5 Maintenance - hinges and locks

Every 3000 miles or three months, the door, hood and trunk lid hinges should be lubricated with a few drops of oil. The door striker plates should also be given a thin coat of white lithium-based grease to reduce wear and ensure free movement.

6 Body repair - minor damage

Below is a list of the equipment and materials necessary to perform the following repair procedures:

Wax, grease and silicone removing solvent
Cloth-backed body tape
Sanding discs
Drill motor with three-inch disc holder
Hand sanding block
Rubber squeegees
Sandpaper
Non-porous mixing palette
Wood paddle or putty knife
Curved tooth body file
Flexible parts repair material

Flexible panels

Note: *The following repair procedure applies to the bumpers, fenders and lower side trim panels, all of which are made of this material.*

1 Remove the damaged panel if necessary or desirable. In most cases, repairs can be carried out with the panel installed.

2 Clean the area(s) to be repaired with a wax, grease and silicone removing solvent applied with a water-dampened cloth.

3 If the damage is structural, that is, if it extends through the panel, clean the backside of the panel area to be repaired as well. Wipe dry.

4 Sand the rear surface about 1-1/2 inches beyond the break.

5 Cut two pieces of fiberglass cloth large enough to overlap the break by about 1-1/2 inches. Cut only to the required length.

6 Mix the adhesive from the flexible parts repair kit according to the instructions included, and apply a layer of the mixture approximately 1/8-inch thick on the backside of the panel. Overlap the break by at least 1-1/2 inches.

7 Apply one piece of fiberglass cloth to the adhesive and cover the cloth with additional adhesive. Apply a second piece of fiberglass cloth to the adhesive and immediately cover the cloth with additional adhesive in sufficient quantity to fill the weave.

8 Allow the repair to cure for 20 to 30 minutes at 60-degrees to 80-degrees F.

9 If necessary, trim the excess repair material at the edge.

10 Remove all of the paint film over and around the area(s) to be repaired. The repair material should not overlap the painted surface.

11 With a drill motor and a sanding disc (or a rotary file), cut a "V" along the break line approximately 1/2-inch wide. Remove all dust and loose particles from the repair area.

12 Mix and apply the repair material. Apply a light coat first over the damaged area; then continue applying material until it reaches a level slightly higher than the surrounding finish.

13 Cure the mixture for 20 to 30 minutes at 60-degrees to 80-degrees F.

14 Roughly establish the contour of the area being repaired with a body file. If low areas or pits remain, mix and apply additional adhesive.

15 Block sand the damaged area with sandpaper to establish the actual contour of the surrounding surface.

16 If desired, the repaired area can be temporarily protected with several light coats of primer. Because of the special paints and techniques required for flexible body panels, it is recommended that the vehicle be taken to a paint shop for completion of the body repair.

Non-flexible panels

17 The non-flexible panels (the roof, side, hood and tailgate panels) are made of a reinforced plastic material. Repair of these panels by the home mechanic should be limited to scratch repair.

18 If the scratch is superficial, lightly rub the scratched area with a fine rubbing compound to remove loose paint and built up wax. Rinse the area with clean water.

19 Apply touch-up paint to the scratch, using a small brush. Continue to apply thin layers of paint until the surface of the paint is level with the surrounding paint. Allow the new paint at least two weeks to harden, then blend it into the surrounding paint by rubbing with a very fine rubbing compound. Finally, apply a coat of wax to the scratch area.

20 If the scratch penetrates all the way through the paint and into the panel material, clean the area as above, then use a rubber or nylon applicator to coat the scratched area with a glaze-type filler. If required, the filler can be mixed with thinner to provide a very thin paste, which is ideal for filling very narrow scratches. Before the glaze filler in the scratch hardens, wrap a piece of small cotton cloth around the tip of a finger. Dip the cloth in thinner and quickly wipe it along the surface of the scratch. This will ensure that the surface of the filler is slightly concave. The scratch can now be painted as described earlier.

7 Body repair - major damage

1 Major damage must be repaired by an auto body/frame repair shop with the necessary welding and hydraulic straightening equipment.

2 If the damage has been serious, it is vital that the structure be checked for proper alignment or the vehicle's handling characteristics may be adversely affected. Other problems, such as excessive tire wear and wear in the driveline and steering may occur.

3 Due to the fact that all of the major body components (hood, fenders, etc.) are separate and replaceable units, any seriously damaged components should be replaced rather than repaired. Sometimes these components can be found in a wrecking yard that specializes in used vehicle components, often at considerable savings over the cost of new parts.

Chapter 11 Body

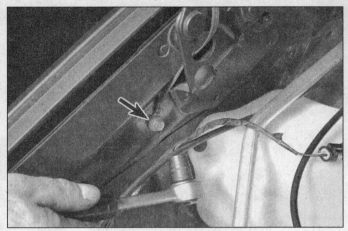

9.3 Remove the bolt (arrow) securing the ground cable to the engine compartment

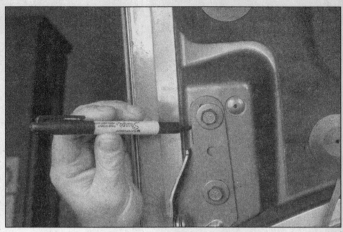

9.4 Clearly mark around both hinge plates

9.11 Screw the hood bumpers in or out to adjust the hood flush with the fenders

10.1 Unclip the cable from the latch assembly by uncoupling the retaining clip

10.3 Remove the two screws securing the hood release lever (arrows)

8 Windshield and fixed glass - replacement

1   Replacement of the windshield and fixed glass requires the use of special fast setting adhesive/caulk materials. These operations should be left to a dealer or a shop specializing in glass work.
2   Windshield-mounted rear view mirror support removal is also best left to experts, as the bond to the glass also requires special tools and adhesives.

9 Hood - removal, installation and adjustment

Refer to illustrations 9.3, 9.4 and 9.11
Note: *The hood is somewhat awkward to remove and install - at least two people should perform this procedure.*

Removal and installation

1   Disconnect the negative cable from the battery (see the "Note" at the beginning of this Chapter).
2   Open the hood and place rags or covers over the fenders to protect them during the removal procedure.
3   Disconnect the ground cable **(see illustration)**.
4   Mark around the hinge-to-hood plate **(see illustration)**.
5   Have an assistant support the weight of the hood and unclip the prop rod.
6   Remove the hinge-to-hood bolts and lift off the hood.
7   Installation is the reverse of removal.

Adjustment

8   Fore-and-aft and side-to-side adjustment of the hood is done by moving the hood in relation to the hinge plate after loosening the hinge-to-hood bolts or nuts.
9   Mark around the entire hinge plate so you can judge the amount of movement.
10   Loosen the bolts or nuts and move the hood into correct alignment. Move it only a little at a time. Tighten the hinge bolts and carefully lower the hood to check the alignment.
11   Finally, adjust the hood bumpers on the radiator support so the hood, when closed, is flush with the fenders **(see illustration)**.
12   The hood latch assembly, as well as the hinges, should be periodically lubricated with white lithium-based grease to prevent sticking and wear.

10 Hood release latch and cable - removal and installation

Refer to illustrations 10.1 and 10.3.

Removal

1   Remove the cable from the latch assembly by uncoupling the release mechanism retaining clip **(see illustration)**.
2   Attach a piece of string or thin wire to the end of the cable and unclip all cable retaining clips.
3   Working in the passenger compartment, remove the hood release lever **(see illustration)**.
4   Remove the cable and grommet rearward, pulling them into the passenger compartment.

Chapter 11 Body

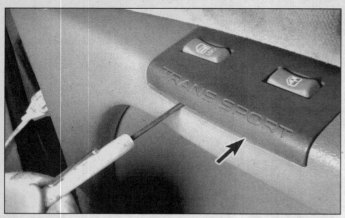

11.2a On models equipped with power windows and locks, gently pry off the armrest control panel, releasing the hidden clips (arrow) . . .

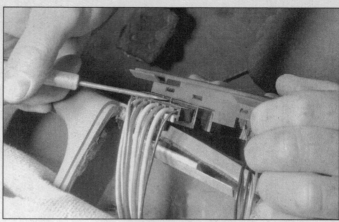

11.2b . . . then pry up the retaining tab and disconnect the electrical connector from the back of the control panel

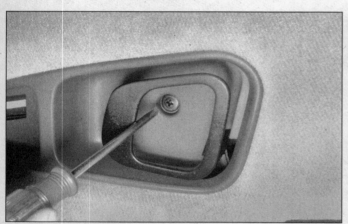

11.2c Remove this screw and pull out the door handle backing plate

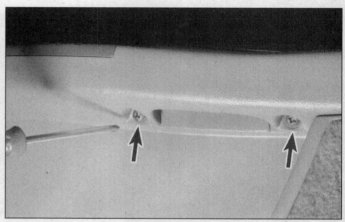

11.2d Remove the screws located either below (early models) or above the door pull

Installation

5 Ensure that the new cable has a grommet attached and then connect the string or wire to the new cable and pull it forward into the engine compartment.
6 Connect the cable and clip it back into its retaining clip.
7 Ensure that the grommet is in place and install the hood release lever.

11 Door trim panels - removal and installation

Refer to illustrations 11.2a through 11.2i, 11.5, 11.6, 11.8, 11.9 and 11.11a through 11.11f

Front door trim panel
Removal

1 On power window equipped models, disconnect the negative cable at the battery (see the "Note" at the beginning of this Chapter).
2 On manual window equipped models, remove the handle by pressing the bearing

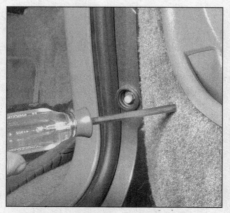

11.2e Locate and remove the hidden screws that attach the lower door trim panel to the door frame. Hidden screws are usually found in front of the door panel pocket, behind the door panel pocket . . .

plate and door trim panel in and, with a piece of hooked wire, pulling off the spring clip. A special tool is available for this purpose but its use is not essential. With the clip removed,

11.2f . . . and at the top right hand corner of the panel carpet

take off the handle and the bearing plate. To remove the trim panel follow illustrations 11.2c through 11.2i, taking care to read the accompanying captions (see illustrations).

Installation

3 Installation is the reverse of the removal procedure.

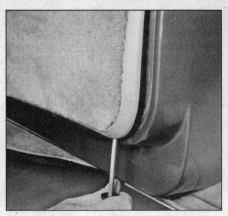

11.2g Gently pry the lower panel free, initially with a small screwdriver (take care to avoid scratching the paint), and then pull it free with your fingers

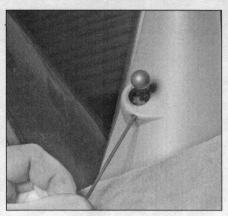

11.2h Use a hex drive to loosen the rear view mirror adjustment knob (if equipped), and gently pull the top trim panel free, starting at the top left hand corner

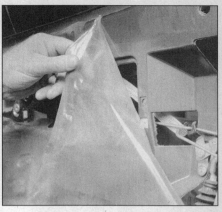

11.2i For access to the inner door components, peel back the protective plastic cover - if you're careful, it can be reused

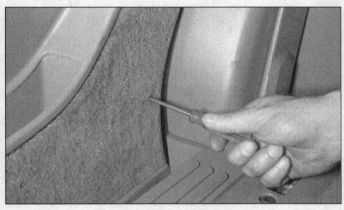

11.5 To remove the sliding door trim panel, remove the hidden screw; on early models, it is located in the lower right hand corner (shown); on later models it is located on the same side but higher up

11.6 Gently pry and pull the panel free of the door trim fasteners

Sliding door trim panel

Removal

4 Remove the screws securing the door handle and pull the backing plate away from the handle.
5 Remove the hidden screw located at the bottom right hand corner of the panel **(see illustration)**.
6 Disconnect the door trim fasteners, although there is a special tool available for this task it is possible to remove it without it **(see illustration)**.
7 Pull the lower panel free.
8 Carefully drill out the rivet securing the garnish to the door frame; it is located in the bottom left corner **(see illustration)**.
9 Gently pry the garnish free **(see illustration)**.

11.8 Drill out the rivet (arrow) - on installation, install a new rivet with a hand riveting tool

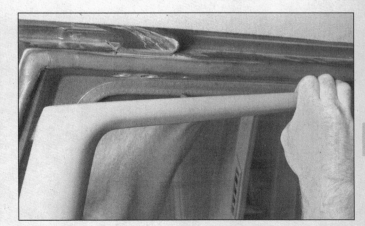

11.9 Starting at the top left, gently pry and pull the panel free

11-6   Chapter 11 Body

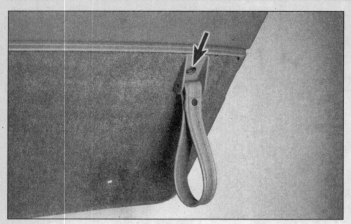

11.11a To remove the liftgate panel, start by removing this screw (arrow) and lifting off the pull handle

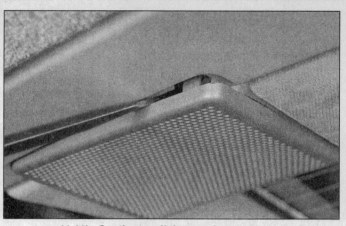

11.11b Gently pry off the speaker covers . . .

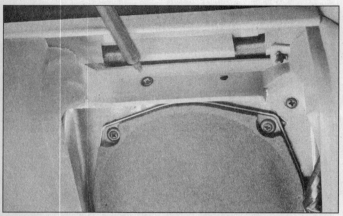

11.11c . . . and remove the screw near the speaker mountings

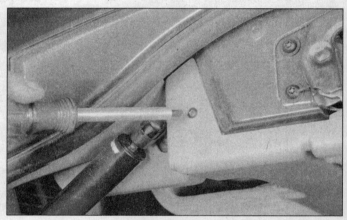

11.11d Remove the panel located near the top end of the liftgate strut on either side of the vehicle

Installation

10 Installation is the reverse of the removal procedure.

Liftgate trim panel

Note: *Because of the awkward size of the trim panel it is recommended that at least two people carry out this procedure.*

Removal

11 Follow **illustrations 11.11a** through **11.11f**. Be sure to stay in order and to read the accompanying captions.

Installation

12 Installation is the reverse of the removal procedure.

12 Side trim panel - removal and installation

Refer to illustrations 12.2a through 12.2g

1 Disconnect the negative cable from the battery (see the "Note" at the beginning of this Chapter).

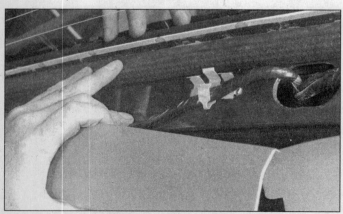

11.11e Gently pull the window surround panel off, starting at the split at the top of the liftgate

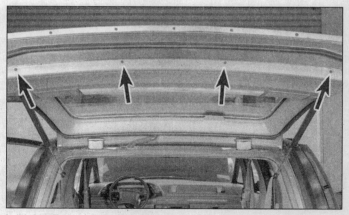

11.11f Remove the four screws from the bottom of the liftgate and lift the panel clear

Chapter 11 Body

11-7

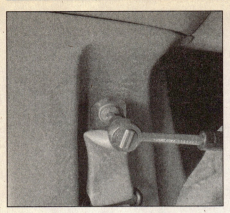

12.2a Pull the plastic covers from the front seat belt retaining bolts and remove them - both the upper and lower bolts are Torx type, so you'll need a special Torx-drive tool

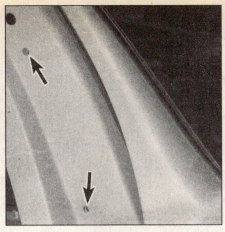

12.2b Remove the panel screws (arrows) from the upper . . .

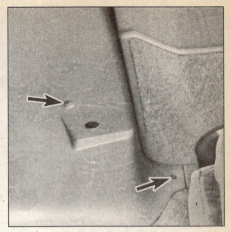

12.2c . . . and lower sections of the door pillar panel

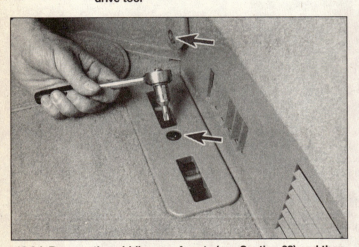

12.2d Remove the middle row of seats (see Section 23) and then remove the anchor cover and the lower middle seat belt retaining bolt - they're both Torx type (arrows)

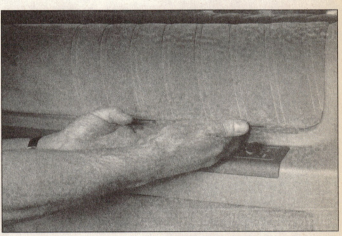

12.2e Carefully remove the padded trim from the top of the panel, which is held in place by retaining clips. On models with rear heating/air conditioning, remove the rear control panel (See Chapter 3, Section 12)

2   Refer to **illustrations 12.2a through 12.2g** for the removal procedure. Be sure to stay in order and to read the accompanying captions **(see illustrations)**.

3   To install the side panel, reverse the removal sequence. Ensure that the seat belt retaining Torx screws are tightened to the torque figure listed in the Specifications at the beginning of this Chapter.

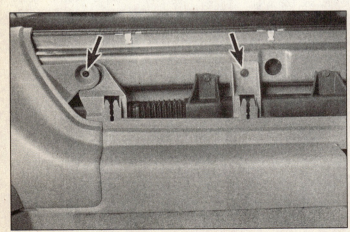

12.2f Remove the upper panel retaining bolts (arrows) - late model shown. Early models have three bolts

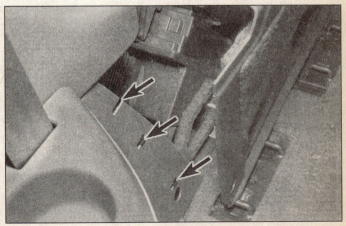

12.2g Unclip the left side retaining clips and pull the panel free

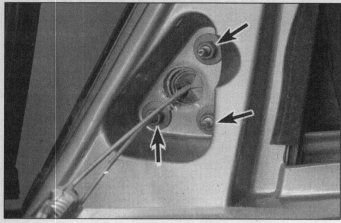

13.3 Remove the mirror retaining nut(s)

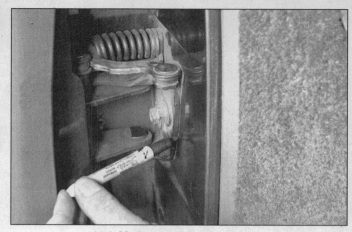

14.4 Mark around the door hinges

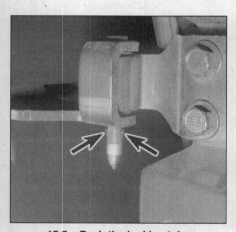

15.5a Push the locking tabs in (arrows) . . .

15.5b . . . with a pair of pliers, and, with a hammer and punch, tap the retaining pin free

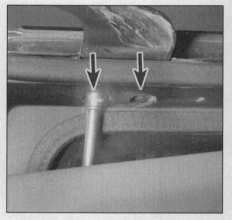

15.6 The upper roller bracket retaining bolts are set in the door frame (arrows)

13 Outside mirror - removal and installation

Refer to illustration 13.3

Removal

1 Disconnect the negative cable from the battery (see the "Note" at the beginning of this Chapter) and remove the front door trim panel (see Section 11).
2 On power mirrors, unplug the electrical connector.
3 Remove the mirror-to-door retaining nut(s) **(see illustration)**.
4 Lift off the mirror assembly.

Installation

5 Installation is the reverse of removal.

14 Front door - removal and installation

Refer to illustration 14.4

1 On doors with power components (power windows and door locks), disconnect the negative cable from the battery (see the "Note" at the beginning of this Chapter).
2 Remove the door trim panel (see Section 11). On doors with power components, unplug all electrical connections from the door (it is a good idea to label all connections to aid the reassembly process) and remove the electrical harness from the door.
3 Open the door all the way and support it on jacks or blocks covered with cloth or pads to prevent damaging the paint.
4 Mark around the door hinges to facilitate realignment during reassembly **(see illustration)**.
5 Have an assistant hold the door, remove the hinge bolts and lift the door away.
6 Install the door by reversing the removal procedure. Tighten the nuts and bolts to the torque specified at the beginning of this Chapter.

15 Sliding door - removal and installation

Refer to illustrations 15.5a, 15.5b and 15.6

Removal

1 Disconnect the negative cable from the battery (see the "Note" at the beginning of this Chapter).
2 Remove the door trim panel (see Section 11).
3 Open the door slightly and support it on jacks or blocks covered with cloth or pads to prevent damaging the paint.
4 Disconnect the cable below the lower roller bracket and remove the cable retaining screw.
5 Remove the center track roller retaining pin **(see illustrations)**.
6 Have an assistant hold the door steady and remove the upper roller bracket from the door **(see illustration)**.
7 Remove the lower roller bracket retaining bolts and lift the door free.

Installation

8 Installation is the reverse of the removal procedure. Be sure to tighten all bolts securely.

16 Liftgate - removal and installation

Refer to illustrations 16.2a through 16.2g
Note: *The liftgate is heavy and somewhat*

Chapter 11 Body

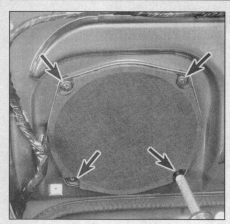

16.2a Remove the four mounting screws from each speaker (arrows) and . . .

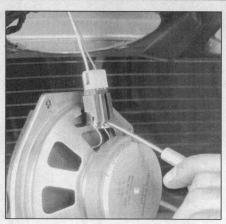

16.2b . . . disconnect the speaker wiring

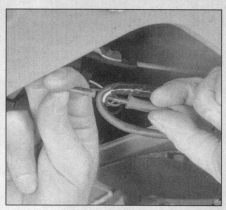

16.2c Pull the rear window washer water supply tube clear of the liftgate frame and separate its connection

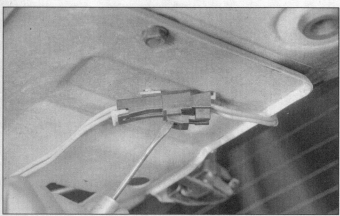

16.2d Disconnect all electrical connections from the door

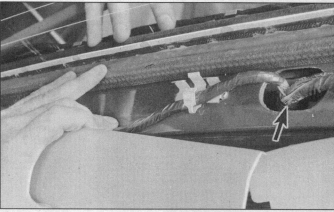

16.2e Do not forget the heated rear window connection, which is hidden in the frame panel at the top of the liftgate

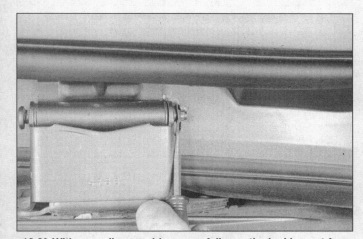

16.2f With a small screwdriver, carefully pry the locking nut free of the hinge pin - the nut may fly out, so be careful not to lose it

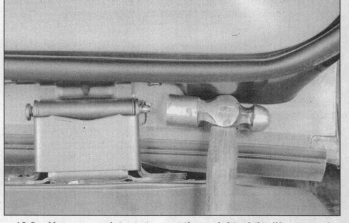

16.2g Have an assistant support the weight of the liftgate and drive the hinge pins out with a hammer and punch

awkward to remove and install - at least two people should perform this procedure.

Removal

1   Disconnect the negative cable from the battery (see the "Note" at the beginning of this Chapter) and remove the liftgate trim panel (see Section 11).

2   Refer to **illustrations 16.2a through 16.2g** for the liftgate removal procedure; study all the illustrations and read all the captions **(see illustrations)**.

Installation

3   Installation is the reverse of the removal procedure.

17 Liftgate strut - replacement

Refer to illustrations 17.1a, 17.1b and 17.1c
Note: *The liftgate is heavy and somewhat awkward to hold - at least two people should perform this procedure.*

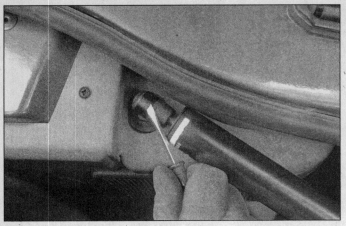

17.1a Open the liftgate fully and unclip the retaining clip from the lower end of the strut

17.1b Ensure that the liftgate is supported and pry the lower end of the strut away from the liftgate, freeing it from the liftgate ball stud

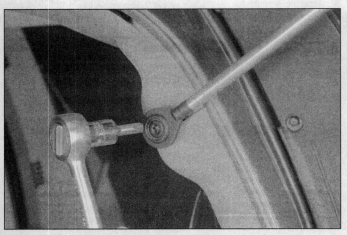

17.1c Remove the Torx screw securing the top of the strut to the door frame, and remove the strut assembly

18.3 Drill out the handle retaining rivets (arrows) - early model shown

Removal

1 Follow the accompanying illustrations, beginning with 17.1a, for the actual strut replacement procedure. Be sure to stay in order and to read the caption under each illustration **(see illustrations)**.

Installation

2 Installation is the reverse of the removal procedure.

18 Door lock and handles - removal and installation

Refer to illustrations 18.3 and 18.12

Front door

Inside handle

1 With the window glass in the full up position, remove the door trim panel (see Section 11).
2 Disconnect the lock rod from the handle.
3 Drill out the rivets securing the handle to the door frame and remove the handle **(see illustration)**.
4 Installation is the reverse of the removal procedure.

Outside handle

5 Remove the door trim panel (see Section 11).
6 Remove the handle-to-door nuts.
7 Disengage the handle from the lock rod and pull the handle free.
8 Reverse the removal procedure to install the handle.

Lock

9 To remove the lock cylinder, remove the door trim panel (see Section 11) and the lock rod from the cylinder.
10 Disengage the cylinder retainer and pull the lock cylinder from the door **(see illustration 18.6)**.
11 From inside the door, disconnect the locking knob rod, inside handle rod and inside handle lock rod from the lock assembly.

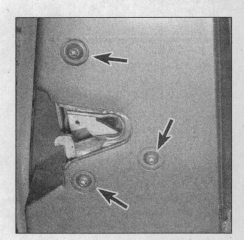

18.12 To remove the lock assembly, unscrew the retaining screws (arrows) at the rear edge of the door

12 Remove the lock assembly retaining screws and pull the assembly free **(see illustration)**.

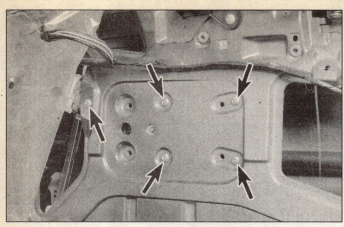

20.4 Drill out the door glass regulator rivets (arrows) - power regulator shown

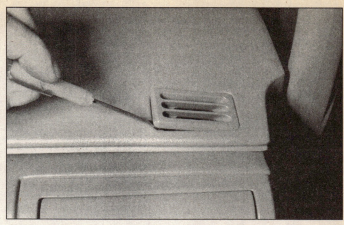

22.1a With a small screwdriver, gently pry the small front air deflector grilles free . . .

Sliding door

Inside handle

13   Remove the sliding door trim panel (see Section 11).
14   Disconnect the control rods from the handle by prying the rod clip out and pushing the rod away from the assembly.
15   Remove the inside handle screws and pull the handle free.
16   Installation is the reverse of the removal procedure.

Outside handle

17   Remove the sliding door trim panel (see Section 11).
18   Disconnect the control rods from the handle by prying the rod clip out and pushing the rod away from the assembly.
19   Remove the retaining bolts or nuts and pull the handle free.
20   Installation is the reverse of the removal procedure.

Lock cylinder

21   Remove the door trim panel (see Section 11).
22   Disconnect the lock cylinder rod by prying the rod clip out and pushing the rod away from the assembly.
23   Carefully remove the lock cylinder retaining clip and pull the cylinder free.
24   Reverse the removal procedure to install the lock cylinder and assembly.

Liftgate lock assembly

25   Disconnect the negative cable from the battery (see the "Note" at the beginning of this Chapter) and remove the liftgate trim panel (see Section 11).
26   Remove the liftgate lock actuator electrical connector (if equipped).
27   Disconnect the lock rods and the lock plate assembly.
28   Remove the wire retainer from the lock plate and drill the lock actuator retaining rivets to free the lock actuator from the liftgate.
29   Installation is the reverse of the removal procedure.

19   Front door window glass - removal and installation

Warning: *Safety glasses and gloves should be worn when performing this procedure.*

1   If the glass being replaced is cracked, criss-cross it with masking tape to help hold the glass together.
2   Remove the door trim panel (see Section 11).
3   Remove the bolts securing the lower run channel and the outer window sealing strip.
4   Remove the window run channel from the door.
5   Disconnect the window from the rear run channel.
6   Pull the window up and clear of the door.
7   Installation is the reverse of removal.

20   Front door window glass regulator - removal and installation

Removal

Refer to illustration 20.4

1   On power window equipped models, disconnect the negative cable at the battery (see the "Note" at the beginning of this Chapter).
2   With the window glass in the full up position, remove the door trim panel and water shield (see Section 11).
3   Secure the window glass in the "Up" position with strong adhesive tape fastened to the glass and wrapped over the door frame (place a rag over the top of the door so the tape doesn't contact the weatherstrip). Do not proceed until you're sure the tape will support the weight of the window.
4   Punch out the center pins of the rivets that secure the window regulator and drill the rivets out with a 3/16-inch drill bit (manual regulators) or a 1/4-inch drill bit (power regulators) **(see illustration)**.
5   On power window equipped models, unplug the electrical connector.
6   Move the regulator until it is disengaged from the sash channel. Lift the regulator from the door.
7   To disconnect the power window motor from the regulator, drill out the motor-to-regulator rivets.

Installation

8   Place the regulator in position in the door and engage it in the sash channel.
9   Secure the regulator to the door using 3/16-inch (manual) or 1/4-inch (power) rivets and a rivet tool. Alternatively use 1/2-inch long bolts of the correct diameter to attach the regulator.
10   Plug in the electrical connector (if equipped).
11   Install the door trim panel. Connect the negative battery cable.

21   Steering column cover - removal and installation

1   Remove the driver's side sound insulation panel (see Section 25).
2   Remove the three screws at the lower edge of the cover.
3   To free the cover from the slots in the lower trim pad carefully pull the cover back and down.
4   Installation is the reverse of the removal procedure.

22   Instrument panel trim - removal and installation

Upper trim pad

Removal

Refer to illustrations 22.1a, through 22.1e

1   To remove the upper trim pad follow illustrations 22.1a through 22.1e. Be sure not to miss any and to read the accompanying captions **(see illustrations)**.

Chapter 11 Body

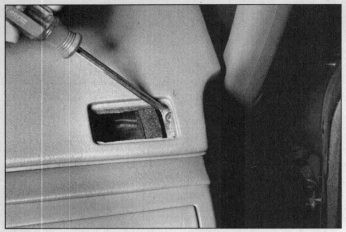

22.1b ... and remove the screw previously hidden by the grille

22.1c Open the glove compartment and remove the screws (arrows) in the lower section of the trim pad

22.1d Remove the two screws (arrows) that attach the upper trim pad to the instrument cluster bezel

22.1e Gently pull the upper trim pad away from the lower trim panel assembly retaining clips (arrows)

Installation

2 Installation is the reverse of the removal procedure.

Lower trim panel

Removal

Refer to illustration 22.9

Note: *Owing to the size and weight of the lower trim panel, the help of an assistant will be required for this procedure.*

3 Disconnect the negative battery cable from the battery (see the "Note" at the beginning of this Chapter).
4 Remove the front passenger seat to improve access to the lower trim components (see Section 23).
5 Remove the upper trim pad (see above procedure).
6 Remove the ashtray assembly (see Section 24) and the left and right sound insulator panels (see Section 25).
7 Working on the left side of the lower trim panel, disconnect and remove the brake release handle and the hood release handle.
8 Remove the lower cowl side panels and disconnect all electrical connectors and the

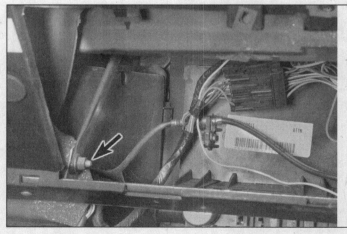

22.9 Remove this left brace nut (arrow)

ground wires connected to the frame. Be sure to clearly label them to jog your memory during reassembly.
9 Remove the outer supports, the left brace and lower the steering column **(see illustration)**. Rest it on the drivers seat - be sure to place a clean cloth over the seat to protect it.

10 Support the lower trim panel and unbolt the center support bracket located behind the stereo/radio support panel. Remove the lower brace.
11 Working in the engine compartment, push the electrical harness through the bulkhead grommet into the passenger compartment.

Chapter 11 Body

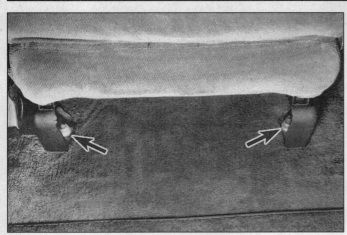

23.3 Remove the rear securing nuts

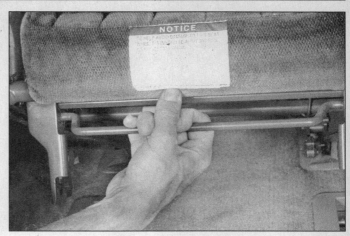

23.8a Pull the rear locking bar up and . . .

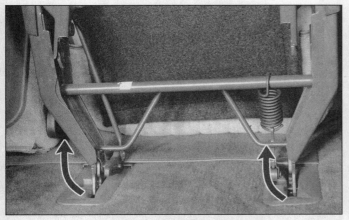

23.8b . . . rotate the seat up and forward

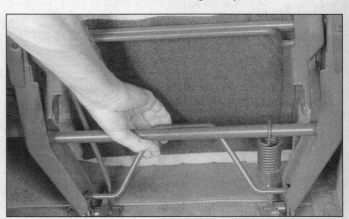

23.9 Pull the front locking bar up to disengage the locking hooks and remove the seat

12 With the help of an assistant remove the lower trim panel.

Installation

13 Installation is the reverse of the removal procedure.

23 Seats - removal and installation

Refer to illustrations 23.3, 23.8a, 23.8b and 23.9

Front seats

Removal

1 Disconnect the negative cable from the battery - driver's power seat only - (see the "Note" at the beginning of this Chapter).
2 Disconnect the electrical connectors (driver's power seat only) and the fuel filler door release cable located under the driver's seat.
3 Remove the rear nuts from the seat anchor plate **(see illustration)**.
4 Remove the front securing nuts and remove the seat.

Installation

5 Installation is the reverse of the removal

procedure. Tighten the seat securing nuts to the torque listed in this Chapter's Specifications.

Second and third row seats

Note: *Remove the second row seats before removing the third row seats.*

Removal

6 Remove the plate hole plugs.
7 Push the backrests fully forward.
8 Fold the seat forward **(see illustrations)**.
9 Disengage the front locking bar and pull the seat free **(see illustration)**.

Installation

10 Installation is the reverse of the removal procedure. Check that the seats are secured firmly to all four seat anchors.

24 Ashtray assembly (1990 through 1992) - removal and installation

Refer to illustrations 24.2, 24.3 and 24.5

1 Disconnect the negative cable from the

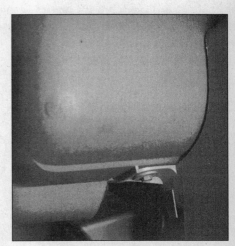

24.2 A screw is located on both sides of the ashtray assembly

battery (see the "Note" at the beginning of this Chapter).
2 Remove the screws securing the assembly to both the passenger's and driver's side insulator panels **(see illustration)**.
3 Remove the rear retaining screws (two)

24.3 The rear screws are located behind the assembly and are accessible from below (arrow) - components removed for clarity

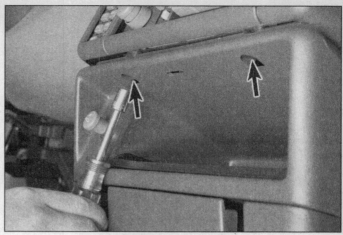

24.5 Remove the hidden nuts (arrows)

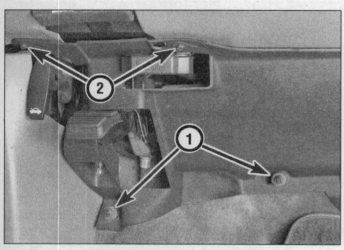

26.1 Driver's side insulator panel
1 Lower panel retaining nuts 2 Panel-to-trim panel bolts

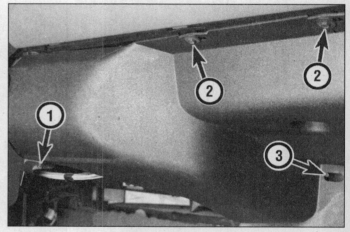

26.4 Passenger's side insulation panel details - early model shown
1 Lower panel screw 3 Lower panel nut
2 Upper panel bolts

located behind the assembly **(see illustration)**.
4 Remove the nut securing the assembly bracket to the floor.
5 Unscrew the hidden nuts at the front of the assembly **(see illustration)**.
6 Lower the assembly and disconnect the storage shelf lamp and the electrical connector from the lighter.
7 Remove the assembly.
8 Installation is the reverse of the removal procedure.

25 Lower extension housing (1993 on) - removal and installation

1 Detach the cable from the negative terminal of the battery.
2 Remove the screws securing the steering column opening filler panel and remove the panel from the retaining clips.
3 Open the glove box door.
4 Remove the upper screws (four), middle screws (two) and lower screws (two) securing the lower extension to the instrument panel.
5 Pull the lower extension away from the instrument panel and disconnect all electrical connectors.
6 Remove the lower extension housing.
7 Installation is the reverse of removal.

26 Sound insulation panels - removal and installation

Refer to illustrations 26.1 and 26.4

Driver's side insulation panel

1 Remove the insulator lower panel retaining nuts, the bolts connecting the panel to the lower trim panel **(see illustration)** and the ashtray assembly (see Section 24).
2 Lift the insulation panel free.
3 Installation is the reverse of the removal procedure.

Passenger's side insulation panel

4 Remove the lower insulator panel nut and the lower panel screw (on earlier models) or screws (on later models) **(see illustration)**.
5 On earlier models, remove the two upper panel bolts **(see illustration 26.4)**. On later models, remove the bolt in the upper right hand corner and the nut in the upper left hand corner of the panel. Installation is the reverse of the removal procedure.

27 Glove box - removal and installation

Refer to illustration 27.5

1 Disconnect the negative cable from the battery (see the "Note" at the beginning of this Chapter).
2 Remove the passenger's side sound insulation panel (see Section 25).
3 To remove the glove box, open it fully and release the door stops by gently lifting them up and out.
4 Remove the lower left hand bolt that was previously hidden behind the sound

Chapter 11 Body

27.5 The screws are located at the top of the glove box (arrows)

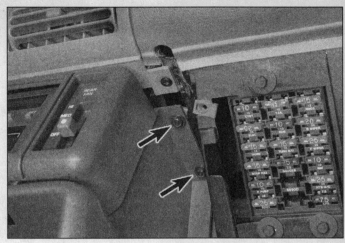

28.2 Remove the bolts (arrows) from either side of the cover panel

insulator.
5 Remove the top screws, including the rubber bumpers **(see illustration)**.
6 Disconnect the glove box lamp and switch connectors and remove the glove box.
7 Installation is the reverse of the removal procedure.

28 Stereo, air conditioning and heater cover panel - removal and installation

Refer to illustration 28.2
1 Remove the steering column cover (see Section 21) and open the glove box.
2 Remove the two bolts from the top of the cover panel, located on both the passenger's and driver's side **(see illustration)**.
3 Installation is the reverse of the removal procedure.

Notes

Chapter 12
Chassis electrical system

Contents

	Section		Section
Air bag - general information and precautions	23	Power door lock system - description and check	22
Bulb replacement	14	Power window system - description and check	21
Circuit breakers - general information	5	Radio and speakers - general information,	
Cruise control system - description and check	20	removal and installation	15
Door glass regulator - removal and installation	See Chapter 11	Radio antenna - removal and installation	14
Door lock assembly - removal and installation	See Chapter 11	Rear window defogger - check and repair	17
Electrical troubleshooting - general information	2	Rear window wiper motor - removal and installation	19
Fuses - general information	3	Relays - general information	6
Fusible links - general information	4	Starter safety switch check	See Chapter 1
General information	1	Turn signal and hazard flashers - check and replacement	7
Headlight bulb - replacement	12	Turn signal switch assembly - removal and installation	8
Headlight switch - removal and installation	10	Windshield wiper motor - removal and installation	18
Headlights - adjustment	13	Wiper/washer switch - removal and installation	11
Ignition key lock cylinder - replacement	9	Wiring diagrams - general information	24
Instrument cluster - removal and installation	16		

Specifications

Bulbs

Front
Headlight	
Outer	9006
Inner	9005
Sidemarker	194NA
Parking/turn signal	3057NA
Underhood lamp	561

Interior
Instrument cluster indicator lights	74
Sliding door stepwell	561
Dome lamp	561
Reading light	
Front	562
Rear	906

Rear
Brake/turn/tail light	
Upper two bulbs	194
Lower two bulbs	2057

Chapter 12 Chassis electrical system

1 General information

The electrical system is a 12-volt, negative ground type. Power for the lights and all electrical accessories is supplied by a lead/acid-type battery which is charged by the alternator.

This Chapter covers repair and service procedures for the various electrical components not associated with the engine. Information on the battery, alternator, ignition system and starter motor can be found in Chapter 5.

It should be noted that when portions of the electrical system are serviced, the negative battery cable should be disconnected from the battery to prevent electrical shorts and/or fires. **Note:** *On models equipped with the Delco Loc II audio system, be sure the lockout feature is turned off before performing any procedure which requires disconnecting the battery.*

2 Electrical troubleshooting - general information

A typical electrical circuit consists of an electrical component, any switches, relays, motors, fuses, fusible links or circuit breakers related to that component and the wiring and electrical connectors that link the component to both the battery and the chassis. To help you pinpoint an electrical circuit problem, wiring diagrams are included at the end of this book.

Before tackling any troublesome electrical circuit, first study the appropriate wiring diagrams to get a complete understanding of what makes up that individual circuit. Trouble spots, for instance, can often be narrowed down by noting if other components related to the circuit are operating properly. If several components or circuits fail at one time, chances are the problem is in a fuse or ground connection, because several circuits are often routed through the same fuse and ground connections.

Electrical problems usually stem from simple causes, such as loose or corroded connections, a blown fuse, a melted fusible link or a bad relay. Visually inspect the condition of all fuses, wires and connections in a problem circuit before troubleshooting it.

If testing instruments are going to be utilized, use the diagrams to plan ahead of time where you will make the necessary connections in order to accurately pinpoint the trouble spot.

The basic tools needed for electrical troubleshooting include a circuit tester or voltmeter (a 12-volt bulb with a set of test leads can also be used), a continuity tester, which includes a bulb, battery and set of test leads, and a jumper wire, preferably with a circuit breaker incorporated, which can be used to bypass electrical components. Before attempting to locate a problem with test instruments, use the wiring diagram(s) to decide where to make the connections.

Voltage checks

Voltage checks should be performed if a circuit is not functioning properly. Connect one lead of a circuit tester to either the negative battery terminal or a known good ground. Connect the other lead to an electrical connector in the circuit being tested, preferably nearest to the battery or fuse. If the bulb of the tester lights, voltage is present, which means that the part of the circuit between the electrical connector and the battery is problem free. Continue checking the rest of the circuit in the same fashion. When you reach a point at which no voltage is present, the problem lies between that point and the last test point with voltage. Most of the time the problem can be traced to a loose connection. **Note:** *Keep in mind that some circuits receive voltage only when the ignition key is in the Accessory or Run position.*

Finding a short

One method of finding shorts in a circuit is to remove the fuse and connect a test light or voltmeter in its place to the fuse terminals. There should be no voltage present in the circuit. Move the wiring harness from side-to-side while watching the test light. If the bulb goes on, there is a short to ground somewhere in that area, probably where the insulation has rubbed through. The same test can be performed on each component in the circuit, even a switch.

Ground check

Perform a ground test to check whether a component is properly grounded. Disconnect the battery and connect one lead of a self-powered test light, known as a continuity tester, to a known good ground. Connect the other lead to the wire or ground connection being tested. If the bulb goes on, the ground is good. If the bulb does not go on, the ground is not good.

Continuity check

A continuity check is done to determine if there are any breaks in a circuit - if it is passing electricity properly. With the circuit off (no power in the circuit), a self-powered continuity tester can be used to check the circuit. Connect the test leads to both ends of the circuit (or to the "power" end and a good ground), and if the test light comes on the circuit is passing current properly. If the light doesn't come on, there is a break somewhere in the circuit. The same procedure can be used to test a switch, by connecting the continuity tester to the switch terminals. With the switch turned On, the test light should come on.

Finding an open circuit

When diagnosing for possible open circuits, it is often difficult to locate them by sight because oxidation or terminal misalignment are hidden by the electrical connectors.

3.1 The passenger compartment fuse block is accessible after lifting out the glove box insert

Merely wiggling an electrical connector on a sensor or in the wiring harness may correct the open circuit condition. Remember this when an open circuit is indicated when troubleshooting a circuit. Intermittent problems may also be caused by oxidized or loose connections.

Electrical troubleshooting is simple if you keep in mind that all electrical circuits are basically electricity running from the battery, through the wires, switches, relays, fuses and fusible links to each electrical component (light bulb, motor, etc.) and to ground, from which it is passed back to the battery. Any electrical problem is an interruption in the flow of electricity to and from the battery.

3 Fuses - general information

Refer to illustrations 3.1 and 3.3

The electrical circuits of the vehicle are protected by a combination of fuses, circuit breakers and fusible links. The fuse block is located under a cover on the left side of the instrument panel **(see illustration)**.

Each of the fuses is designed to protect a specific circuit, and the various circuits are identified on the fuse panel itself.

Miniaturized fuses are employed in the fuse block. These compact fuses, with blade terminal design, allow fingertip removal and replacement. If an electrical component fails, always check the fuse first. A blown fuse is easily identified through the clear plastic body. Visually inspect the element for evidence of damage **(see illustration)**. If a continuity check is called for, the blade terminal tips are exposed in the fuse body.

Be sure to replace blown fuses with the correct type. Fuses of different ratings are physically interchangeable, but only fuses of the proper rating should be used. Replacing a fuse with one of a higher or lower value than specified is not recommended. Each electrical circuit needs a specific amount of protection. The amperage value of each fuse is

Chapter 12 Chassis electrical system

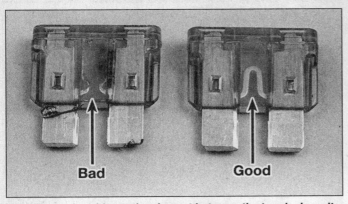

3.3 When a fuse blows, the element between the terminals melts - the fuse on the left is blown, the fuse on the right is good

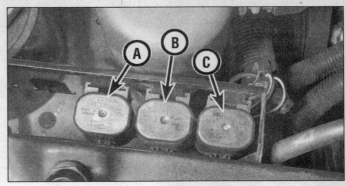

6.2a Typical 3.1L engine relays

- A Fuel pump relay
- B Cooling fan relay
- C Air conditioning compressor relay

molded into the fuse body.

If the replacement fuse immediately fails, don't replace it again until the cause of the problem is isolated and corrected. In most cases, the cause will be a short circuit in the wiring caused by a broken or deteriorated wire.

4 Fusible links - general information

Some circuits are protected by fusible links. Fusible links are used in circuits which are not ordinarily fused, such as the ignition circuit.

Although the fusible links appear to be a heavier gauge than the wire they are protecting, the appearance is due to the thick insulation. All fusible links are four wire gauges smaller than the wire they are designed to protect.

Fusible links cannot be repaired, but a new link of the same size wire can be put in its place. The procedure is as follows:

a) Disconnect the negative cable from the battery.
b) Disconnect the fusible link from the wiring harness.

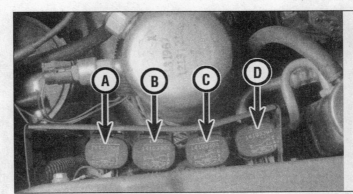

c) Cut the damaged fusible link out of the wiring just behind the electrical connector.
d) Strip the insulation back approximately 1/2-inch.
e) Position the electrical connector on the new fusible link and crimp it into place.
f) Use rosin core solder at each end of the new link to obtain a good solder joint.
g) Use plenty of electrical tape around the soldered joint. No wires should be exposed.
h) Connect the battery ground cable. Test the circuit for proper operation.

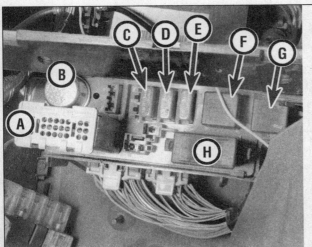

6.2c Typical convenience center components

- A Audio alarm module
- B Hazard warning flasher
- C Power door lock circuit breaker
- D Power window circuit breaker
- E Rear window defogger circuit breaker
- F Blower motor relay
- G Horn relay
- H Rear window defogger relay

6.2b Typical 3800 engine relays

- A Fuel pump relay
- B Air conditioning compressor relay
- C Cooling puller fan relay
- D Cooling pusher fan relay

5 Circuit breakers - general information

Circuit breakers protect components such as power windows, power door locks and headlights. Some circuit breakers are located in the convenience center (see Section 6).

On some models the circuit breaker resets itself automatically, so an electrical overload in a circuit breaker protected system will cause the circuit to fail momentarily, then come back on. If the circuit doesn't come back on, check it immediately. Once the condition is corrected, the circuit breaker will resume its normal function. Some circuit breakers must be reset manually.

6 Relays - general information

Refer to illustrations 6.2a and 6.2b

Several electrical accessories in the vehicle use relays to transmit the electrical signal to the component. If the relay is defective, that component will not operate properly.

The various relays are grouped together in several locations. Some relays are grouped together in the engine compartment **(see illustrations)**. Other relays controlling function inside the vehicles are found in the convenience center which is located under the

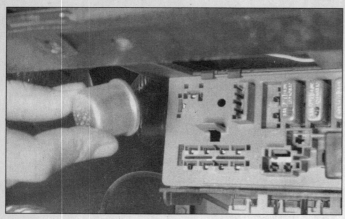

7.9 Grasp the hazard flasher unit and pull it straight out of the convenience center

8.4 Remove the hazard warning knob screw (arrow) and the knob

8.5 Remove the cancel cam assembly

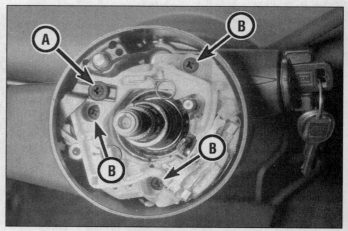

8.6 Remove the turn signal lever screw (A), followed by the switch screws (B)

right side of the dashboard behind the sound insulator panel **(see illustration)**.

If a faulty relay is suspected, it can be removed and tested by a dealer service department or a repair shop. Defective relays must be replaced as a unit.

7 Turn signal and hazard flashers - check and replacement

Refer to illustration 7.9

Turn signal flasher

1 The turn signal flasher, a small canister-shaped unit located under the dash, in a clip to the left of the steering column, flashes the turn signals.
2 When the flasher unit is functioning properly, an audible click can be heard during its operation. If the turn signals fail on one side or the other and the flasher unit does not make its characteristic clicking sound, a faulty turn signal bulb is indicated.
3 If both turn signals fail to blink, the problem may be due to a blown fuse, a faulty flasher unit, a broken switch or a loose or open connection. If a quick check of the fuse box indicates that the turn signal fuse has blown, check the wiring for a short before installing a new fuse.
4 To replace the flasher, simply unplug it and pull it out of the clip.
5 Make sure that the replacement unit is identical to the original. Compare the old one to the new one before installing it.
6 Installation is the reverse of removal.

Hazard flasher

7 The hazard flasher, a small canister-shaped unit, located in the convenience center, flashes all four turn signals simultaneously when activated.
8 The hazard flasher is checked in a fashion similar to the turn signal flasher (see Steps 2 and 3).
9 To replace the hazard flasher, pull it from the convenience center **(see illustration)**.
10 Make sure the replacement unit is identical to the one it replaces. Compare the old one to the new one before installing it.
11 Installation is the reverse of removal.

8 Turn signal switch assembly - removal and installation

Refer to illustrations 8.4, 8.5, 8.6, 8.8a and 8.8b

1 Detach the cable from the negative battery terminal.
2 Remove the steering wheel (see Chapter 10, Section 21).
3 Use a small screwdriver to pry the retaining ring out of the groove in the steering column and remove the lock plate. It may be necessary to use a lock plate removal tool to depress the lock plate for access to the ring.
4 Remove the hazard warning knob **(see illustration)**.
5 Remove the turn signal cancel cam **(see illustration)**.
6 Place the turn signal lever in the right turn position, remove the screw and detach the lever, then remove the turn signal switch mounting screws **(see illustration)**.
7 Remove the left under-dash panel below the steering column.
8 Locate the turn signal switch electrical connector and unplug it **(see illustration)**.

Chapter 12 Chassis electrical system

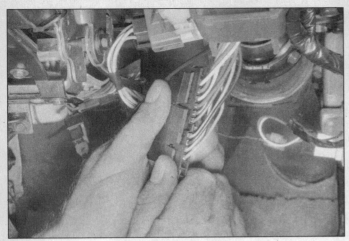

8.8a The turn signal switch electrical connector is located under the dash near the steering column – unplug it as shown, then . . .

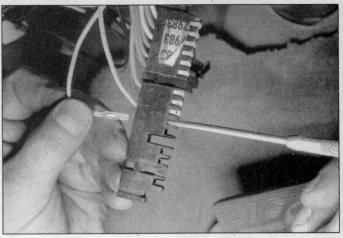

8.8b . . . detach the turn signal switch wires

9.5 Lift off the buzzer switch and clip with needle nose pliers

9.7 The lock is held in place by a Torx-head screw (arrow)

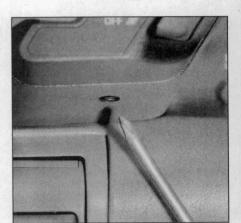

10.2 Use a Phillips head screwdriver to remove the retaining screw at the bottom of the switch housing

Detach the wires to the switch assembly from the back of the electrical connector **(see illustration)**.

9 Pull the wiring harness and electrical connector up through the steering column and remove the switch assembly.

10 Installation is the reverse of removal.

9 Ignition key lock cylinder - replacement

Refer to illustrations 9.5 and 9.7

1 Detach the cable from the negative battery terminal.
2 Remove the steering wheel (see Chapter 10, Section 21).
3 Remove the turn signal switch assembly (see Section 8).
4 Make sure the key is removed from the lock cylinder.
5 Using needle nose pliers, remove the buzzer switch **(see illustration)**.
6 Insert the key and place the lock cylinder in the Lock position.

7 Remove the lock cylinder retaining screw **(see illustration)**.
8 Remove the lock cylinder.
9 Installation is the reverse of removal.

10 Headlight switch - removal and installation

Refer to illustrations 10.2 and 10.4

1 Detach the cable from the negative battery terminal.
2 Remove the retaining screw at the bottom of the cluster housing **(see illustration)**.
3 Pry carefully at the upper outer corner of the switch with a small screwdriver, detach the retaining clips and rotate the switch out of the cluster housing.
4 Withdraw the switch from the opening and detach the electrical connector with a small screwdriver **(see illustration)**.
5 Plug in the electrical connector, insert the tabs and rotate the switch into position, then press in until the clips engage. Install the retaining screw.

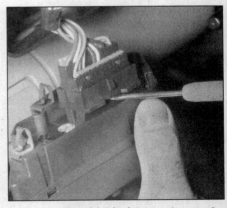

10.4 Pry the electrical connector apart with a small screwdriver

11 Wiper/washer switch - removal and installation

Refer to illustrations 11.2 and 11.3

1 Detach the cable from the negative battery terminal.

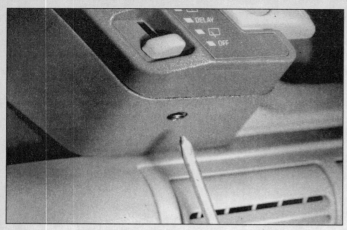

11.2 The wiper/washer is held in place at the bottom by a Phillips head screw

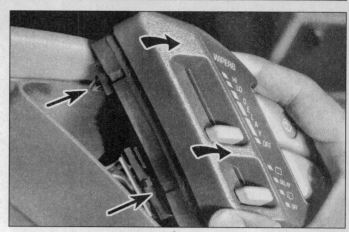

11.3 Pry at the corner of the switch to detach the clips (arrows) that rotate the switch out of the housing

12.2 Unscrew the two plastic screws and rotate the headlight housing forward for access to the bulb

12.3 Rotate the bulb holder counterclockwise and lift it out of the housing

12.4 Hold the electrical connector securely and pull the bulb straight out

2 Remove the retaining screw at the bottom of the cluster housing **(see illustration)**.
3 Pry carefully at the upper outer corner of the switch with a small screwdriver, detach the retaining clips and rotate the switch out of the cluster housing **(see illustration)**.
4 Withdraw the switch from the opening and detach the electrical connector with a small screwdriver.
5 Plug in the electrical connector, insert the tabs and rotate the switch into position, then press in until the clips engage. Install the retaining screw.

12 Headlight bulb - replacement

Refer to illustrations 12.2, 12.3 and 12.4
Warning: *Halogen gas filled bulbs are under pressure and may shatter if the surface is scratched or the bulb is dropped. Wear eye protection and handle the bulbs carefully, grasping only the base whenever possible. Do not touch the surface of the bulb with your fingers because the oil from your skin could cause it to overheat and fail prematurely. If you do touch the bulb surface, clean it with rubbing alcohol.*
1 Detach the cable from the negative terminal of the battery.
2 Remove the two plastic screws, then rotate the headlight housing forward for access to the bulbs **(see illustration)**.
3 Grasp the bulb holder assembly securely, turn it counterclockwise and pull the bulb holder assembly out of the headlight **(see illustration)**.
4 Pull the bulb straight out of the bulb holder **(see illustration)**.
5 Installation is the reverse of removal.

13 Headlights - adjustment

Refer to illustrations 13.1a, 13.1b and 13.2
Note: *The headlights must be aimed correctly. If adjusted incorrectly they could blind the driver of an oncoming vehicle and cause a serious accident or seriously reduce your ability to see the road. The headlights should be checked for proper aim every 12 months and any time a new headlight is installed or front end body work is performed. It should be emphasized that the following procedure is only an interim step which will provide tem-*

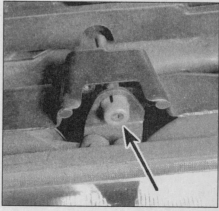

13.1a The headlight vertical adjustment screw is located at the top of the housing (arrow)

Chapter 12 Chassis electrical system

13.1b A special Torx-head tool will be required for making horizontal headlight adjustments

porary adjustment until the headlights can be adjusted by a properly equipped shop.

1   Headlights have two spring loaded adjusting screws, one on the top controlling up-and-down movement and one on the side controlling left-and-right movement **(see illustrations)**.

2   There are several methods of adjusting the headlights. The simplest method requires a blank wall 25-feet in front of the vehicle and a level floor **(see illustration)**.

3   Position masking tape vertically on the wall in reference to the vehicle centerline and the centerlines of both headlights.

4   Position a horizontal tape line in reference to the centerline of all the headlights.
Note: It may be easier to position the tape on the wall with the vehicle parked only a few inches away.

5   Adjustment should be made with the vehicle sitting level, the gas tank half-full and no unusually heavy load in the vehicle.

6   Starting with the low beam adjustment, position the high intensity zone so it's two inches below the horizontal line and two inches to the right of the headlight vertical line. Adjustment is made by turning the top adjusting screw clockwise to raise the beam and counterclockwise to lower the beam. The adjusting screw on the side should be used in the same manner to move the beam left or right.

7   With the high beams on, the high intensity zone should be vertically centered with the exact center just below the horizontal line. **Note:** It may not be possible to position the headlight aim exactly for both high and low beams. If a compromise must be made, keep in mind that the low beams are the most used and have the greatest effect on driver safety.

8   Have the headlights adjusted by a dealer service department or service station at the earliest opportunity.

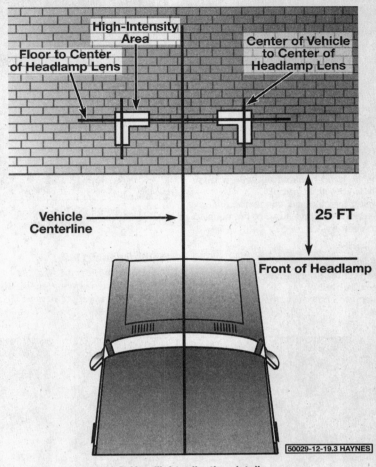

13.2 Headlight adjusting details

14 Bulb replacement

Front

Side marker light

Refer to illustrations 14.1, 14.2, 14.3a and 14.3b

1   Tilt the headlight housing forward for access (Section 12), then remove the side marker housing nuts **(see illustration)**.

2   Slide the light housing forward, then pull it out for access to the bulb holder **(see illustration)**.

3   Detach the bulb holder by turning it

14.1 Side marker light housing retaining nut locations (arrow)

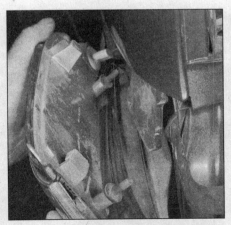

14.2 Slide the housing forward to detach it from the fender, then pull it back for access to the bulb holder

counterclockwise, then remove the bulb by pulling it straight out (see illustrations).

Park/turn signal

Refer to illustrations 14.4 and 14.5

4   Reach up under the front bumper, rotate the bulb holder counterclockwise and withdraw it from the housing (see illustration).
5   Pull the bulb straight out of the holder (see illustration).

Interior

Stepwell lamp

Refer to illustrations 14.6, 14.7a and 14.7b

6   Pry the lamp housing out with a small screwdriver (see illustration).
7   Unplug the electrical connector, unhook the bulb end and rotate it out of the housing (see illustrations).

Dome light

Refer to illustrations 14.8 and 14.9

8   Use a small screwdriver to pry off the lens (see illustration).
9   Detach the bulb from the contacts (see illustration).

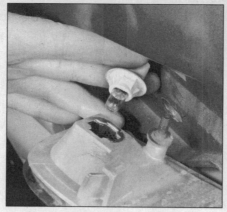

14.3a Pull the bulb holder out of the housing

Dome/reading light

Refer to illustrations 14.10 and 14.11

10   On these models, remove the lens and bulb, then use pliers to remove the push nuts and lower the light housing (see illustration).

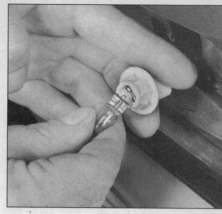

14.3b The bulb is inserted into the holder

11   Detach the bulb holder from the housing, then remove the bulb by pushing it in and turning counterclockwise (see illustration).

Instrument panel light

Refer to illustration 14.12

12   To gain access to the instrument panel

14.4 Reach up under the front bumper, grasp the bulb holder and pull it out of the park/turn signal housing

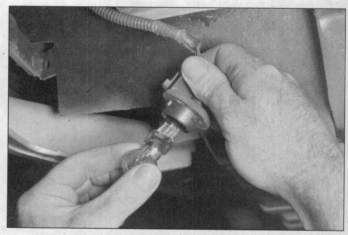

14.5 Grasp the bulb and pull it straight out of the holder

14.6 Use a small screwdriver to pry the lamp out

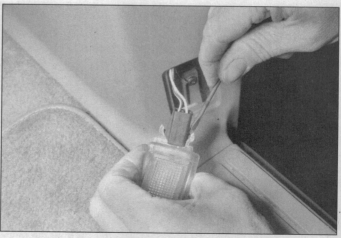

14.7a Disconnect the plug with a screwdriver

Chapter 12 Chassis electrical system

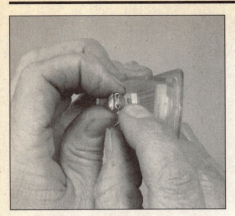

14.7b The end of the bulb fits over the hook in the housing

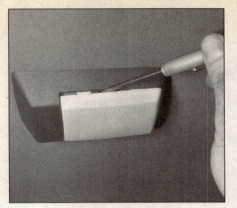

14.8 Pry the lens off with a small screwdriver

14.9 Use a small screwdriver to carefully detach the hook on the end of the bulb

lights, the instrument cluster will have to be removed first (see Section 16). To remove a bulb holder, rotate it counterclockwise and pull it out of the cluster **(see illustration)**. To remove the bulb from the holder, simply pull it straight out.

Rear
Tail lights
Refer to illustrations 14.13, 14.14a, 14.14b, 14.15a, 14.15b, 14.16a and 14.16b

13 Remove the two screws from the side of the housing **(see illustration)**.

14 Detach the two clips and rotate the housing out for access to the bulb holders

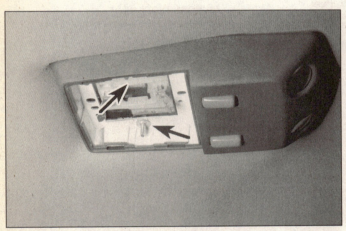

14.10 Use pliers to remove the pushnuts (arrows) and lower the housing

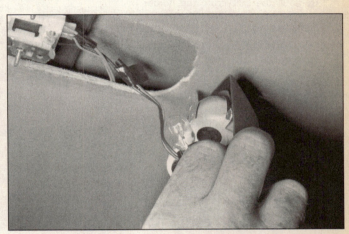

14.11 Detach the holder from the housing, then pull the bulb out

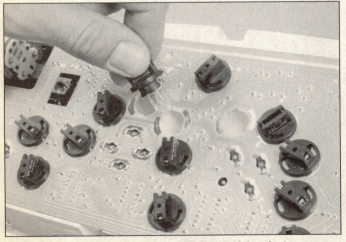

14.12 Rotate the bulb and lift it out of the cluster

14.13 You'll need a Phillips-head screwdriver to remove the two tail light housing bulbs (one shown here)

Chapter 12 Chassis electrical system

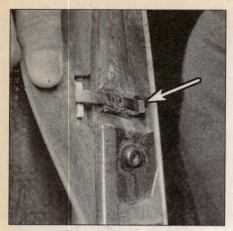

14.14a Release both of the housing clips (arrow)

14.14b Rotate the light housing away from the body

14.15a On some bulbs, you'll have to depress the clip, then turn the holder counterclockwise to remove it

(see illustrations).
15   Turn the bulb holder counterclockwise and withdraw it from the housing **(see illustrations)**.
16   Remove the bulb from the holder by pulling straight out **(see illustrations)**.

Backup lights
Refer to illustrations 14.18, 14.19 and 14.20
17   Remove the rear license plate.
18   Remove the two screws and lift the light housing away **(see illustration)**.
19   Rotate the bulb holder counterclockwise

and withdraw it from the housing **(see illustration)**.
20   Pull the bulb out of the holder **(see illustration)**.

License plate light
21   Remove the license plate.

14.15b The smaller bulb holders can be simply twisted and removed

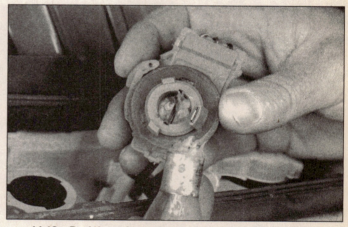

14.16a Push in and turn the larger bulbs to remove them

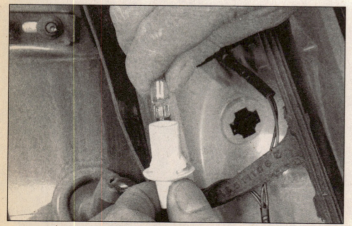

14.16b Pull the smaller bulbs straight out of the holder

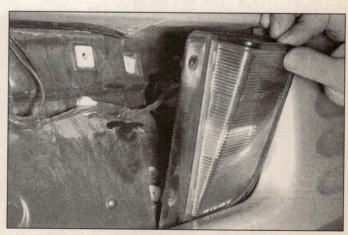

14.18 Remove the screws and detach the backup light housing

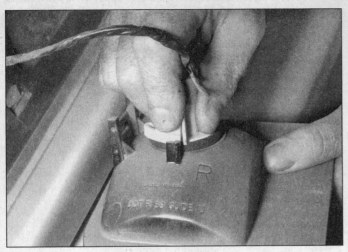

14.19 Turn the bulb holder counterclockwise and lift it out of the housing

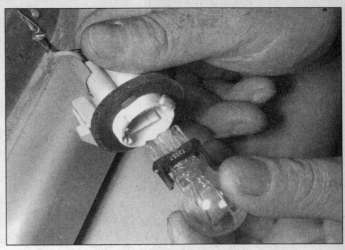

14.20 The bulb pulls straight out of the holder

22 Remove the two Torx-head screws and detach the housing.
23 Withdraw the holder from the housing, then pull the bulb out.

15 Radio and speakers - general information, removal and installation

Delco Loc II general information

1 Some audio systems in the models covered by this manual feature the Delco Loc II anti-theft feature. In this system, the owner can program a secret code into the radio, which will automatically scramble if the radio is deprived of battery power, rendering it inoperative.
2 Before beginning any procedure which requires the battery to be disconnected or the radio to be removed, make sure you know the secret code stored in the radio. It must be entered after the battery is reconnected.
3 The radio can also be unlocked before battery power is disconnected by entering the secret code, at which time the display on the radio will momentarily indicate " - " then the clock will be displayed.
4 If you don't know your secret code or encounter difficulty in reactivating the radio, consult your dealer service department.

Radio

Refer to illustrations 15.7a, 15.7b, 15.8a and 15.8b
5 Detach the cable from the negative terminal of the battery.
6 Remove the instrument panel trim plate (Chapter 11, Section 22).
7 Remove two Torx-head screws at the top of the radio and the two bolts at the bottom **(see illustrations)**.

15.7a Use a Torx-head screwdriver to remove the two screws securing the top of the radio

8 Pull the radio out and disconnect the electrical connection and antenna lead **(see illustrations)**.

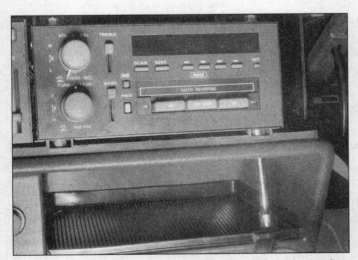

15.7b Remove the two bolts at the top of the radio with a socket and extension

15.8a Depress the clips and detach the electrical connector from the radio

15.8b Pull the antenna lead straight out of the electrical connector in the back of the radio

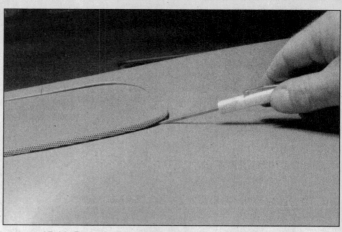

15.11 Detach the speaker cover by prying carefully around the edges

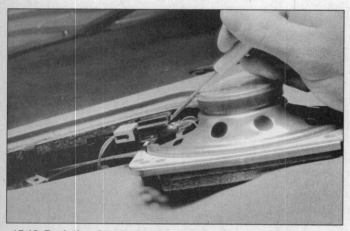

15.12 Push the electrical connector off with a small screwdriver

17.4 When measuring the voltage at the rear window defogger grid, wrap a piece of aluminum foil around the positive probe of the voltmeter and press the foil against the wire with your finger

9 Remove the radio from the dash.
10 Installation is the reverse of removal.

Speakers

Refer to illustrations 15.11 and 15.12

11 Pry off the speaker cover with a small screwdriver **(see illustration)**.
12 Remove the screws, detach the speaker and unplug the electrical connector **(see illustration)**.
13 Installation is the reverse of removal.

16 Instrument cluster - removal and installation

1 Detach the cable from the negative battery terminal.
2 Remove the headlight and wiper/washer switches (see Sections 10- and 11).
3 Remove instrument panel trim pad (Chapter 11, Section 22).
4 Remove the four screws, detach the instrument cluster from the electrical wiring harness connector and lift it from the dash.
5 Installation is the reverse of removal.

17 Rear defogger (electric grid type) - check and repair

Refer to illustrations 17.4 and 17.11

1 This option consists of a rear window with a number of horizontal elements baked into the glass surface during the glass forming operation.
2 Small breaks in the element can be successfully repaired without removing the rear window.
3 To test the grids for proper operation, turn on the system.
4 Ground one lead of a test light and carefully touch the other lead to each element line **(see illustration)**.
5 The brilliance of the test light should increase as the lead is moved across the element from right to left. If the test light glows brightly at both ends of the lines, check for a loose ground wire. All of the lines should be checked in at least two places.
6 To repair a break in a line, it is recommended that a repair kit specifically for this purpose be purchased from a GM dealer parts department or auto parts store.

Included in the repair kit will be a decal, a container of silver plastic and hardener, a mixing stick and instructions.
7 To repair a break, first turn off the system and allow it to de-energize for a few minutes.
8 Lightly buff the element area with fine steel wool then clean it thoroughly with rubbing alcohol.
9 Use the decal supplied in the repair kit or apply strips of electrician's tape above and below the area to be repaired. The space between the pieces of tape should be the same width as the existing lines. This can be checked from outside the vehicle. Press the tape tightly against the glass to prevent seepage.
10 Mix the hardener and silver plastic thoroughly.
11 Using the wood spatula, apply the silver plastic mixture between the pieces of tape, overlapping the undamaged area slightly on either end **(see illustration)**.
12 Carefully remove the decal or tape and apply a constant stream of hot air directly to the repaired area. A heat gun set at 500 to 700-degrees F is recommended. Hold the

Chapter 12 Chassis electrical system

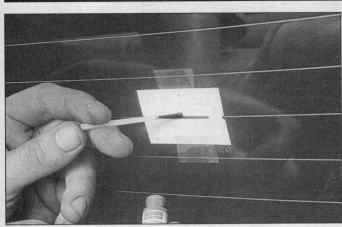

17.11 To use a defogger repair kit, apply masking tape to the inside of the window at the damaged area, then brush on the special conductive coating

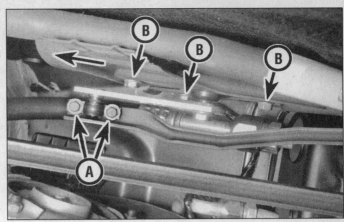

18.10 Loosen the socket bolts (A), and detach the crank arm, then remove the bolts (B) and slide the motor out of the slot in the direction shown (1991 and later models)

19.2 Detach the clip from the wiper arm, then detach the washer hose

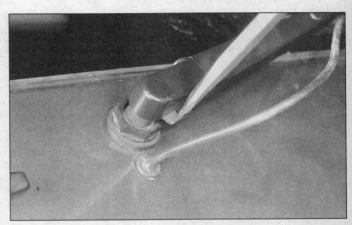

19.3 Use a screwdriver to pull out the release lever, then remove the wiper arm from the spindle

gun one inch from the glass for two minutes.
13   If the new element appears off color, tincture of iodine can be used to clean the repair and bring it back to the proper color. This mixture should not remain on the repair for more than 30 seconds.
14   Although the defogger is now fully operational, the repaired area should not be disturbed for at least 24 hours.

18 Windshield wiper motor - removal and installation

Refer to illustration 18.10
1   Raise the hood.
2   Detach the cable from the negative terminal of the battery.

1990 models

3   Remove the nuts and detach the windshield wiper arms.
4   Remove the air cleaner and unplug the electrical connector from the wiper motor.
5   Remove the two wiper housing bolts using a 55 mm Torx-head tool and the three module frame-to-body nuts with a 10 mm socket. Lower the module and lift it from the engine compartment.
6   Remove the two crank arm socket bolts sufficiently to allow the socket to be disconnected from the ball on the crank arm.
7   Remove the three bolts and detach the wiper motor from the module.
8   Installation is the reverse of removal.

1991 on

9   Unplug the electrical connector from the wiper motor.
10   Loosen the two socket bolts, detach the crank arm from the linkage, then remove the motor retaining bolts and slide the motor out of the slot in the module frame, lowering it into the engine compartment (see illustration).
11   Installation is the reverse of removal.

19 Rear window wiper motor - removal and installation

Refer to illustrations 19.2, 19.3, 19.4 and 19.6
1   Detach the cable from the negative terminal of the battery.
2   Use a small screwdriver to disconnect

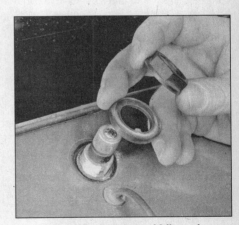

19.4 Wiper motor spindle and washer details

the washer hose from the wiper arm (see illustration).
3   Pull out the release clip and detach the wiper arm from the motor spindle (see illustration).
4   Remove the wiper motor nut and washer, noting the direction in which the washer faces (see illustration).

19.6 Lower the motor through the liftgate access hole

5   Open the liftgate and remove the door trim panel (Chapter 11, Section 11).
6   Working through the access hole unplug the wiper motor electrical connector. Disconnect the tailgate latch rod. Remove the bolts, detach the motor and lower it through the access hole (see illustration).
7   Installation is the reverse of removal.

20 Cruise control system - description and check

The cruise control system maintains vehicle speed with a vacuum actuated servo motor located in the engine compartment, which is connected to the throttle linkage by a cable. The system consists of the servo motor, clutch switch, brake switch, control switches, a relay and associated vacuum hoses.

Because of the complexity of the cruise control system and the special tools and techniques required for diagnosis, repair should be left to a dealer service department or a repair shop. However, it is possible for the home mechanic to make simple checks of the wiring and vacuum connections for minor faults which can be easily repaired. These include:

a) *Inspect the cruise control actuating switches for broken wires and loose connections.*
b) *Check the cruise control fuse.*
c) *The cruise control system is operated by vacuum so it's critical that all vacuum switches, hoses and connections are secure. Check the hoses in the engine compartment for tight connections, cracks and obvious vacuum leaks.*

21 Power window system - description and check

The power window system operates the electric motors mounted in the doors which lower and raise the windows. The system consists of the control switches, the motors and glass lift mechanisms (regulators), and associated wiring.

Because of the complexity of the power window system, repair should be left to a dealer service department or other repair shop. However, it is possible for the home mechanic to make simple checks of the wiring connections and motors for minor faults which can be easily repaired. These include:

a) *Inspect the power window actuating switches for broken wires and loose connections.*
b) *Check the power window fuse/and or circuit breaker.*
c) *Remove the door panel(s) and check the power window motor wires to see if they're loose or damaged. Inspect the glass mechanisms for damage which could cause binding.*

22 Power door lock system - description and check

The power door lock system operates the door lock actuators mounted in each door. The system consists of the switches, actuators and associated wiring. Because of the complexity of the power door lock system, repair should be left to a dealer service department or other repair shop. However, it is possible for the home mechanic to make simple checks of the wiring connections and actuators for minor faults which can be easily repaired. These include:

a) *Check the system fuse and/or circuit breaker.*
b) *Check the switch wires for damage and loose connections. Check the switches for continuity.*
c) *Remove the door panel(s) and check the actuator wiring connections to see if they're loose or damaged. Inspect the actuator rods (if equipped) to make sure they aren't bent or damaged. Inspect the actuator wiring for damaged or loose connections. The actuator can be checked by applying battery power momentarily. A discernible click indicates that the solenoid is operating properly.*

23 Airbag - general information and precautions

General information

1   Some later models are equipped with a Supplemental Inflatable Restraint (SIR) system, more commonly known as an airbag. This system is designed to protect the driver from serious injury in the event of a head-on or frontal collision. It consists of an airbag module in the center of the steering wheel, two discriminating (crash) sensors mounted at the front and the interior of the vehicle, an arming sensor located behind the glove box, an SIR coil assembly mounted under the steering wheel and a diagnostic module located inside the passenger compartment under the dash.

Airbag module and SIR coil

2   The airbag inflator module contains a housing incorporating the cushion (airbag) and inflator unit, mounted in the center of the steering wheel. The inflator assembly is mounted on the back of the housing over a hole through which gas is expelled, inflating the bag almost instantaneously when an electrical signal is sent from the system. The airbag steering column coil assembly is mounted on the steering column under the steering wheel and carries this signal to the module. The coil assembly can transmit an electrical signal regardless of steering wheel position.

Sensors

3   The system has three sensors: a forward discriminating sensor mounted on the radiator support and a passenger compartment discriminating sensor mounted under the front of the center console. These sensors are basically pressure sensitive switches that complete an electrical circuit during an impact of sufficient G force. The electrical signal from these sensors is sent to the arming sensor which then completes the circuit and inflates the airbag.

Diagnostic Energy Reserve Module (DERM)

4   The DERM supplies the current to the airbag system in the event of the collision, even if battery power is cut off. It checks this system every time the vehicle is started, causing the "AIR BAG" light to go on then off, if the system is operating properly. If there is a fault in the system, the light will go on and stay on and the DERM will store fault codes indicating the nature of the fault. If the AIR BAG light goes on and stays on, the vehicle should be taken to your dealer immediately for service.

Operation

5   For the airbag(s) to deploy, an impact of sufficient force must occur within 30-degrees of the vehicle centerline. When this condition

Chapter 12 Chassis electrical system

occurs, the circuit to the airbag inflator is closed and the airbag inflates. If the battery is destroyed by the impact, or is too low to power the inflator, a back-up power supply inside the diagnostic/energy reserve module supplies current to the airbag.

Self-diagnosis system

6 A self-diagnosis circuit in the module displays a light when the ignition switch is turned to the On position. If the system is operating normally, the light should go out after seven flashes. If the light doesn't come on, or doesn't go out after seven flashes, or if it comes on while you're driving the vehicle, there's a malfunction in the SIR system. Have it inspected and repaired as soon as possible. Do not attempt to troubleshoot or service the SIR system yourself. Even a small mistake could cause the SIR system to malfunction when you need it.

Servicing components near the SIR system

7 Nevertheless, there are times when you need to remove the steering wheel, radio or service other components on or near the instrument panel or at the front of the vehicle. At these times, you'll be working around components and wiring harnesses for the SIR system. SIR system wiring is easy to identify; they're all covered by a bright yellow conduit. Do not unplug the connectors for the SIR system wiring, except to disable the system. And do not use electrical test equipment on the SIR system wiring. **Always disable the SIR system before working near the SIR system components or related wiring.**

Disabling the SIR system

8 Turn the steering wheel to the straight ahead position, place the ignition switch key in the Lock position and remove the key. Remove the airbag fuse from the fuse block (see Section 3). It's also a good idea to disconnect the cable from the negative terminal of the battery, although this is not actually specified by the manufacturer. **Caution:** *If the vehicle is equipped with a Delco Loc II audio system, make sure you have the correct activation code before disconnecting the battery. See the information at the front of this manual for the radio re-activation procedure.*

9 Remove the steering column covers and the left sound insulator panel below the instrument panel (see Chapter 11).

10 Unplug the yellow Connector Position Assurance (CPA) steering column harness connector.

Enabling the SIR system

11 After you've disabled the airbag and performed the necessary service, plug in the steering column CPA connectors. Reinstall the steering column lower trim panel, and the sound insulator panel.

12 Install the airbag fuse. Connect the negative battery terminal.

Removing, centering and installing the SIR coil

13 Anytime some part of the steering system is disassembled for service or replacement, the steering column should be immobilized to ensure that the SIR coil doesn't become uncentered (moved). This can occur, for example, if the steering column is separated from the steering gear, or if the centering spring is pushed down, allowing the hub to rotate while the coil is removed from the steering column. If the coil becomes accidentally uncentered, re-center it as follows before reassembling the steering system.

14 Make sure that the wheels are pointed straight ahead.

15 Remove the coil assembly snap-ring and remove the coil assembly.

16 Holding the coil assembly with its bottom side facing up, depress the spring lock and rotate the hub in the direction of the arrow until it stops (the arrow is on the back of the coil assembly). The coil ribbon should be wound up snug against the center hub.

17 Rotate the coil hub in the opposite direction two and three-quarters turns, then release the spring lock. The coil is now centered.

18 Install the SIR coil and secure it with the snap-ring. The tab will be at the top and the marks aligned when the coil is properly installed.

24 Wiring diagrams - general information

Since it isn't possible to include all wiring diagrams for every year covered by this manual, the following diagrams are those that are typical and most commonly needed.

Prior to troubleshooting any circuit, check the fuse and circuit breakers (if equipped) to make sure they're in good condition. Make sure the battery is properly charged and check the cable connections (see Chapter 1, Section 8).

When checking a circuit, make sure that all electrical connectors are clean, with no broken or loose terminals. When unplugging an electrical connector, do not pull on the wires. Pull only on the connector housings themselves.

12-16 Chapter 12 Chassis electrical system

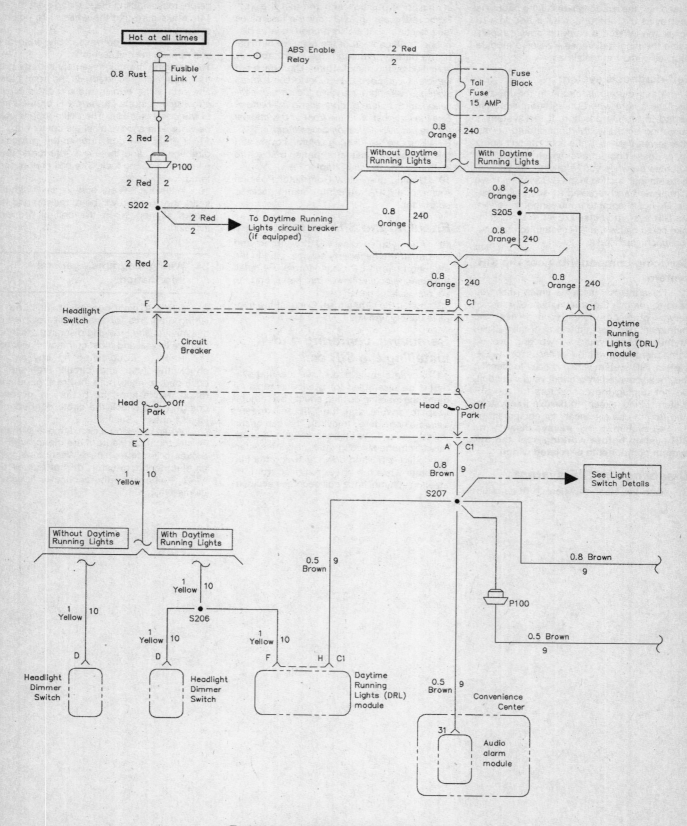

Typical light switch wiring diagram (1 of 3)

Chapter 12 Chassis electrical system

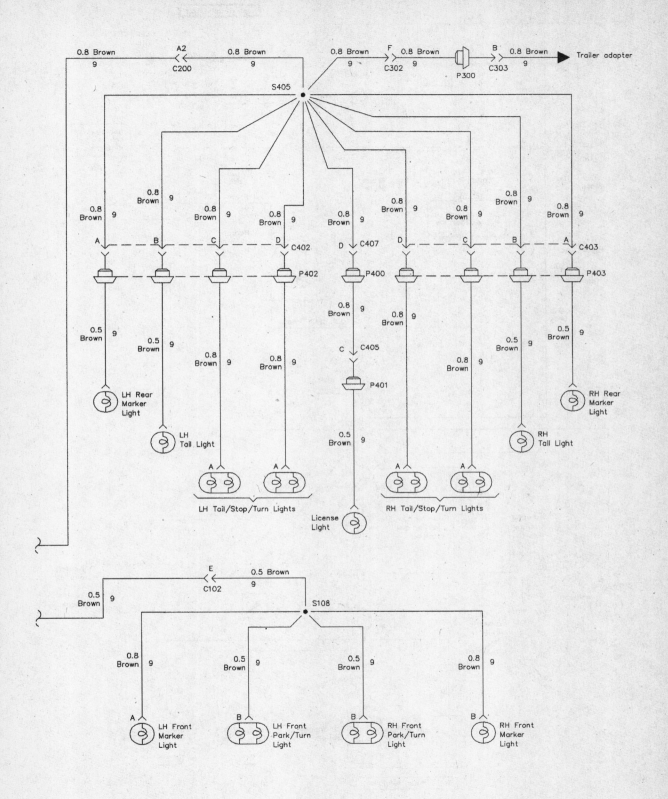

Typical light switch wiring diagram (2 of 3)

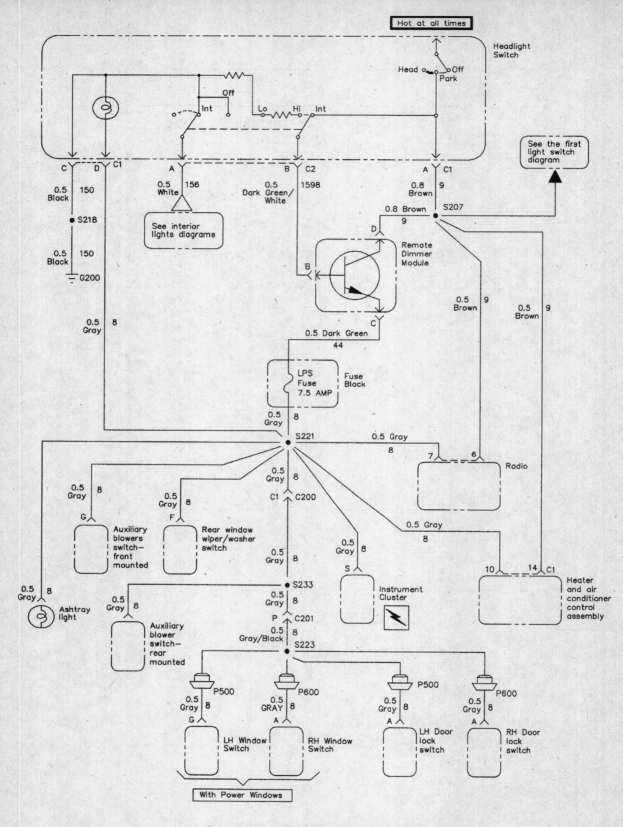

Typical light switch wiring diagram (3 of 3)

Chapter 12 Chassis electrical system

12-19

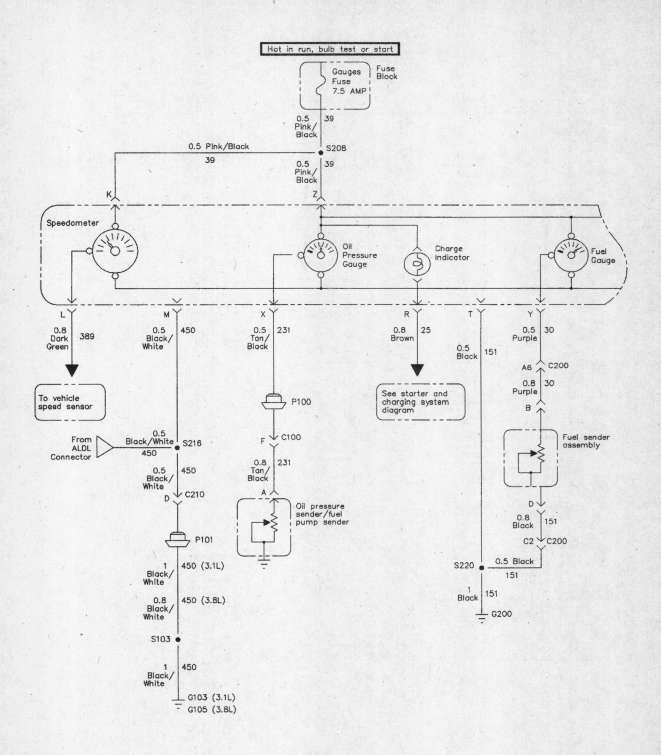

Typical instrument panel wiring diagram (1 of 3)

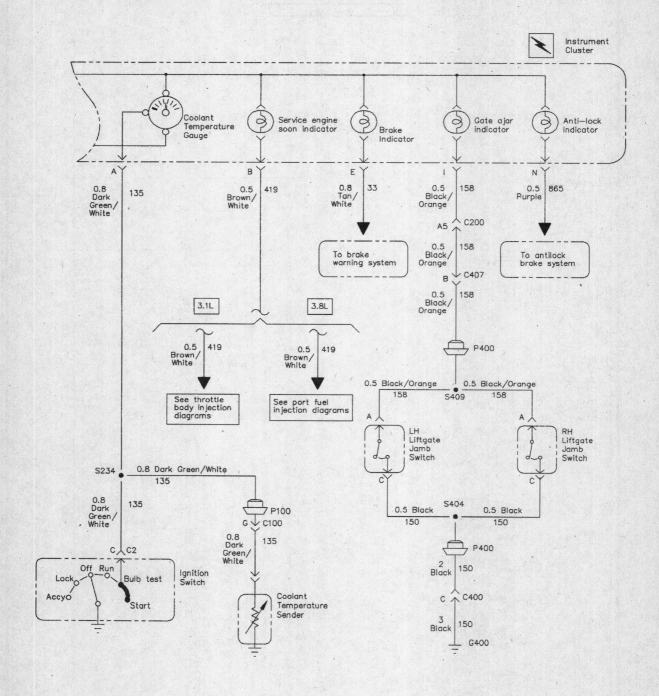

Typical instrument panel wiring diagram (2 of 3)

Chapter 12 Chassis electrical system

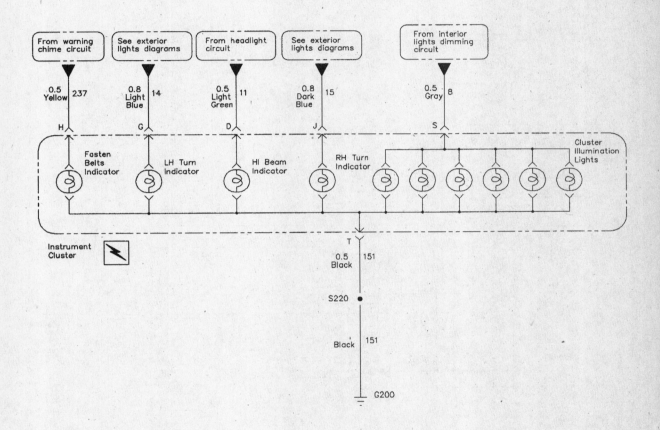

Typical instrument panel wiring diagram (3 of 3)

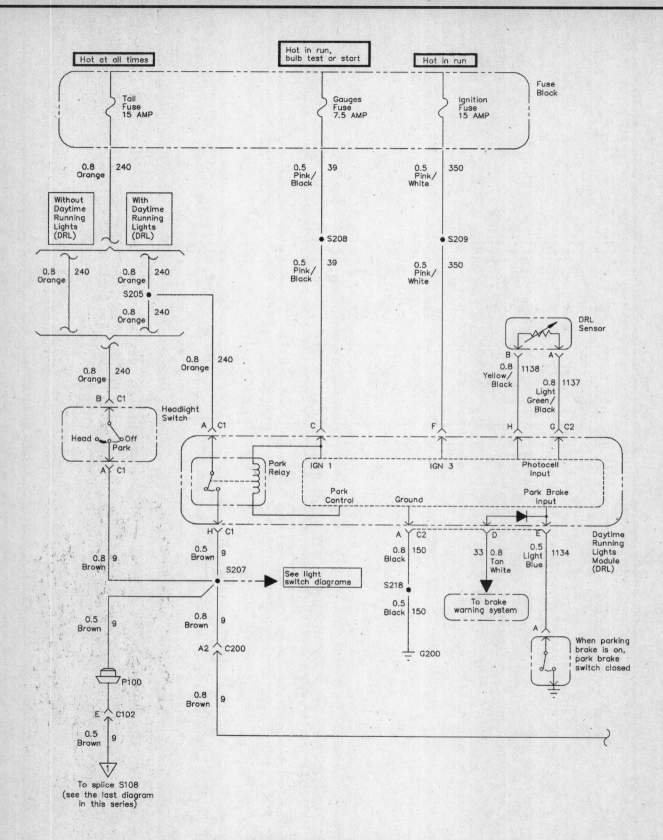

Typical exterior lights wiring diagram (1 of 5)

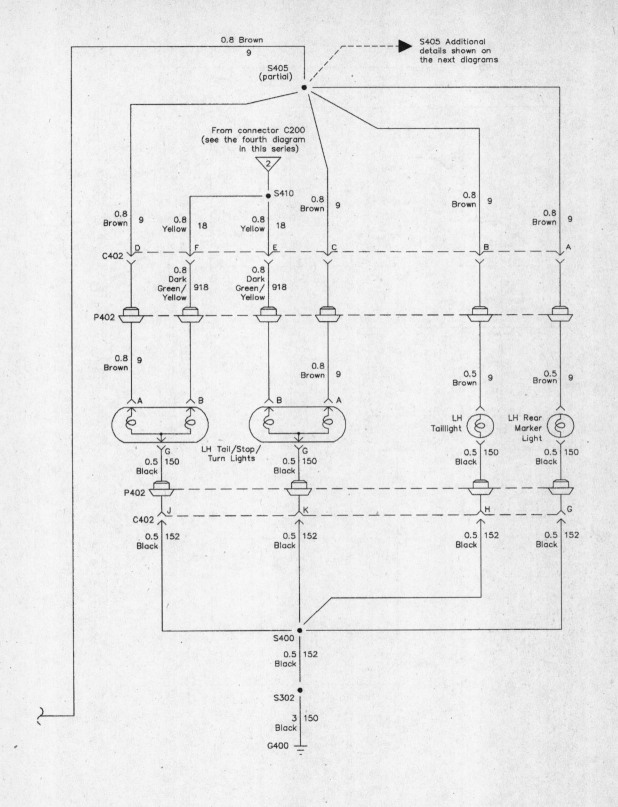

Typical exterior lights wiring diagram (2 of 5)

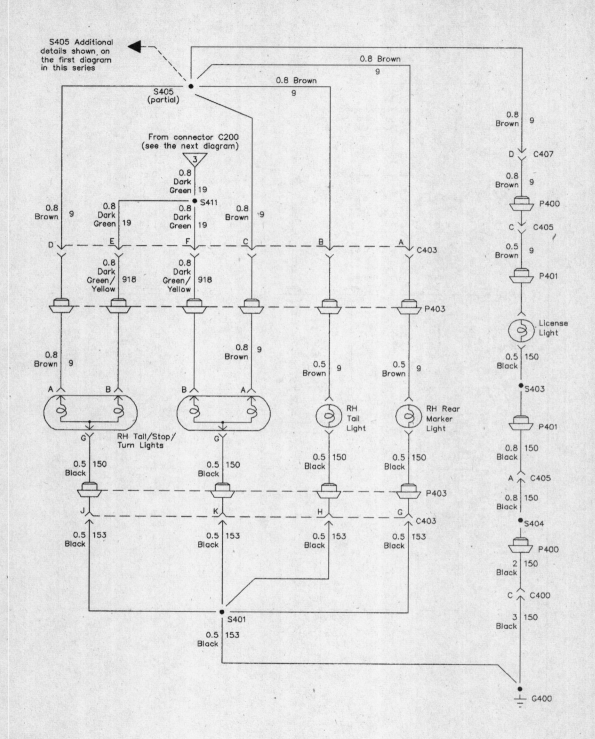

Typical exterior lights wiring diagram (3 of 5)

Chapter 12 Chassis electrical system

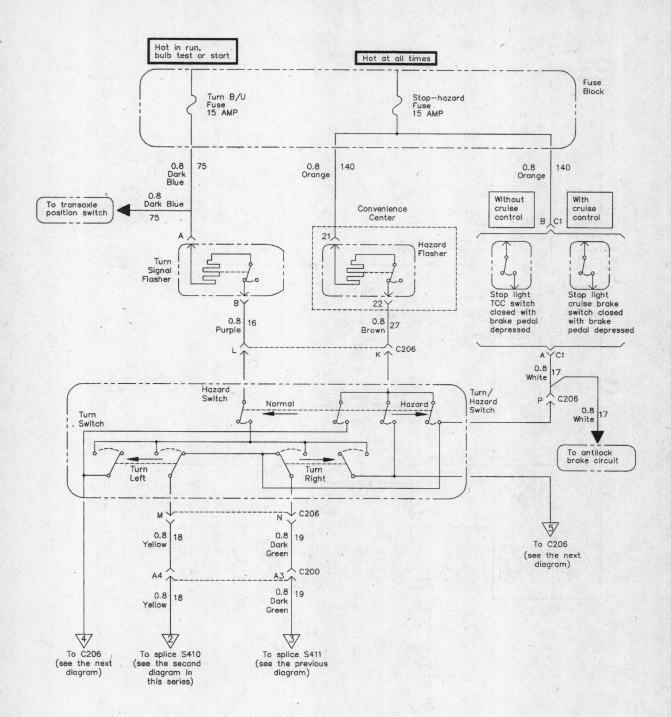

Typical exterior lights wiring diagram (4 of 5)

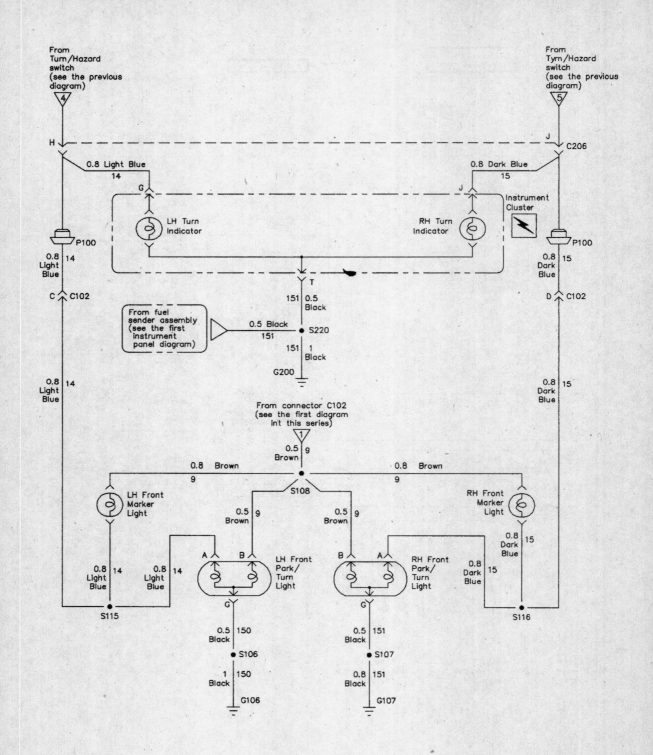

Typical exterior lights wiring diagram (5 of 5)

Chapter 12 Chassis electrical system

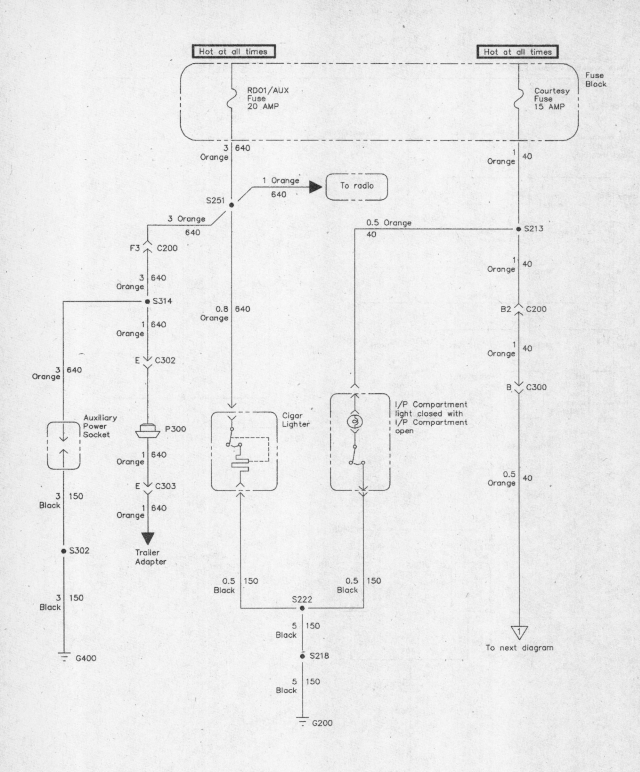

Typical interior lights wiring diagram (1 of 4)

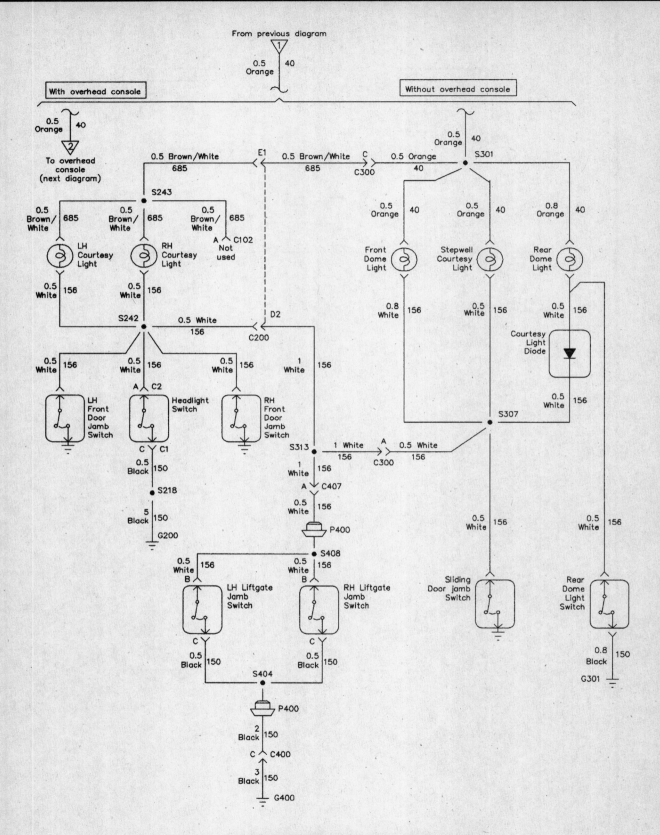

Typical interior lights wiring diagram (2 of 4)

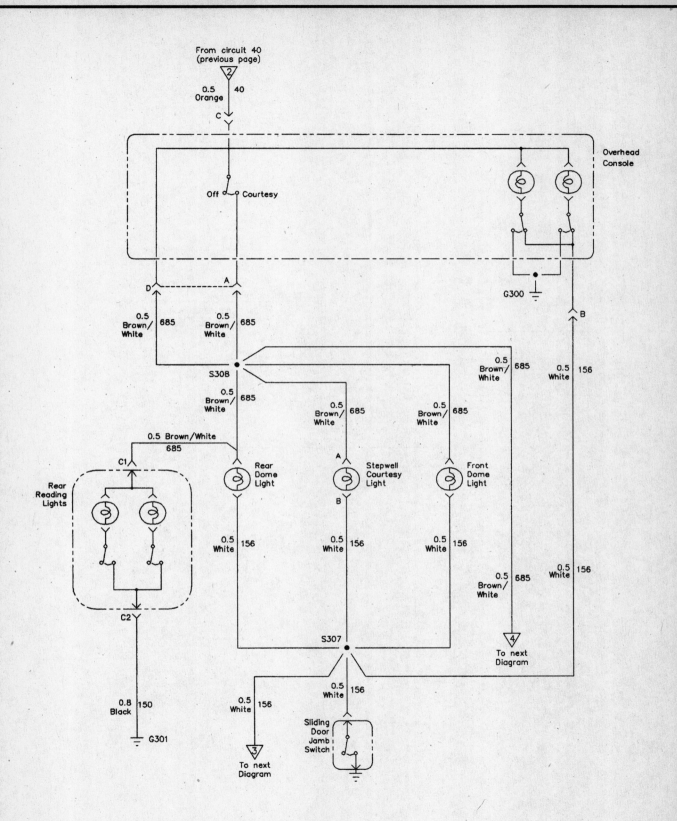

Typical interior lights wiring diagram (3 of 4)

12-30 Chapter 12 Chassis electrical system

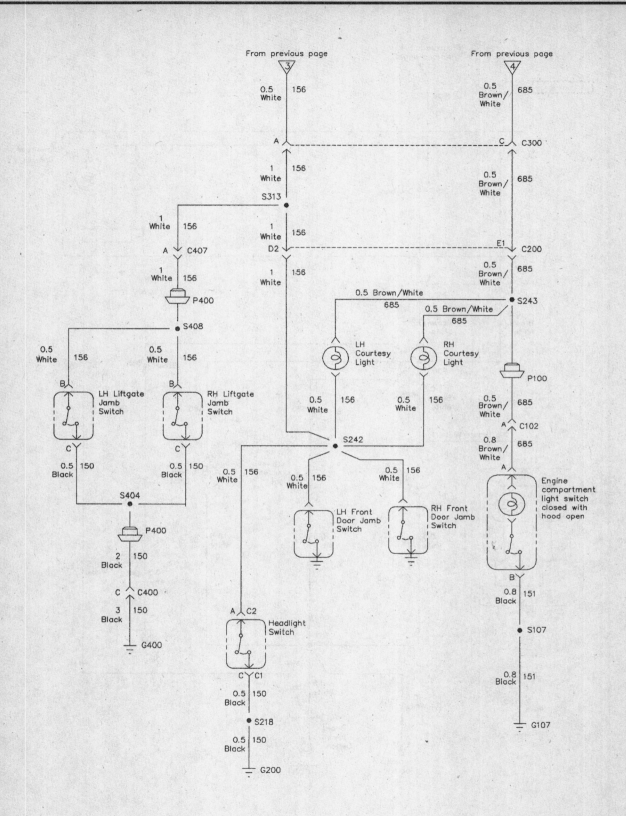

Typical interior lights wiring diagram (4 of 4)

Chapter 12 Chassis electrical system

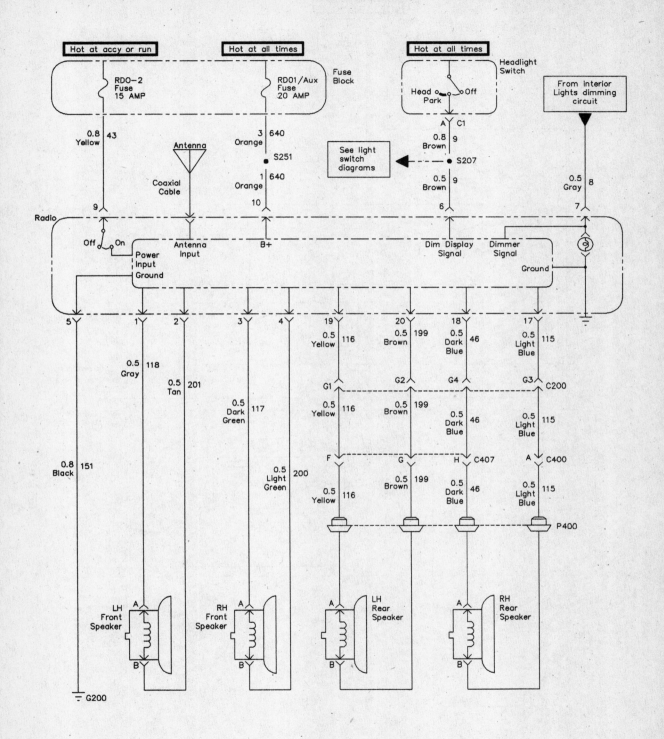

Typical radio wiring diagram

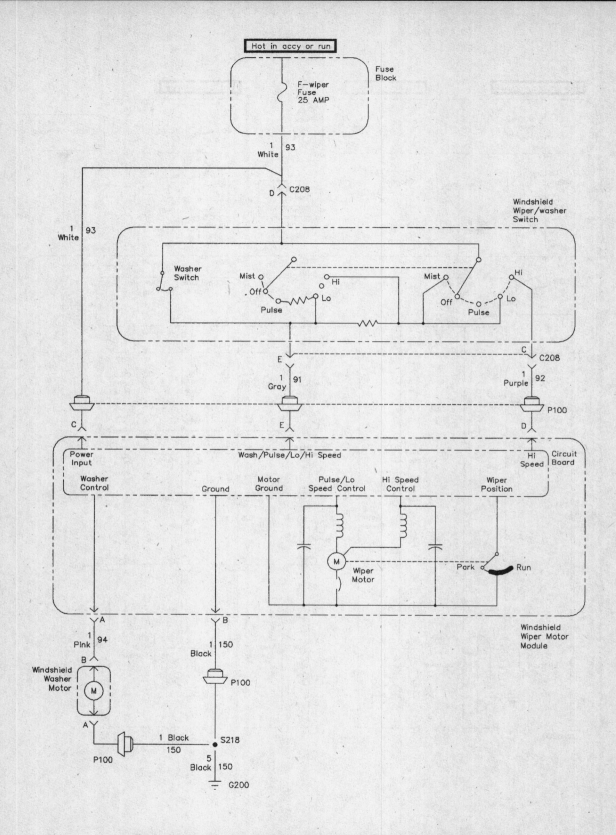

Typical front wiper/washer wiring diagram (pulse-type washers shown)

Chapter 12 Chassis electrical system

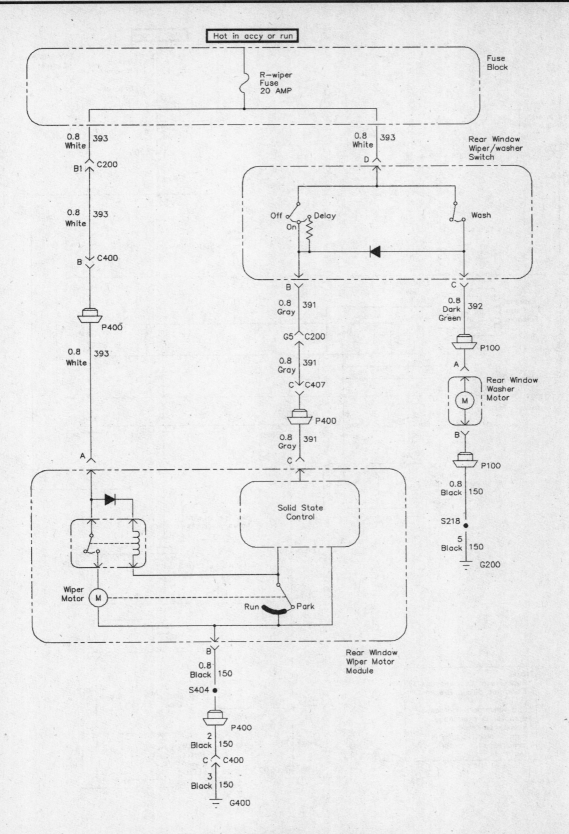

Typical rear wiper/washer wiring diagram

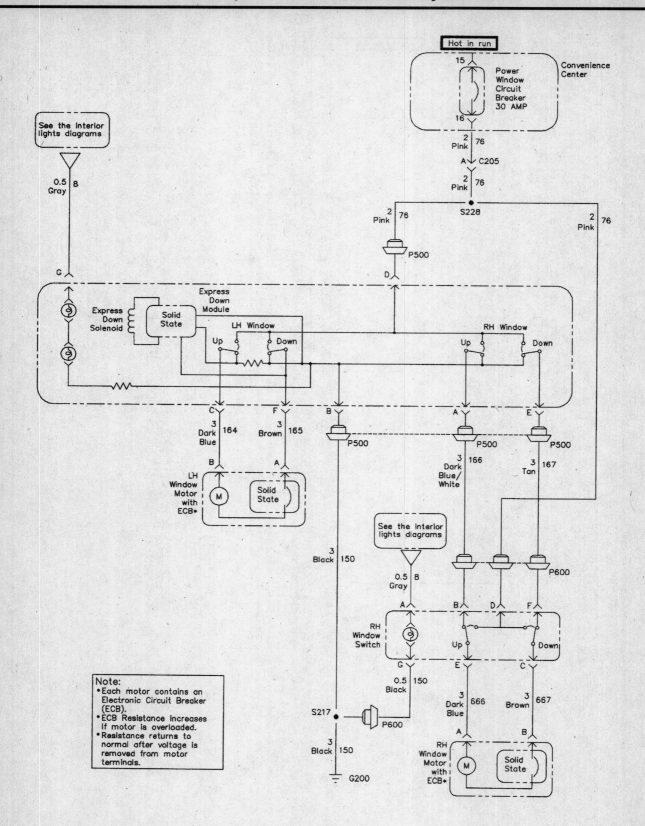

Typical power window wiring diagram

Chapter 12 Chassis electrical system

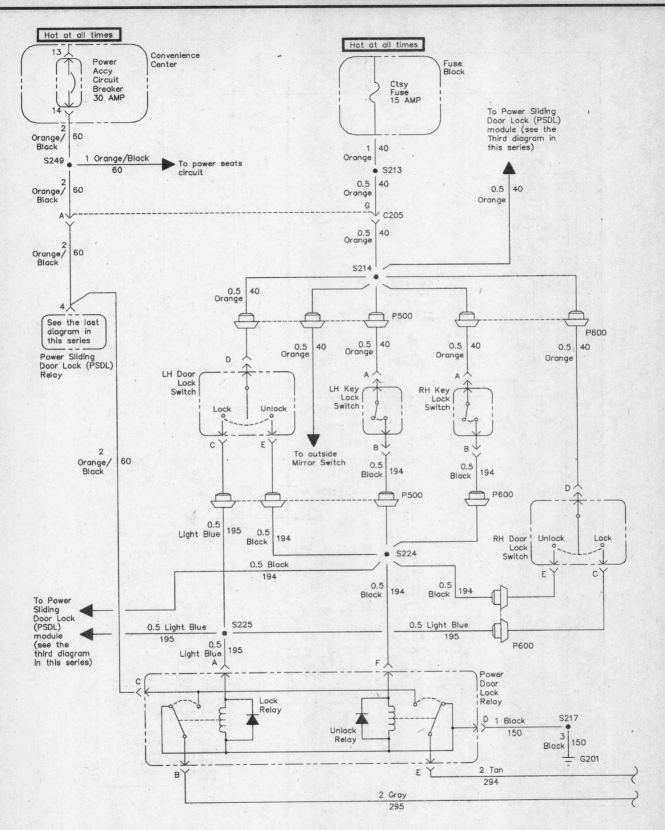

Typical power door lock wiring diagram - front doors and liftgate - (1 of 2)

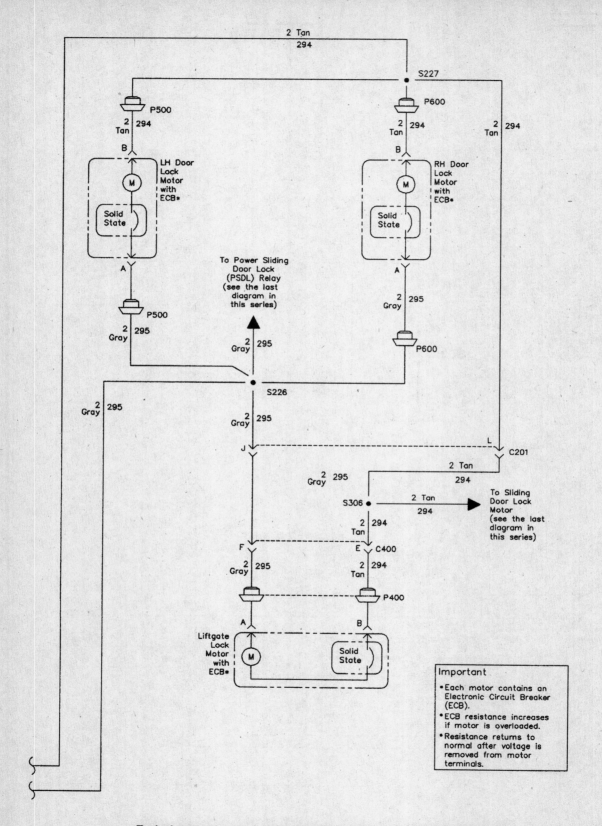

Typical power door lock wiring diagram - front doors and liftgate - (2 of 2)

Chapter 12 Chassis electrical system

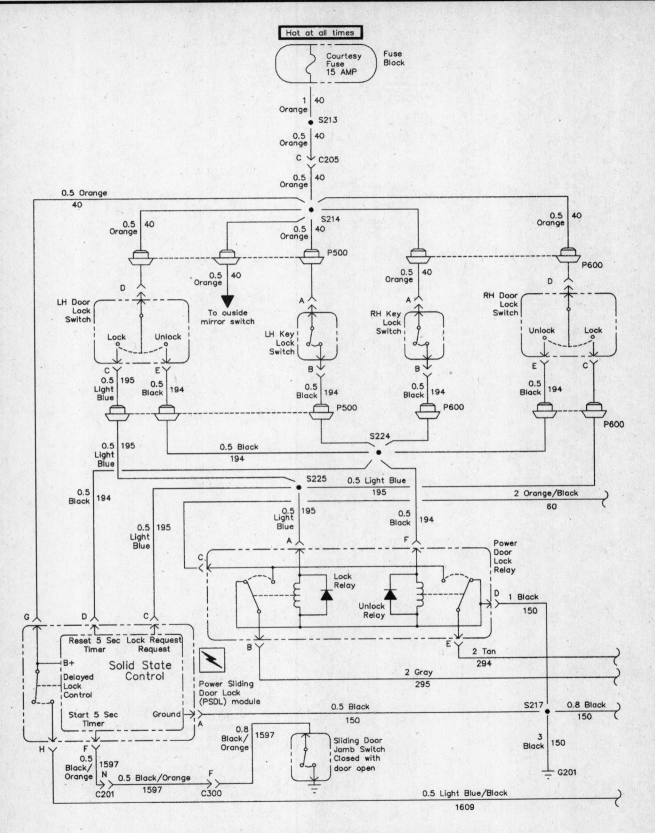

Typical power door lock wiring diagram - sliding door - (1 of 2)

12-38 Chapter 12 Chassis electrical system

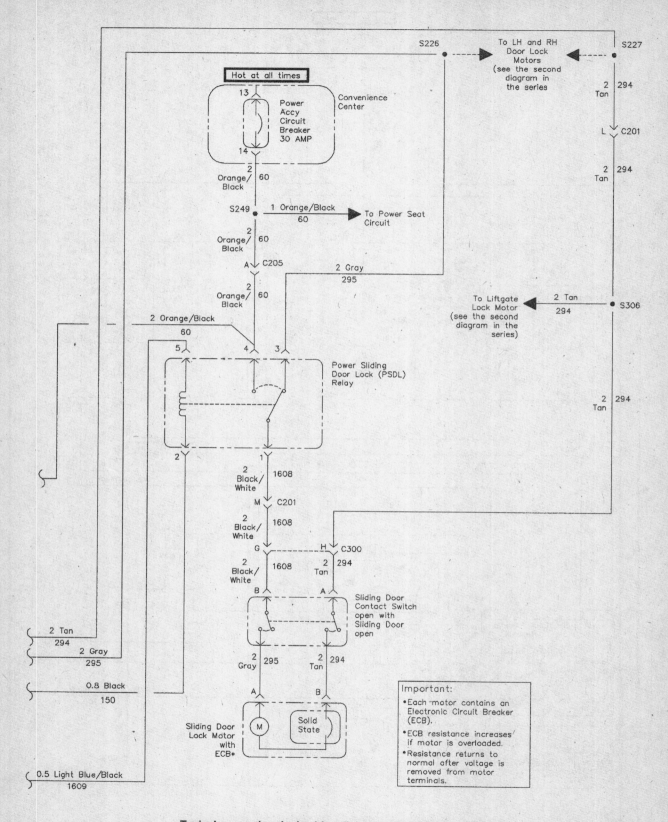

Typical power door lock wiring diagram - sliding door - (2 of 2)

Chapter 12 Chassis electrical system

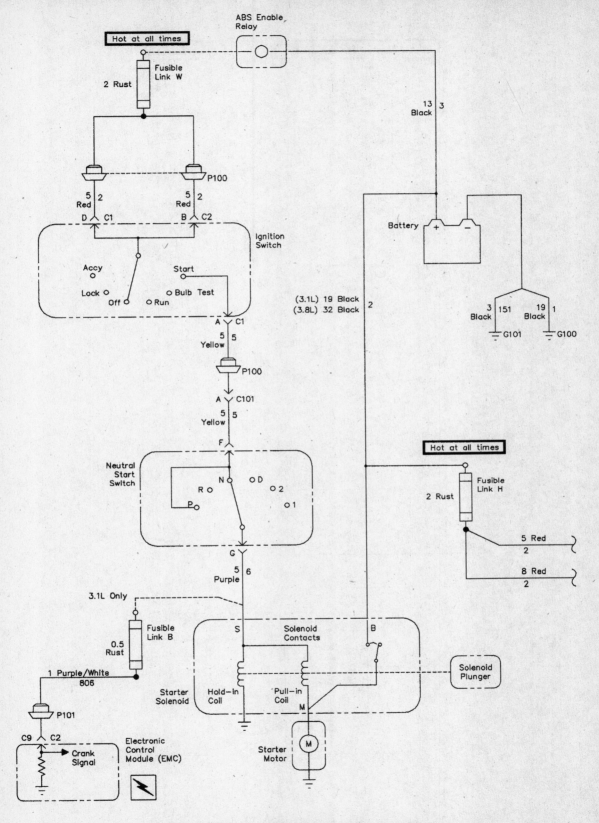

Typical starting and charging systems wiring diagram (1 of 2)

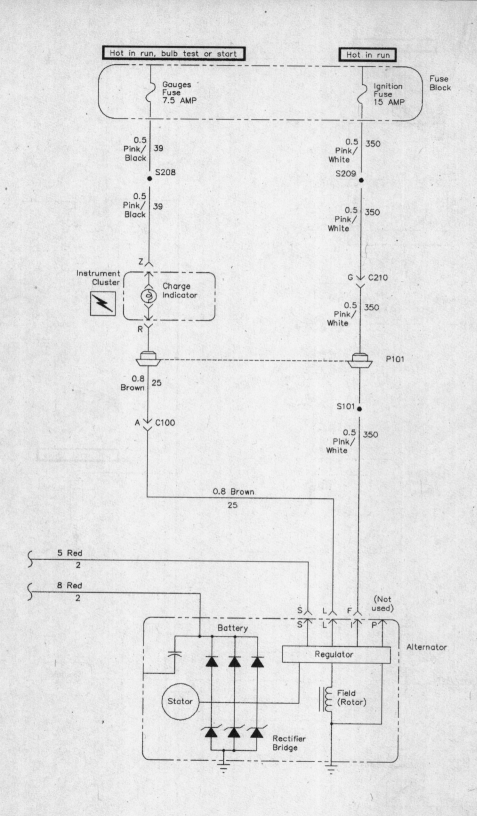

Typical starting and charging systems wiring diagram (2 of 2)

Chapter 12 Chassis electrical system

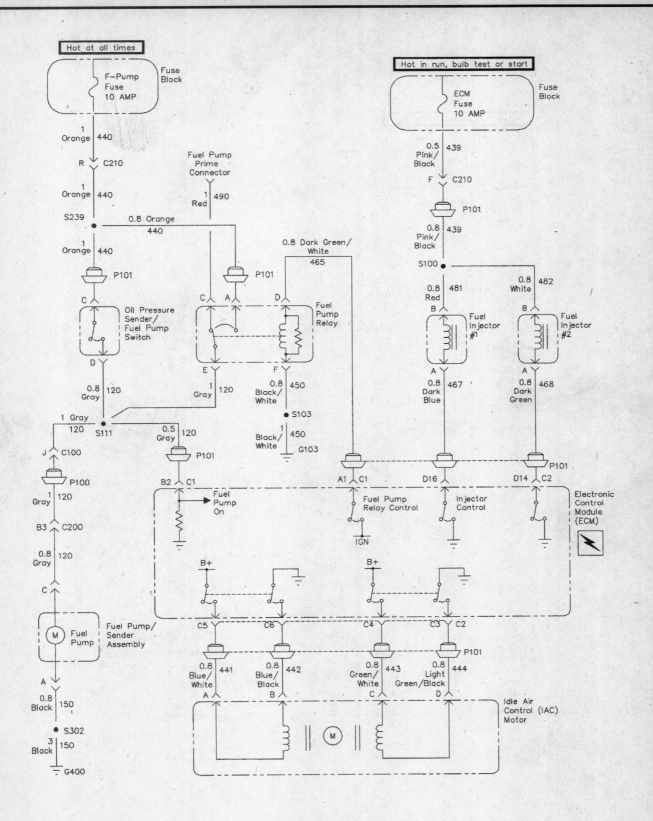

Typical throttle body fuel injection (TBI) wiring diagram - 3.1L engine - (1 of 3)

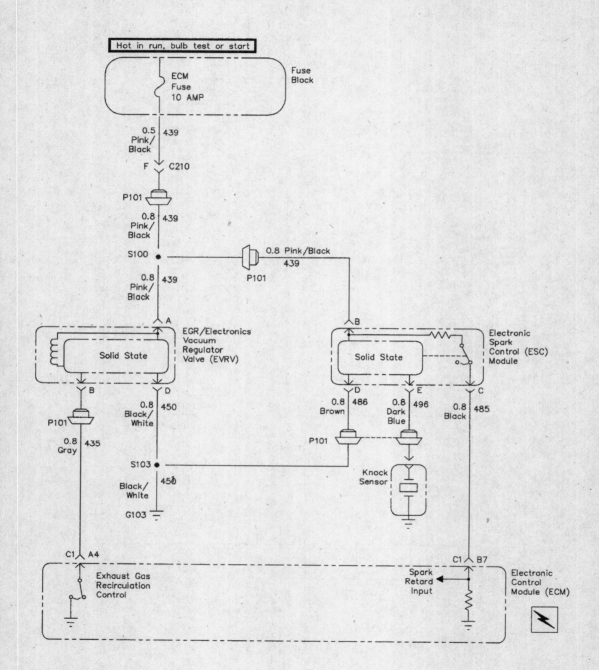

Typical throttle body fuel injection (TBI) wiring diagram - 3.1L engine - (2 of 3)

Chapter 12 Chassis electrical system

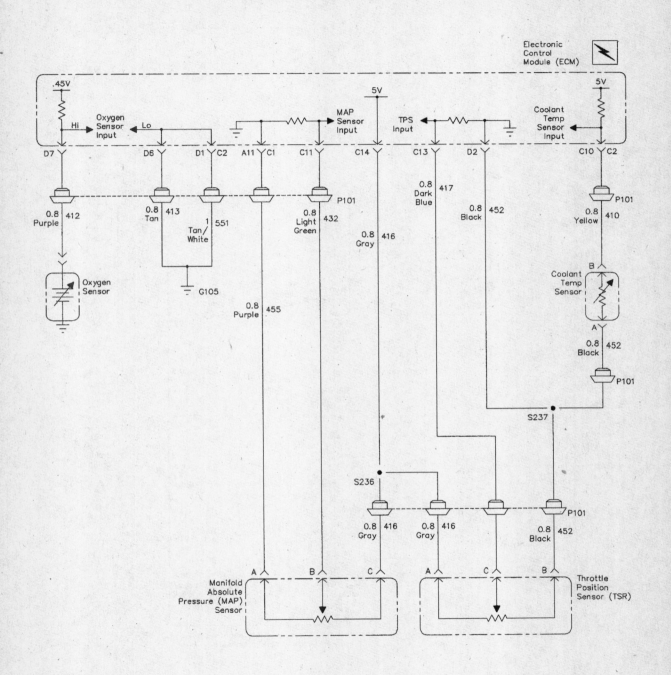

Typical throttle body fuel injection (TBI) wiring diagram - 3.1L engine - (3 of 3)

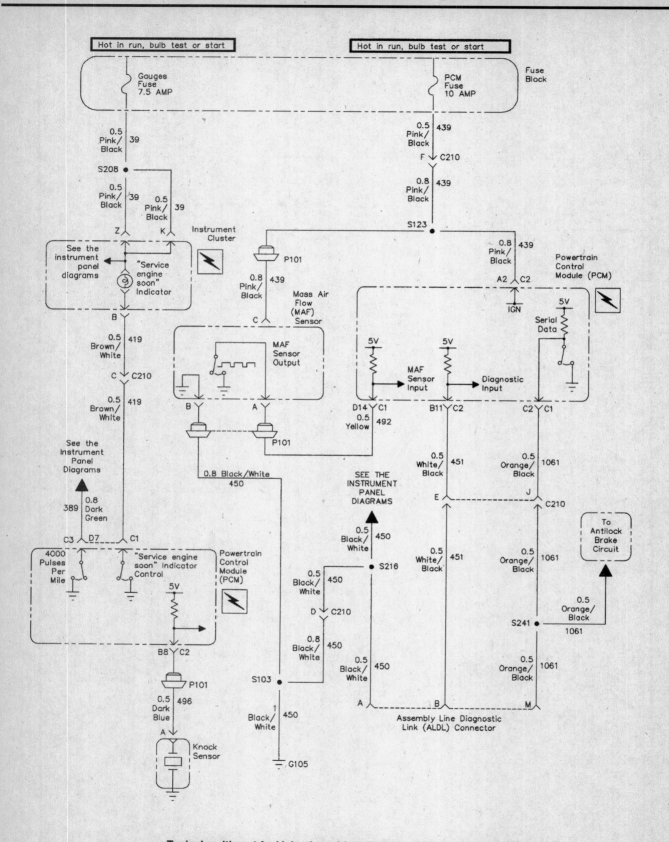

Typical multi-port fuel injection wiring diagram - 3800 engine - (1 of 4)

Chapter 12 Chassis electrical system

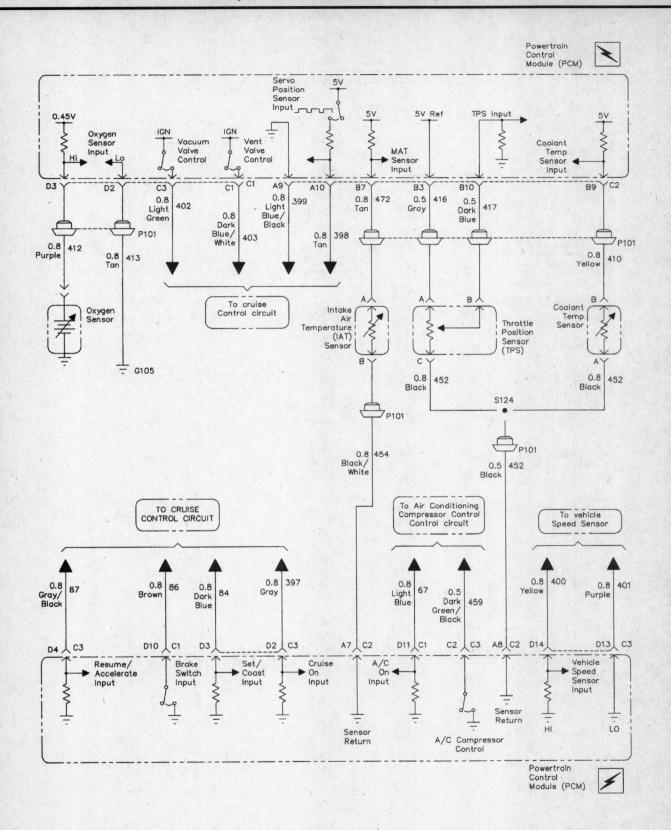

Typical multi-port fuel injection wiring diagram - 3800 engine - (2 of 4)

12-46 Chapter 12 Chassis electrical system

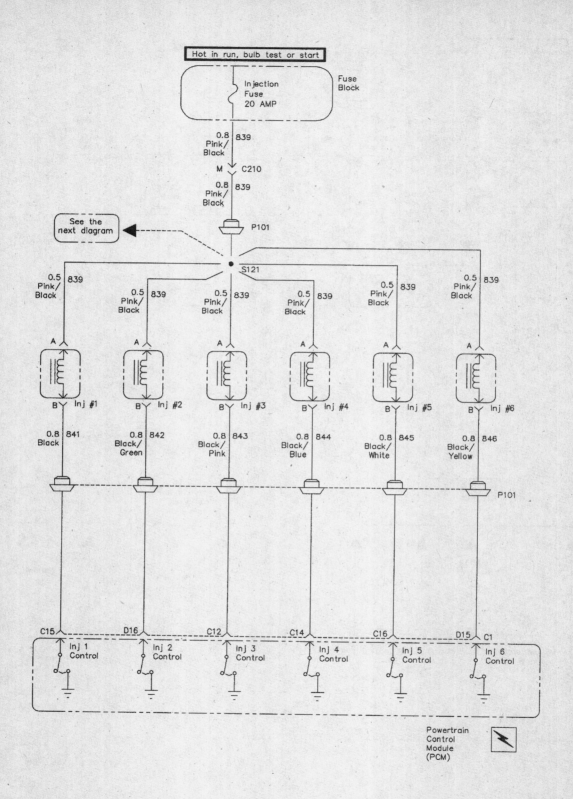

Typical multi-port fuel injection wiring diagram - 3800 engine - (3 of 4)

Chapter 12 Chassis electrical system

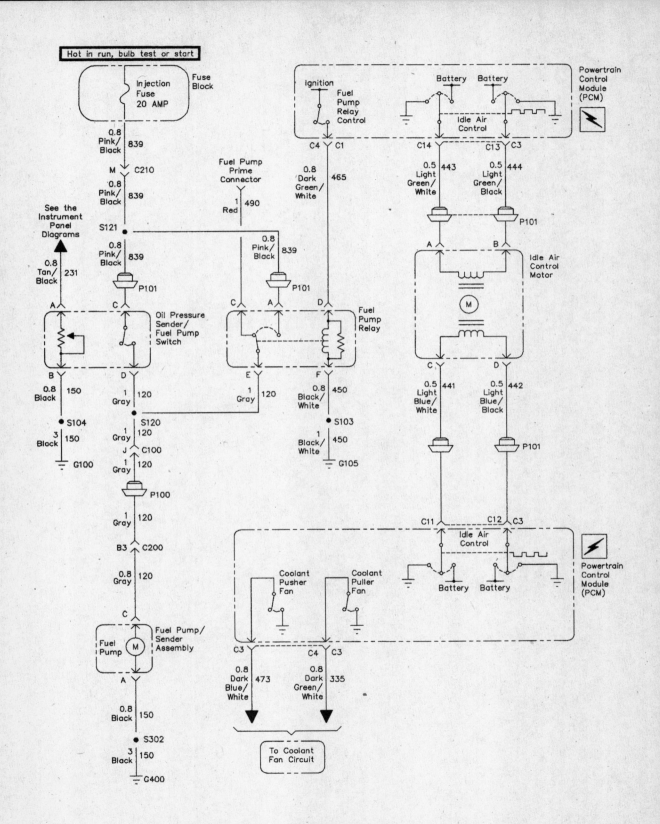

Typical multi-port fuel injection wiring diagram - 3800 engine - (4 of 4)

Notes

Index

A

About this manual, 0-5
Acknowledgements, 0-2
Air bag, general information, 12-14
Air cleaner housing, removal and installation, 4-6
Air conditioning
    accumulator, removal and installation, 3-13
    compressor, removal and installation, 3-14
    condenser, removal and installation, 3-15
    expansion (orifice) tube, removal and installation, 3-17
    system, check and maintenance, 3-13
Air filter replacement, 1-18
Alignment, general information, 10-17
Alternator, removal and installation, 5-9
Antifreeze, general information, 3-2
Ashtray assembly (1990 through 1992), removal and installation, 11-13
Automatic transaxle, 7-1 through 7-6
    diagnosis, general, 7-2
    differential oil seals, replacement, 7-5
    fluid and filter change, 1-22
    fluid level check, 1-10
    general information, 7-2
    mount, check and replacement, 7-5
    neutral start switch, replacement and adjustment, 7-4
    park/lock cable, removal and installation, 7-5
    removal and installation, 7-6
    shift cable, replacement and adjustment, 7-4
    throttle valve (TV) cable, replacement and adjustment, 7-3
Automotive chemicals and lubricants, 0-14
Axle assembly, rear, removal and installation, 10-11

B

Balljoint, check and replacement, 10-8
Battery
    cables, check and replacement, 5-2
    check, maintenance and charging, 1-11
    emergency jump starting, 0-13
    removal and installation, 5-2
Body, 11-1 through 11-16
    ashtray assembly (1990 through 1992), removal and installation, 11-13
    door
        lock and handles, removal and installation, 11-10
        trim panels, removal and installation, 11-4
    general information, 11-1
    glove box, removal and installation, 11-14
    hinges and locks, maintenance, 11-2
    hood release latch and cable, removal and installation, 11-3
    hood, removal, installation and adjustment, 11-3
    instrument panel trim, removal and installation, 11-11
    liftgate, removal, installation and adjustment (hardtop models only), 11-8
    liftgate strut, replacement, 11-9
    lower extension housing (1993 on), removal and installation, 11-14
    maintenance, 11-2
    outside mirror, removal and installation, 11-8
    repair
        major damage, 11-2
        minor damage, 11-2
    seats, removal and installation, 11-13
    side trim panel, removal and installation, 11-6

sound insulation panels, removal and installation, 11-14
steering column cover, removal and installation, 11-11
stereo, air conditioning and heater cover panel, removal and installation, 11-15
upholstery and carpets, maintenance, 11-2
vinyl trim, maintenance, 11-2
windshield and fixed glass, replacement, 11-3

Booster battery (jump) starting, 0-13

Brakes, 9-1 through 9-18
brake disc, inspection, removal and installation, 9-5
brake light switch, removal, installation and adjustment, 9-17
check, 1-16
disc brake
caliper, removal, overhaul and installation, 9-4
pads, replacement, 9-2
general information, 9-2
hoses and lines, inspection and replacement, 9-15
master cylinder, removal, overhaul and installation, 9-12
parking brake
adjustment, 9-16
cables, removal and installation, 9-16
power brake booster, inspection, removal and installation, 9-17
rear brake shoes, inspection and replacement, 9-7
rear wheel cylinder, removal, overhaul and installation, 9-11
shoes, rear, inspection and replacement, 9-7
system bleeding, 9-15

Bulb replacement, 12-7

Buying parts, 0-7

C

Camshaft, 2C-23

Camshaft, balance shaft and bearings, removal and inspection, 2C-12

Catalytic converter, 6-20

Charging system
check, 5-9
general information and precautions, 5-8
precautions, 5-9

Chassis electrical system, 12-1 through 12-48
air bag, general information, 12-14
bulb replacement, 12-7
circuit breakers, general information, 12-3
cruise control system, description and check, 12-14
door lock system, power, description and check, 12-14
electrical troubleshooting, general information, 12-2
fuses, general information, 12-2
fusible links, general information, 12-3
general information, 12-2
headlight bulb, replacement, 12-6
headlight switch, removal and installation, 12-5
headlights, adjustment, 12-6
ignition key lock cylinder, replacement, 12-5
instrument cluster, removal and installation, 12-12
radio and speakers, general information, removal and installation, 12-11
rear window wiper motor, removal and installation, 12-13
relays, general information, 12-3
turn signal
and hazard flashers, check and replacement, 12-4
switch assembly, removal and installation, 12-4
window
defogger, rear, check and repair, 12-12
system, power, description and check, 12-14
windshield wiper motor, removal and installation, 12-13
wiper/washer switch, removal and installation, 12-5
wiring diagrams, general information, 12-15

Chassis lubrication, 1-21

Circuit breakers, general information, 12-3

Coil springs, rear, removal, inspection and installation, 10-10

Computer Command Control (CCC) system and trouble codes, 6-2

Control arm, removal and installation, 10-5

Conversion Factors, 0-15

Coolant
reservoir, removal and installation, 3-7
temperature sending unit, check and replacement, 3-8

Cooling, heating and air conditioning systems, 3-1 through 3-18
accumulator, removal and installation, 3-13
air conditioning system, check and maintenance, 3-13
antifreeze, general information, 3-2
compressor, removal and installation, 3-14
condenser, removal and installation, 3-15
coolant
reservoir, removal and installation, 3-7
temperature sending unit, check and replacement, 3-8
engine cooling fan, check and replacement, 3-3
evaporator core(s), removal and installation, 3-16
expansion (orifice) tube, removal and installation, 3-17
heater and air conditioner
blower motor, removal and installation, 3-9
control assembly, removal and installation, 3-12
heater core(s), removal and installation, 3-11
radiator, removal and installation, 3-5
thermostat, check and replacement, 3-2
water pump
check, 3-7
removal and installation, 3-7

Cooling system
check, 1-12
servicing (draining, flushing and refilling), 1-24

Crankshaft
inspection, 2C-19
installation and main bearing oil clearance check, 2C-22
removal, 2C-14

Crankshaft front oil seal, replacement
3.1L V6 engine, 2A-11
3800 V6 engine, 2B-9

Cruise control system, description and check, 12-14

Cylinder honing, 2C-17

Cylinder compression check, 2C-4

Cylinder head
cleaning and inspection, 2C-10

Index

disassembly, 2C-9
reassembly, 2C-12
removal and installation
   3.1L V6 engine, 2A-8
   3800 V6 engine, 2B-7

D

Differential oil seals, replacement, 7-5
Disc brake
  caliper, removal, overhaul and installation, 9-4
  pads, replacement, 9-2
Distributor (3.1L engine), removal and installation, 5-3
Door
  lock and handles, removal and installation, 11-10
  removal and installation
    front, 11-8
    sliding, 11-8
  trim panels, removal and installation, 11-4
Door lock system, power, description and check, 12-14
Door window glass
  front, removal and installation, 11-11
  regulator, front, removal and installation, 11-11
Driveaxles, 8-1 through 8-10
  boot
    check, 1-15
    replacement and Constant Velocity (CV) joint
      overhaul, 8-3
  general information, 8-1
  removal and installation, 8-3
Drivebelt check and replacement, 1-19
Driveplate, removal, 2C-25

E

Electrical troubleshooting, general information, 12-2
Electronic Control Module (ECM) PROM (3.1L engine)/Powertrain Control Module (PCM) MEM/CAL (3800 engine), 6-8
Electronic fuel injection system, general information, 4-7
Electronic Level Control (ELC), general information, 10-11
Electronic Spark Control (ESC) system (1990 through 1992) or Knock Sensor (KS) system (1993 on), 6-15
Electronic Spark Timing (EST), 6-15
Emissions control systems, 6-1 through 6-20
  catalytic converter, 6-20
  Computer Command Control (CCC) system and trouble codes, 6-2
  Electronic Control Module (ECM) PROM (3.1L engine)/Power-train Control Module (PCM) MEM/CAL (3800 engine), 6-8
  Electronic Spark Control (ESC) system (1990 through 1992) or Knock Sensor (KS) system (1993 on), 6-15
  Electronic Spark Timing (EST), 6-15
  Evaporative Emission Control System (EECS), 6-17
  Exhaust Gas Recirculation (EGR) system (all 3.1L engines and 1993 and later 3800 engines), 6-16
  general information, 6-1
  information sensors, 6-10
    camshaft sensor (3800 engine), 6-14
    crankshaft sensor (3800 engine), 6-14
    Intake Air Temperature (IAT) Sensor (3800 engine), 6-14
    knock sensor, 6-15
    Manifold Absolute Pressure (MAP) sensor (3.1L engine), 6-11
    Mass Air Flow Sensor (MAF) (3800 engine), 6-13
    oxygen sensor, 6-11
    Park/Neutral switch, 6-12
    Throttle Position Sensor (TPS), 6-12
    Vehicle Speed Sensor (VSS), 6-13
  Positive Crankcase Ventilation (PCV) system, 6-18
  Thermostatic air cleaner (THERMAC) (3.1L engine), 6-18
  Transmission Converter Clutch (TCC), 6-20
Engine cooling fan, check and replacement, 3-3
Engine electrical systems, 5-1 through 5-10
  alternator, removal and installation, 5-9
  battery
    cables, check and replacement, 5-2
    emergency jump starting, 0-13
    removal and installation, 5-2
  charging system
    check, 5-9
    general information and precautions, 5-8
    precautions, 5-9
  distributor (3.1L engine), removal and installation, 5-3
  ignition
    coil, removal, testing and installation, 5-7
    module and distributorless ignition system, check and replacement, 5-4
    pick-up coil (3.1L engine), check and replacement, 5-6
    system, check, 5-2
  starter motor
    removal and installation, 5-10
    testing in vehicle, 5-9
  starter solenoid, removal and installation, 5-10
  starting system, general information, 5-9
Engine oil and filter change, 1-14
Engines
  3.1L V6 engine, 2A-1 through 2A-14
    crankshaft front oil seal, replacement, 2A-11
    cylinder head(s), removal and installation, 2A-8
    exhaust manifolds, removal and installation, 2A-7
    hydraulic lifters, removal, inspection and installation, 2A-9
    intake manifold, removal and installation, 2A-4
    mounts, check and replacement, 2A-14
    oil pan, removal and installation, 2A-13
    oil pump, removal and installation, 2A-14
    repair operations possible with the engine in the vehicle, 2A-2
    rocker arms and pushrods, removal, inspection and installation, 2A-5
    timing chain
      and sprockets, inspection, removal and installation, 2A-12
      cover, removal and installation, 2A-12

Top Dead Center (TDC) for number one piston, locating, 2A-2
valve covers, removal and installation, 2A-3
valve springs, retainers and seals, replacement, 2A-6
vibration damper, removal and installation, 2A-11
3800 V6 engine, 2B-1 through 2B-12
crankshaft front oil seal, replacement, 2B-9
cylinder heads, removal and installation, 2B-7
exhaust manifolds, removal and installation, 2B-7
intake manifold, removal and installation, 2B-5
mount, check and replacement, 2B-12
oil filter adapter and pressure regulator valve, removal and installation, 2B-10
oil pan, removal and installation, 2B-12
oil pump
pick-up tube and screen assembly, removal and installation, 2B-12
removal, inspection and installation, 2B-10
repair operations possible with the engine in the vehicle, 2B-2
rocker arms and pushrods, removal, inspection and installation, 2B-3
timing chain and sprockets, removal and installation, 2B-11
timing chain cover, removal and installation, 2B-9
Top Dead Center (TDC) for number one piston, locating, 2B-2
valve covers, removal and installation, 2B-3
valve lifters, removal, inspection and installation, 2B-5
valve springs, retainers and seals, replacement, 2B-4
vibration damper, removal and installation, 2B-8
General engine overhaul procedures, 2C-1 through 2C-26
block
cleaning, 2C-15
inspection, 2C-16
camshaft
and balance shaft, installation, 2C-23
balance shaft and bearings, removal and inspection, 2C-12
crankshaft
inspection, 2C-19
installation and main bearing oil clearance check, 2C-22
removal, 2C-14
cylinder compression check, 2C-4
cylinder head
disassembly, 2C-9
cleaning and inspection, 2C-10
reassembly, 2C-12
cylinder honing, 2C-17
driveplate, removal and installation, 2C-25
engine overhaul
disassembly sequence, 2C-8
general information, 2C-4
reassembly sequence, 2C-20
engine rebuilding alternatives, 2C-8
engine, removal and installation, 2C-5
engine removal, methods and precautions, 2C-5
initial start-up and break-in after overhaul, 2C-26
main and connecting rod bearings, inspection, 2C-19
piston rings, installation, 2C-20
pistons and connecting rods
installation and rod bearing oil clearance check, 2C-24
removal, 2C-13
rear main oil seal, replacement, 2C-23
valves, servicing, 2C-12
Evaporative Emission Control System (EECS), 6-17
Evaporative emissions control system check, 1-25
Evaporator core(s), removal and installation, 3-16
Exhaust Gas Recirculation (EGR) system (all 3.1L engines and 1993 and later 3800 engines), 6-16
Exhaust manifolds, removal and installation
3.1L V6 engine, 2A-7
3800 V6 engine, 2B-7
Exhaust system
check, 1-16
servicing, general information, 4-14

F

Fluid level checks, 1-7
automatic transaxle, 1-10
battery electrolyte, 1-8
brake fluid, 1-8
engine coolant, 1-8
engine oil, 1-7
power steering, 1-11
windshield washer fluid, 1-8
Fuel and exhaust systems, 4-1 through 4-14
air cleaner housing, removal and installation, 4-6
electronic fuel injection system, general information, 4-7
exhaust system servicing, general information, 4-14
filter replacement, 1-20
fuel injection system, check, 4-8
fuel pressure relief procedure, 4-2
injection system, check, 4-8
lines and fittings, repair and replacement, 4-3
Model 220 Throttle Body Injection (TBI) (3.1L engine), removal, overhaul and installation, 4-8
Port Fuel Injection (PFI) (3800 engine), component removal and installation, 4-12
pump/fuel pressure, testing, 4-2
pump, removal and installation, 4-5
system check, 1-17
tank
cleaning and repair, general information, 4-5
removal and installation, 4-4
throttle cable, removal and installation, 4-7
Fuses, general information, 12-2
Fusible links, general information, 12-3

G

Gear boots, steering, replacement, 10-14
Gear, steering, removal and installation, 10-12

Index

General engine overhaul procedures, 2C-1 through 2C-26
  block
    cleaning, 2C-15
    inspection, 2C-16
  camshaft
    and balance shaft, installation, 2C-23
    balance shaft and bearings, removal and
      inspection, 2C-12
  crankshaft
    installation and main bearing oil clearance
      check, 2C-22
    inspection, 2C-19
    removal, 2C-14
  cylinder head
    cleaning and inspection, 2C-10
    disassembly, 2C-9
    reassembly, 2C-12
  cylinder honing, 2C-17
  driveplate, removal and installation, 2C-25
  engine overhaul
    disassembly sequence, 2C-8
    general information, 2C-4
    reassembly sequence, 2C-20
  engine rebuilding alternatives, 2C-8
  engine removal, methods and precautions, 2C-5
  initial start-up and break-in after overhaul, 2C-26
  main and connecting rod bearings, inspection, 2C-19
  piston rings, installation, 2C-20
  pistons and connecting rods
    inspection, 2C-17
    installation and rod bearing oil clearance check, 2C-24
    removal, 2C-13
  rear main oil seal, replacement, 2C-23
  valves, servicing, 2C-12
Glove box, removal and installation, 11-14

H

Headlight bulb, replacement, 12-6
Headlight switch, removal and installation, 12-5
Headlights, adjustment, 12-6
Heater and air conditioner
  blower motor, removal and installation, 3-9
  control assembly, removal and installation, 3-12
Heater core(s), removal and installation, 3-11
Hinges and locks, maintenance, 11-2
Hood release latch and cable, removal and installation, 11-3
Hood, removal, installation and adjustment, 11-3
Hub and wheel bearing assembly, removal and installation
  front, 10-6
  rear, 10-10
Hydraulic lifters, removal, inspection and installation, 3.1L V6 engine, 2A-9

I

Ignition
  coil, removal, testing and installation, 5-7
  key lock cylinder, replacement, 12-5
  module and distributorless ignition system, check and replacement, 5-4
  pick-up coil (3.1L engine), check and replacement, 5-6
  system, check, 5-2
  timing check and adjustment, 1-27
Information sensors, 6-10
  Camshaft sensor (3800 engine), 6-14
  Crankshaft sensor (3800 engine), 6-14
  Intake Air Temperature (IAT) Sensor (3800 engine), 6-14
  knock sensor, 6-15
  Manifold Absolute Pressure (MAP) sensor (3.1L engine), 6-11
  Mass Air Flow Sensor (MAF) (3800 engine), 6-13
  oxygen sensor, 6-11
  Park/Neutral switch, 6-12
  Throttle Position Sensor (TPS), 6-12
  Vehicle Speed Sensor (VSS), 6-13
Initial start-up and break-in after overhaul, 2C-26
Instrument cluster, removal and installation, 12-12
Instrument panel trim, removal and installation, 11-11
Intake manifold, removal and installation
  3.1L V6 engine, 2A-4
  3800 V6 engine, 2B-5
Introduction to the Chevrolet Lumina, Oldsmobile Silhouette and Pontiac Trans Sport, 0-5

J

Jacking and towing, 0-13

K

Knock sensor, 6-15

L

Liftgate, removal, installation and adjustment (hardtop models only), 11-8
Liftgate strut, replacement, 11-9
Lower extension housing (1993 on), removal and installation, 11-14

M

Main and connecting rod bearings, 2C-19
Maintenance
  schedule, 1-6
  techniques, tools and working facilities, 0-7
Master cylinder, removal, overhaul and installation, 9-12

Index

Model 220 Throttle Body Injection (TBI) (3.1L engine), removal, overhaul and installation, 4-8
Mounts, engine, check and replacement
    3.1L V6 engine, 2A-14
    3800 V6 engine, 2B-12

N

Neutral start switch, replacement and adjustment, 7-4

O

Oil filter adapter and pressure regulator valve, removal and installation, 3800 V6 engine, 2B-10
Oil pan, removal and installation
    3.1L V6 engine, 2A-13
    3800 V6 engine, 2B-12
Oil pump pick-up tube and screen assembly, removal and installation, 3800 V6 engine, 2B-12
Oil pump, removal and installation
    3.1L V6 engine, 2A-14
    3800 V6 engine, 2B-10
Outside mirror, removal and installation, 11-8
Oxygen sensor, 6-11

P

Park/lock cable, removal and installation, 7-5
Parking brake
    adjustment, 9-16
    cables, removal and installation, 9-16
Piston rings, installation, 2C-20
Pistons and connecting rods
    inspection, 2C-17
    installation and rod bearing oil clearance check, 2C-24
    removal, 2C-13
Port Fuel Injection (PFI) (3800 engine), component removal and installation, 4-12
Positive Crankcase Ventilation (PCV) system, 6-18
Positive Crankcase Ventilation (PCV) valve check and replacement, 1-24
Power brake booster, inspection, removal and installation, 9-17
Power steering
    fluid level check, 1-11
    pump, removal and installation, 10-14
    system, bleeding, 10-15

R

Radiator, removal and installation, 3-5
Radio and speakers, general information, removal and installation, 12-11
Rear main oil seal, replacement, 2C-23
Relays, general information, 12-3
Repair operations possible with the engine in the vehicle
    3.1L V6 engine, 2A-2
    3800 V6 engine, 2B-2
Rocker arms and pushrods, removal, inspection and installation
    3.1L V6 engine, 2A-5
    3800 V6 engine, 2B-3

S

Safety first!, 0-16
Seat belt check, 1-20
Seats, removal and installation, 11-13
Shift cable, replacement and adjustment, 7-4
Shock absorbers, rear, removal, inspection and installation, 10-8
Side trim panel, removal and installation, 11-6
Sound insulation panels, removal and installation, 11-14
Spare tire and jack check, 1-20
Spark plug
    replacement, 1-25
    wire, distributor cap and rotor check and replacement, 1-26
Stabilizer bar and bushings, front, removal and installation, 10-5
Starter motor
    removal and installation, 5-10
    testing in vehicle, 5-9
Starter safety switch check, 1-20
Starter solenoid, removal and installation, 5-10
Starting system, general information, 5-9
Steering column cover, removal and installation, 11-11
Steering knuckle and hub, removal and installation, 10-7
Steering system
    alignment, general information, 10-17
    general information, 10-12
    power steering
        pump, removal and installation, 10-14
        system, bleeding, 10-15
    steering gear boots, replacement, 10-14
    steering gear, removal and installation, 10-12
    steering wheel, removal and installation, 10-15
    tie-rod ends, removal and installation, 10-12
    wheels and tires, general information, 10-16
Steering wheel, removal and installation, 10-15
Stereo, air conditioning and heater cover panel, removal and installation, 11-15
Strut
    assembly, removal and installation, 10-4
    replacement, 10-4
Suspension and steering check, 1-16
Suspension and steering systems, 10-1 through 10-18
    axle assembly, rear, removal and installation, 10-11
    balljoint, check and replacement, 10-8
    check, 1-16
    control arm, removal and installation, 10-5
    Electronic Level Control (ELC), general information, 10-11

hub and wheel bearing assembly, removal and installation
   front, 10-6
   rear, 10-10
rear coil springs, removal, inspection and installation, 10-10
rear shock absorbers, removal, inspection and installation, 10-8
stabilizer bar and bushings, front, removal and installation, 10-5
steering knuckle and hub, removal and installation, 10-7
strut assembly, removal and installation, 10-4
strut, replacement, 10-4
track bar assembly, removal and installation, 10-9

T

Thermostat, check and replacement, 3-2
Thermostatic air cleaner (THERMAC) (3.1L engine), 6-18
Thermostatic air cleaner check (3.1L engine only), 1-23
Throttle Body Injection (TBI) mounting bolt torque check (3.1L engine only), 1-19
Throttle cable, removal and installation, 4-7
Throttle valve (TV) cable, replacement and adjustment, 7-3
Tie-rod ends, removal and installation, 10-12
Timing chain and sprockets, inspection, removal and installation
   3.1L V6 engine, 2A-12
   3800 V6 engine, 2B-11
Timing chain cover, removal and installation
   3.1L V6 engine, 2A-12
   3800 V6 engine, 2B-9
Tire and tire pressure checks, 1-9
Tire rotation, 1-16
Tools, 0-9
Top Dead Center (TDC) for number one piston, locating
   3.1L V6 engine, 2A-2
   3800 V6 engine, 2B-2
Track bar assembly, removal and installation, 10-9
Transaxle, automatic, 7-1 through 7-6
   diagnosis, general, 7-2
   differential oil seals, replacement, 7-5
   fluid and filter change, 1-22
   fluid level check, 1-10
   general information, 7-2
   mount, check and replacement, 7-5
   neutral start switch, replacement and adjustment, 7-4
   park/lock cable, removal and installation, 7-5
   removal and installation, 7-6
   shift cable, replacement and adjustment, 7-4
   throttle valve (TV) cable, replacement and adjustment, 7-3

Transmission Converter Clutch (TCC), 6-20
Troubleshooting, 0-17
Tune-up and routine maintenance, 1-1 through 1-28
Tune-up general information, 1-7
Turn signal
   and hazard flashers, check and replacement, 12-4
   switch assembly, removal and installation, 12-4

U

Underhood hose check and replacement, 1-13
Upholstery and carpets, maintenance, 11-2

V

Valve covers, removal and installation
   3.1L V6 engine, 2A-3
   3800 V6 engine, 2B-3
Valve lifters, removal, inspection and installation, 3800 V6 engine, 2B-5
Valve springs, retainers and seals, replacement
   3.1L V6 engine, 2A-6
   3800 V6 engine, 2B-4
Valves, servicing, 2C-12
Vehicle identification numbers, 0-6
Vibration damper, removal and installation
   3.1L V6 engine, 2A-11
   3800 V6 engine, 2B-8
Vinyl trim, maintenance, 11-2

W

Water pump
   check, 3-7
   removal and installation, 3-7
Wheel cylinder, rear, removal, overhaul and installation, 9-11
Wheels and tires, general information, 10-16
Window
   defogger, rear, check and repair, 12-12
   system, power, description and check, 12-14
   wiper motor, rear, removal and installation, 12-13
Windshield
   and fixed glass, replacement, 11-3
   wiper motor, removal and installation, 12-13
Wiper blade inspection and replacement, 1-13
Wiper/washer switch, removal and installation, 12-5
Wiring diagrams, general information, 12-15
Working facilities, 0-12

Haynes Automotive Manuals

NOTE: New manuals are added to this list on a periodic basis. If you do not see a listing for your vehicle, consult your local Haynes dealer for the latest product information.

ACURA
- 12020 Integra '86 thru '89 & Legend '86 thru '90
- 12021 Integra '90 thru '93 & Legend '91 thru '95

AMC
- Jeep CJ - see JEEP (50020)
- 14020 Concord/Hornet/Gremlin/Spirit '70 thru '83
- 14025 (Renault) Alliance & Encore '83 thru '87

AUDI
- 15020 4000 all models '80 thru '87
- 15025 5000 all models '77 thru '83
- 15026 5000 all models '84 thru '88

AUSTIN
- Healey Sprite - see MG Midget (66015)

BMW
- *18020 3/5 Series '82 thru '92
- *18021 3 Series including Z3 models '92 thru '98
- 18025 320i all 4 cyl models '75 thru '83
- 18050 1500 thru 2002 except Turbo '59 thru '77

BUICK
- Century (FWD) - see GM (38005)
- *19020 Buick, Oldsmobile & Pontiac Full-size (Front wheel drive) '85 thru '00
- 19025 Buick Oldsmobile & Pontiac Full-size (Rear wheel drive) '70 thru '90
- 19030 Mid-size Regal & Century '74 thru '87
- Regal - see GENERAL MOTORS (38010)
- Skyhawk - see GM (38030)
- Skylark - see GM (38020, 38025)
- Somerset - see GENERAL MOTORS (38025)

CADILLAC
- 21030 Cadillac Rear Wheel Drive '70 thru '93
- Cimarron, Eldorado & Seville - see GM (38015, 38030, 38031)

CHEVROLET
- 10305 Chevrolet Engine Overhaul Manual
- *24010 Astro & GMC Safari Mini-vans '85 thru '98
- 24015 Camaro V8 all models '70 thru '81
- 24016 Camaro all models '82 thru '92
- Cavalier - see GM (38015)
- Celebrity - see GM (38005)
- *24017 Camaro & Firebird '93 thru '00
- 24020 Chevelle, Malibu, El Camino '69 thru '87
- 24024 Chevette & Pontiac T1000 '76 thru '87
- Citation - see GENERAL MOTORS (38020)
- 24032 Corsica/Beretta all models '87 thru '96
- 24040 Corvette all V8 models '68 thru '82
- *24041 Corvette all models '84 thru '96
- 24045 Full-size Sedans Caprice, Impala, Biscayne, Bel Air & Wagons '69 thru '90
- 24046 Impala SS & Caprice and Buick Roadmaster '91 thru '96
- Lumina '90 thru '94 - see GM (38010)
- *24048 Lumina & Monte Carlo '95 thru '01
- Lumina APV - see GM (38035)
- 24050 Luv Pick-up all 2WD & 4WD '72 thru '82
- Malibu - see GM (38026)
- 24055 Monte Carlo all models '70 thru '88
- Monte Carlo '95 thru '01 - see LUMINA
- 24059 Nova all V8 models '69 thru '79
- 24060 Nova/Geo Prizm '85 thru '92
- 24064 Pick-ups '67 thru '87 - Chevrolet & GMC, all V8 & in-line 6 cyl, 2WD & 4WD '67 thru '87; Suburbans, Blazers & Jimmys '67 thru '91
- 24065 Pick-ups '88 thru '98 - Chevrolet & GMC, all full-size models '88 thru '98; C/K Classic '99 & '00; Blazer & Jimmy '92 thru '94; Suburban '92 thru '99; Tahoe & Yukon '95 thru '99
- *24066 Pick-ups '99 thru '01 - Chevrolet Silverado & GMC Sierra '99 thru '01; Suburban/Tahoe/Yukon/Yukon XL '00 & '01
- 24070 S-10 & GMC S-15 Pick-ups '82 thru '93
- *24071 S-10, Gmc S-15 & Jimmy '94 thru '01
- *24075 Sprint & Geo Metro '85 thru '94
- *24080 Vans - Chevrolet & GMC '68 thru '96

CHRYSLER
- 10310 Chrysler Engine Overhaul Manual
- 25015 Chrysler Cirrus, Dodge Stratus, Plymouth Breeze, '95 thru '98
- 25020 Full-size Front-Wheel Drive '88 thru '93
- K-Cars - see DODGE Aries (30008)
- Laser - see DODGE Daytona (30030)
- 25025 Chrysler LHS, Concorde & New Yorker, Dodge Intrepid, Eagle Vision, '93 thru '97
- *25026 Chrysler LHS, Concorde, 300M, Dodge Intrepid '98 thru '01
- 25030 Chrysler/Plym. Mid-size front wheel drive '82 thru '95
- Rear-wheel Drive - see DODGE (30050)
- *25040 Chrysler Sebring/Dodge Avenger '95 thru '02

DATSUN
- 28005 200SX all models '80 thru '83
- 28007 B-210 all models '73 thru '78
- 28009 210 all models '79 thru '82
- 28012 240Z, 260Z & 280Z Coupe '70 thru '78
- 28014 280ZX Coupe & 2+2 '79 thru '83
- 300ZX - see NISSAN (72010)
- 28016 310 all models '78 thru '82
- 28018 510 & PL521 Pick-up '68 thru '73
- 28020 510 all models '78 thru '81
- 28022 620 Series Pick-up all models '73 thru '79
- 720 Series Pick-up - NISSAN (72030)
- 28025 810/Maxima all gas models '77 thru '84

DODGE
- 400 & 600 - see CHRYSLER (25030)
- 30008 Aries & Plymouth Reliant '81 thru '89
- 30010 Caravan & Ply. Voyager '84 thru '95
- *30011 Caravan & Ply. Voyager '96 thru '99
- 30012 Challenger/Plymouth Saporro '78 thru '83
- Challenger '67-'76 - see DART (30025)
- 30016 Colt/Plymouth Champ '78 thru '87
- 30020 Dakota Pick-ups all models '87 thru '96
- *30021 Durango '98 & '99, Dakota '97 thru '99
- 30025 Dart, Challenger/Plymouth Barracuda & Valiant 6 cyl models '67 thru '76
- 30030 Daytona & Chrysler Laser '84 thru '89
- Intrepid - see Chrysler (25025, 25026)
- *30034 Dodge & Plymouth Neon '95 thru '99
- *30035 Omni & Plymouth Horizon '78 thru '90
- 30040 Pick-ups all full-size models '74 thru '93
- *30041 Pick-ups all full-size models '94 thru '01
- *30045 Ram 50/D50 Pick-ups & Raider and Plymouth Arrow Pick-ups '79 thru '93
- 30050 Dodge/Ply./Chrysler RWD '71 thru '89
- *30055 Shadow/Plymouth Sundance '87 thru '94
- 30060 Spirit & Plymouth Acclaim '89 thru '95
- *30065 Vans - Dodge & Plymouth '71 thru '99

EAGLE
- Talon - see MITSUBISHI (68030, 68031)
- Vision - see CHRYSLER (25025)

FIAT
- 34010 124 Sport Coupe & Spider '68 thru '78
- 34025 X1/9 all models '74 thru '80

FORD
- 10355 Ford Automatic Transmission Overhaul
- 10320 Ford Engine Overhaul Manual
- 36004 Aerostar Mini-vans '86 thru '97
- Aspire - see FORD Festiva (36030)
- 36006 Contour/Mercury Mystique '95 thru '00
- 36008 Courier Pick-up all models '72 thru '82
- *36012 Crown Victoria & Mercury Grand Marquis '88 thru '00
- 36016 Escort/Mercury Lynx '81 thru '90
- 36020 Escort/Mercury Tracer '91 thru '00
- Expedition - see FORD Pick-up (36059)
- *36024 Explorer & Mazda Navajo '91 thru '01
- 36028 Fairmont & Mercury Zephyr '78 thru '83
- 36030 Festiva & Aspire '88 thru '97
- 36032 Fiesta all models '77 thru '80
- 36034 Focus all models '00 and '01
- 36036 Ford & Mercury Full-size '75 thru '87
- 36040 Granada & Mercury Monarch '75 thru '80
- 36044 Ford & Mercury Mid-size '75 thru '86
- 36048 Mustang V8 all models '64-1/2 thru '73
- 36049 Mustang II 4 cyl, V6 & V8 '74 thru '78
- 36050 Mustang & Mercury Capri '79 thru '86
- *36051 Mustang all models '94 thru '00
- 36054 Pick-ups and Bronco '73 thru '79
- 36058 Pick-ups and Bronco '80 thru '96
- *36059 Pick-ups, Expedition & Lincoln Navigator '97 thru '99
- *36060 Super Duty Pick-up, Excursion '97 thru '02
- 36062 Pinto & Mercury Bobcat '75 thru '80
- 36066 Probe all models '89 thru '92
- *36070 Ranger/Bronco II gas models '83 thru '92
- *36071 Ford Ranger '93 thru '00 & Mazda Pick-ups '94 thru '00
- 36074 Taurus & Mercury Sable '86 thru '95
- *36075 Taurus & Mercury Sable '96 thru '01
- 36078 Tempo & Mercury Topaz '84 thru '94
- 36082 Thunderbird/Mercury Cougar '83 thru '88
- 36086 Thunderbird/Mercury Cougar '89 thru '97
- 36090 Vans all V8 Econoline models '69 thru '91
- 36094 Vans full size '92 thru '01
- *36097 Windstar Mini-van '95 thru '01

GENERAL MOTORS
- 10360 GM Automatic Transmission Overhaul
- 38005 Buick Century, Chevrolet Celebrity, Olds Cutlass Ciera & Pontiac 6000 '82 thru '96
- *38010 Buick Regal, Chevrolet Lumina, Oldsmobile Cutlass Supreme & Pontiac Grand Prix front wheel drive '88 thru '99
- 38015 Buick Skyhawk, Cadillac Cimarron, Chevrolet Cavalier, Oldsmobile Firenza Pontiac J-2000 & Sunbird '82 thru '94
- *38016 Chevrolet Cavalier/Pontiac Sunfire '95 thru '00
- 38020 Buick Skylark, Chevrolet Citation, Olds Omega, Pontiac Phoenix '80 thru '85
- 38025 Buick Skylark & Somerset, Olds Achieva, Calais & Pontiac Grand Am '85 thru '98
- *38026 Chevrolet Malibu, Olds Alero & Cutlass, Pontiac Grand Am '97 thru '00
- 38030 Cadillac Eldorado & Oldsmobile Toronado '71 thru '85, Seville '80 thru '85, Buick Riviera '79 thru '85
- *38031 Cadillac Eldorado & Seville '86 thru '91, Deville & Buick Riviera '86 thru '93, Fleetwood & Olds Toronado '86 thru '92
- *38035 Chevrolet Lumina APV, Oldsmobile Silhouette & Pontiac Trans Sport '90 thru '95
- *38036 Chevrolet Venture, Olds Silhouette, Pontiac Trans Sport & Montana '97 thru '01
- General Motors Full-size Rear-wheel Drive - see BUICK (19025)

GEO
- Metro - see CHEVROLET Sprint (24075)
- Prizm - see CHEVROLET (24060) or TOYOTA (92036)
- 40030 Storm all models '90 thru '93
- Tracker - see SUZUKI Samurai (90010)

GMC
- Vans & Pick-ups - see CHEVROLET

HONDA
- 42010 Accord CVCC all models '76 thru '83
- 42011 Accord all models '84 thru '89
- 42012 Accord all models '90 thru '93
- 42013 Accord all models '94 thru '97
- *42014 Accord all models '98 and '99
- 42020 Civic 1200 all models '73 thru '79
- 42021 Civic 1300 & 1500 CVCC '80 thru '83
- 42022 Civic 1500 CVCC all models '75 thru '79
- 42023 Civic all models '84 thru '91
- 42024 Civic & del Sol '92 thru '95
- *42025 Civic '96 thru '00, CR-V '97 thru '00, Acura Integra '94 thru '00
- Passport - see ISUZU Rodeo (47017)
- *42040 Prelude CVCC all models '79 thru '89

HYUNDAI
- *43010 Elantra all models '96 thru '01
- 43015 Excel & Accent all models '86 thru '98

ISUZU
- Hombre - see CHEVROLET S-10 (24071)
- *47017 Rodeo '91 thru '97, Amigo '89 thru '94, Honda Passport '95 thru '97
- 47020 Trooper '84 thru '91, Pick-up '81 thru '93

JAGUAR
- 49010 XJ6 all 6 cyl models '68 thru '86
- 49011 XJ6 all models '88 thru '94
- 49015 XJ12 & XJS all 12 cyl models '72 thru '85

JEEP
- 50010 Cherokee, Comanche & Wagoneer Limited all models '84 thru '00
- 50020 CJ all models '49 thru '86
- *50025 Grand Cherokee all models '93 thru '00
- 50029 Grand Wagoneer & Pick-up '72 thru '91
- *50030 Wrangler all models '87 thru '00

LEXUS
- ES 300 - see TOYOTA Camry (92007)

LINCOLN
- Navigator - see FORD Pick-up (36059)
- *59010 Rear Wheel Drive all models '70 thru '01

MAZDA
- 61010 GLC (rear wheel drive) '77 thru '83
- 61011 GLC (front wheel drive) '81 thru '85
- *61015 323 & Protegé '90 thru '00
- *61016 MX-5 Miata '90 thru '97
- 61020 MPV all models '89 thru '94
- Navajo - see FORD Explorer (36024)
- 61030 Pick-ups '72 thru '93
- Pick-ups '94 on - see Ford (36071)
- 61035 RX-7 all models '79 thru '85
- 61036 RX-7 all models '86 thru '91
- 61040 626 (rear wheel drive) '79 thru '82
- *61041 626 & MX-6 (front wheel drive) '83 thru '91
- *61042 626 '93 thru '01, MX-6/Ford Probe '93 thru '97

MERCEDES-BENZ
- 63012 123 Series Diesel '76 thru '85
- 63015 190 Series 4-cyl gas models, '84 thru '88
- 63020 230, 250 & 280 6 cyl sohc '68 thru '72
- 63025 280 123 Series gas models '77 thru '81
- 63030 350 & 450 all models '71 thru '80

MERCURY
- 64200 Villager & Nissan Quest '93 thru '01
- All other titles, see FORD listing.

MG
- 66010 MGB Roadster & GT Coupe '62 thru '80
- 66015 MG Midget & Austin Healey Sprite Roadster '58 thru '80

MITSUBISHI
- 68020 Cordia, Tredia, Galant, Precis & Mirage '83 thru '93
- 68030 Eclipse, Eagle Talon & Plymouth Laser '90 thru '94
- *68031 Eclipse '95 thru '01, Eagle Talon '95 thru '98
- 68040 Pick-up '83 thru '96, Montero '83 thru '93

NISSAN
- 72010 300ZX all models incl. Turbo '84 thru '89
- 72015 Altima all models '93 thru '01
- 72020 Maxima all models '85 thru '92
- *72021 Maxima all models '93 thru '01
- 72030 Pick-ups '80 thru '97, Pathfinder '87 thru '95
- *72031 Frontier Pick-up '98 thru '01, Xterra '00 & '01, Pathfinder '96 thru '01
- 72040 Pulsar all models '83 thru '86
- 72050 Sentra all models '82 thru '94
- 72051 Sentra & 200SX all models '95 thru '99
- 72060 Stanza all models '82 thru '90

OLDSMOBILE
- *73015 Cutlass '74 thru '88
- For other OLDSMOBILE titles, see BUICK, CHEVROLET or GM listings.

PLYMOUTH
- For PLYMOUTH titles, see DODGE.

PONTIAC
- 79008 Fiero all models '84 thru '88
- 79018 Firebird V8 models except Turbo '70 thru '81
- 79019 Firebird all models '82 thru '92
- 79040 Mid-size Rear-wheel Drive '70 thru '87
- For other PONTIAC titles, see BUICK, CHEVROLET or GM listings.

PORSCHE
- 80020 911 Coupe & Targa models '65 thru '89
- 80025 914 all 4 cyl models '69 thru '76
- 80030 924 all models incl. Turbo '76 thru '82
- 80035 944 all models incl. Turbo '83 thru '89

RENAULT
- Alliance, Encore - see AMC (14020)

SAAB
- *84010 900 including Turbo '79 thru '88

SATURN
- *87010 Saturn all models '91 thru '99

SUBARU
- 89002 1100, 1300, 1400 & 1600 '71 thru '79
- 89003 1600 & 1800 2WD & 4WD '80 thru '94

SUZUKI
- *90010 Samurai/Sidekick/Geo Tracker '86 thru '01

TOYOTA
- 92005 Camry all models '83 thru '91
- 92006 Camry all models '92 thru '96
- *92007 Camry/Avalon/Solara/Lexus ES 300 '97 thru '01
- 92015 Celica Rear Wheel Drive '71 thru '85
- 92020 Celica Front Wheel Drive '86 thru '99
- 92025 Celica Supra all models '79 thru '92
- 92030 Corolla all models '75 thru '79
- 92032 Corolla rear wheel drive models '80 thru '87
- 92035 Corolla front wheel drive models '84 thru '92
- *92036 Corolla & Geo Prizm '93 thru '01
- 92040 Corolla Tercel all models '80 thru '82
- 92045 Corona all models '74 thru '82
- 92050 Cressida all models '78 thru '82
- 92055 Land Cruiser FJ40/43/45/55 '68 thru '82
- *92056 Land Cruiser FJ60/62/80/FZJ80 '80 thru '96
- 92065 MR2 all models '85 thru '87
- 92070 Pick-up all models '69 thru '78
- 92075 Pick-up all models '79 thru '95
- 92076 Tacoma '95 thru '00, 4Runner '96 thru '00, T100 '93 thru '98
- 92080 Previa all models '91 thru '95
- *92082 RAV4 all models '96 thru '02
- 92085 Tercel all models '87 thru '94

TRIUMPH
- 94007 Spitfire all models '62 thru '81
- 94010 TR7 all models '75 thru '81

VW
- 96008 Beetle & Karmann Ghia '54 thru '79
- *96009 New Beetle '98 thru '00
- 96016 Rabbit, Jetta, Scirocco, & Pick-up gas models '74 thru '91 & Convertible '80 thru '92
- 96017 Golf & Jetta '93 thru '97
- *96018 Golf & Jetta '98 thru '01
- 96020 Rabbit, Jetta, Pick-up diesel '77 thru '84
- *96023 Passat '98 thru '01, Audi A4 '96 thru '01
- 96030 Transporter 1600 all models '68 thru '79
- 96035 Transporter 1700, 1800, 2000 '72 thru '79
- 96040 Type 3 1500 & 1600 '63 thru '73
- 96045 Vanagon air-cooled models '80 thru '83

VOLVO
- 97010 120, 130 Series & 1800 Sports '61 thru '73
- 97015 140 Series all models '66 thru '74
- 97020 240 Series all models '76 thru '93
- 97025 260 Series all models '75 thru '82
- 97040 740 & 760 Series all models '82 thru '88

TECHBOOK MANUALS
- 10205 Automotive Computer Codes
- 10210 Automotive Emissions Control Manual
- 10215 Fuel Injection Manual, 1978 thru 1985
- 10220 Fuel Injection Manual, 1986 thru 1999
- 10225 Holley Carburetor Manual
- 10230 Rochester Carburetor Manual
- 10240 Weber/Zenith/Stromberg/SU Carburetor
- 10305 Chevrolet Engine Overhaul Manual
- 10310 Chrysler Engine Overhaul Manual
- 10320 Ford Engine Overhaul Manual
- 10330 GM and Ford Diesel Engine Repair
- 10340 Small Engine Repair Manual
- 10345 Suspension, Steering & Driveline
- 10355 Ford Automatic Transmission Overhaul
- 10360 GM Automatic Transmission Overhaul
- 10405 Automotive Body Repair & Painting
- 10410 Automotive Brake Manual
- 10415 Automotive Detailing Manual
- 10420 Automotive Eelectrical Manual
- 10425 Automotive Heating & Air Conditioning
- 10430 Automotive Reference Dictionary
- 10435 Automotive Tools Manual
- 10440 Used Car Buying Guide
- 10445 Welding Manual
- 10450 ATV Basics

SPANISH MANUALS
- 98903 Reparación de Carrocería & Pintura
- 98905 Códigos Automotrices de la Computadora
- 98910 Frenos Automotriz
- 98915 Inyección de Combustible 1986 al 1999
- 99040 Chevrolet & GMC Camionetas '67 al '87
- 99041 Chevrolet & GMC Camionetas '88 al '98
- 99042 Chevrolet Camionetas Cerradas '68 al '95
- 99055 Dodge Caravan/Ply. Voyager '84 al '95
- 99075 Ford Camionetas y Bronco '80 al '94
- 99077 Ford Camionetas Cerradas '69 al '91
- 99083 Ford Modelos de Tamaño Grande '75 al '87
- 99088 Ford Modelos de Tamaño Mediano '75 al '86
- 99091 Ford Taurus & Mercury Sable '86 al '95
- 99095 GM Modelos de Tamaño Grande '70 al '90
- 99100 GM Modelos de Tamaño Mediano '70 al '88
- 99110 Nissan Camionetas '80 al '96, Pathfinder '87 al '95
- 99118 Nissan Sentra '82 al '94
- 99125 Toyota Camionetas y 4-Runner '79 al '95

* Listings shown with an asterisk (*) indicate model coverage as of this printing. These titles will be periodically updated to include later model years - consult your Haynes dealer for more information.

Nearly 100 Haynes motorcycle manuals also available

7-02

Haynes North America, Inc., 861 Lawrence Drive, Newbury Park, CA 91320 • (805) 498-6703